UNDERSTAND

CRIME

ANALYZING THE GEOGRAPHY OF CRIME

UNDERSTANDING CRIME

ANALYZING THE GEOGRAPHY OF CRIME

Spencer Chainey

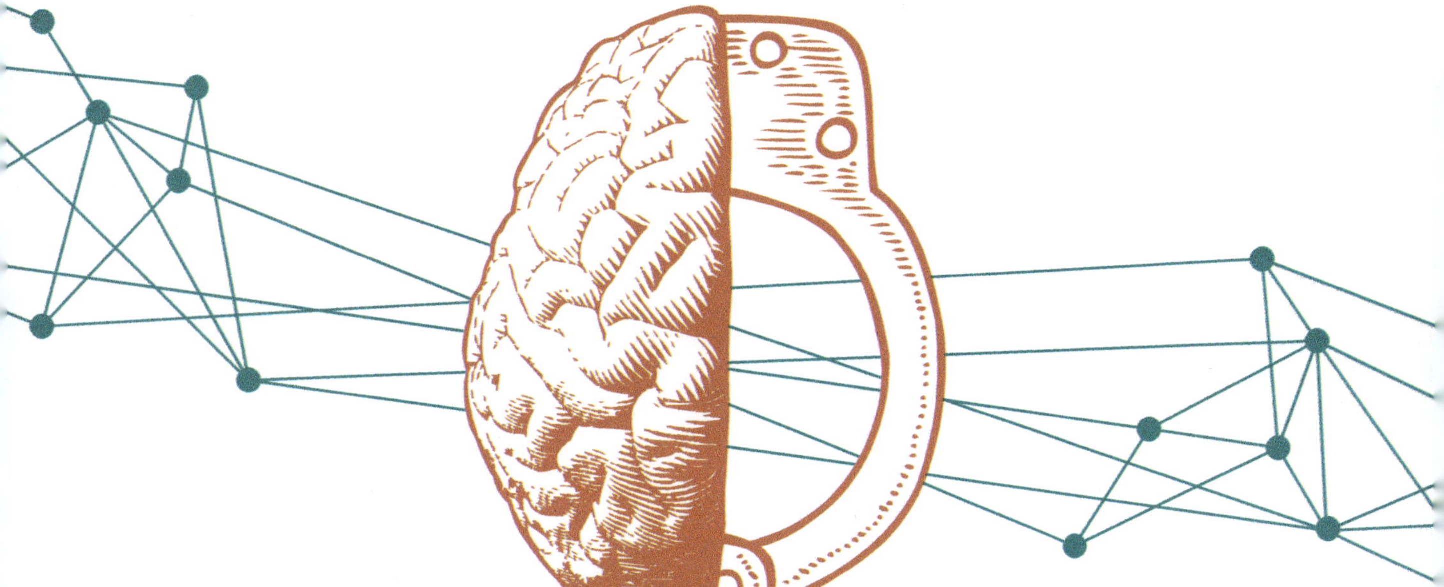

Esri Press
REDLANDS | CALIFORNIA

Esri Press, 380 New York Street, Redlands, California 92373-8100

ISBN: 9781589485846
Library of Congress Control Number: 2020943962

For purchasing and distribution options (both domestic and international), please visit esripress.esri.com.

Contents

Preface vii

Acknowledgments ix

1. The key theories underpinning the geographic analysis of crime 1
2. Statistical measures for spatial patterns, and the journey to crime 21
3. Hot spot analysis: Examining spatial concentrations and clusters of crime 37
4. The temporal analysis of crime 65
5. Risky facilities, and repeat and near repeat victimization 83
6. Persistent, emerging, and dispersed spatial patterns of crime 113
7. Geographic profiling analysis: Supporting criminal investigation 131
8. Applying spatial regression analysis to crime data 161
9. Mapping and analyzing crime risk 225
10. Measuring performance and evaluating impact 247

Index 279

Preface

There are many books about crime analysis, GIS and law enforcement, and crime mapping that discuss techniques on the geographic analysis of crime, but none of these books fully explain how to analyze the geography of crime. There are also many journal articles, if you are fortunate enough to have subscribed access, that include studies on the geographic analysis of crime, many of which introduce new techniques for examining patterns of crime. But often, to make sense of these techniques, you need a PhD in mathematics or statistics, or skills in writing code to apply the techniques to your own crime data. This means we are not taking advantage of a fantastic range of analysis techniques that can help us better understand crime, improve how we study the geography of crime, and improve the quality of the services we provide to our colleagues and our communities.

Improving how we analyze crime is what motivated me to write this book. This book provides a detailed description, explanation, and illustration of geographic analysis techniques that can help us better understand crime. It aims to be comprehensive in scope. It aims to be accessible, regardless of your expertise, experience, and the type of GIS you use. This is not an ArcGIS book. It is purposely more generic so that analysts, students, researchers, police officers, public safety officials, and so on can use the contents of the book to help better examine the geography of crime. The book does cover some technical topics. My aim was not to dumb them down, but to "dissect" them, describe them in plain English so they are easier to understand, and so you can be more confident in how to apply them. Because if you do, you will better understand crime.

This book is not necessarily meant to be read cover to cover, but instead aims to act as a guide that you dip into when needed, to help you apply techniques for a specific task, or help you better understand how certain techniques work. A good place to start is to skim through the pages so you become familiar with the range of content the book covers, so you will hopefully come to regard it as *that reference source* you can refer to repeatedly. The content includes new and old techniques, and ones you may have heard of but have not been demystified for you to be confident in using. Techniques that you may be familiar with have been stripped back to illustrate the role they play in modern geographic crime analysis. New techniques described in this book offer that contemporary edge, and may become the *classics* of the future. And techniques that you often have not been confident in using because of their complexity have been broken down, explained in plain English, and illustrated with numerous examples to help you replicate these techniques with your own data.

The chapters follow a logical order. Chapter 1 describes the main theories that relate to the geography of crime. Do not skip this chapter just because it is about theory, and not about techniques. To understand crime, and to interpret the patterns we observe, we must draw from the theory behind crime. All techniques (or at least the good ones) require theory to inform their design, with the theory also helping us interpret the results that the technique generates. If we don't consider theoretical principles in the interpretation of our results, what we state about our results is merely descriptive. We want to explain *why* the crime patterns exist so we can better recommend what can be done to counter the behaviors, circumstances, and conditions that lead to their formation. To do so, we draw from the theory in our interpretation of the results. The first chapter brings all the main theories that relate to the geographic analysis of crime into one place.

Beginning with chapter 2, we get into the techniques, starting with chapters on ways to examine spatial and temporal patterns of crime, then on analyzing crime with the use of other data (such as demographic, land use, and socioeconomic data), and finishing with ways to evaluate the impact of geographically targeted interventions.

Each chapter is organized as such: a description of the objectives of the chapter, a list of the key learning points, an introduction, the chapter content, a summary, and references. Some chapters also include a list of further resources. Where possible, I have not only cited (out of necessity) the key references, but also attempted to cite as many free-to-access references as possible. In each chapter, there are also *thought boxes*, which may contain an opinion piece, supplementary information, or other things that are useful to consider in the application of techniques. These elements are designed to help you pause for thought as you progress through each chapter. Sometimes they offer an analogy to help you better understand a theoretical principle or technical matter, or they may offer additional information that helps you consider another way to use a technique. The book includes numerous examples, from all corners of the world, using real data. These examples are designed to help you replicate the techniques described so you can compare your results with those from the example, and they can help you interpret your results.

This book is not designed as a workbook, showing you which button to click in ArcGIS, or any other software, because the aim is to make the book generic to whatever software is used. However, you should find that enough detail is provided in each chapter to enable you to replicate any technique using ArcGIS, or other software. In places, I refer to specific tools that are available in ArcGIS because they provide the best way to illustrate the application of a technique. I do, though, describe how the technique works, so that those of you who do not use ArcGIS can attempt to replicate the functionality of the specific ArcGIS tool. In places, I refer to other software tools to illustrate the application of a technique.

I hope this book helps you improve your analysis of the geography of crime.

Acknowledgments

Thanks to my wife, Victoria, and my two sons, Oscar and Isaac, who are partners in everything I do. I also would like to thank all those researchers, students, analysts, police, public safety, and citizen security personnel I work with across the world. The main reasons I do what I do are because of the many great people I work with and the support I get from my family.

I wrote a lot of this book during the peak of the coronavirus disease 2019 (COVID-19) pandemic. All of us were affected in some way. This book is dedicated to the frontline workers, in health, policing, other public services, and businesses, who helped keep us going and went beyond the call of duty. We can take inspiration from what they did, to help us improve what we can do.

Spencer Chainey
London, England

CHAPTER 1

The key theories underpinning the geographic analysis of crime

Objectives

In this chapter, we introduce the main theories that underpin the geographic analysis of crime. It is useful to be familiar with these theoretical principles, as they can help us understand the spatial and temporal patterns we observe in our data. In each of the following chapters, we will draw from these theoretical principles to illustrate how they can help us interpret crime patterns. The main theories discussed are social disorganization, collective efficacy and social capital, the routine-activity perspective, rational choice, crime pattern theory, the least-effort principle, crime generators, and crime attractors.

Key learning points

- Recognize the value of theory in interpreting the patterns observed in the analysis of data.
- Become familiar with key theoretical principles for interpreting patterns in crime data.
- Be able to draw from these theoretical principles in the chapters that follow to help determine how analysis techniques can be used to better understand patterns of crime.

Introduction

The starting point for using any type of geographic analysis output is to consider the theoretical principles that can be used to interpret what the analysis outputs are telling us. If we can interpret the patterns in terms of how offenders have behaved, we can then use the results to help determine the types of activities to implement that may counter or prevent this type of behavior. Several theories relating to the geography of crime have been developed over the years. In this chapter, we bring these theories together to help explain why spatial and temporal patterns of crime are not random.

The theoretical perspectives that are introduced in this chapter include meso-geographic explanations for crime (i.e., at the neighborhood level) and micro-geographic explanations of crime (i.e., at the street, or specific location, level). By drawing together the principal geographic theories of crime, we create a foundation that permits a continual assessment throughout the book on how different analysis techniques can be used in a practical manner for informing operational, investigative, and strategic policing and crime prevention activity.

If analysis outputs of crime are to be used effectively, the interpretation of the analysis outputs must be based on clear theoretical principles. If it is not clear why a crime issue is likely to be present, it makes it difficult to determine what can be done to effectively tackle the crime issue. If the crime patterns can be interpreted effectively, this in turn will help identify the specific activities that could be introduced to counter the observed patterns. Theory can help us interpret crime patterns.

As a starting point, a useful theoretical principle to recognize is that crime has an inherently geographic quality. When a crime occurs, it happens at a place with a geographic location (Chainey and Ratcliffe 2005). This can be observed in Brantingham and Brantingham's (1981) description of the four dimensions of every crime:

- The legal dimension (a law must be broken).
- The victim dimension (someone or something must be targeted).
- The offender dimension (someone must do the crime).
- The spatial dimension (the crime must happen somewhere).

Crimes do not occur randomly. If crimes were random occurrences that had an equal chance of happening anywhere at any time, there would be little point in attempting to observe patterns and consider what could be done to address crime issues.

The main theoretical discipline that underpins the geography of crime is the practical subset of mainstream criminology known as *environmental criminology.* Environmental criminology involves the study of criminal activity and victimization, and how factors of space influence the behavior of offenders and the victimization of people or other types of targets (Bottoms and Wiles 2002). The next section in this chapter begins by describing how the importance of this geographic influence on people came to be recognized, before examining the geographic dynamics of offenders and the interaction of offenders and victims in space. The progression in theoretical development that has taken place over time also helps illustrate the evolution and relationship from a meso-level consideration of crime toward micro-level geographic explanations of crime. We also illustrate that these theories tend not to work in isolation but typically in a complementary manner.

Researchers long have known there is variation in the spatial arrangement of crime. Although spatial studies of crime have been recorded for nearly 200 years, many key research periods have punctuated the history. Although these periods have overlapped across time, for convenience they can be thought of, as defined by Chainey and Ratcliffe (2005), as three distinct schools of thought: the Cartographic School, the Chicago School, and the GIS School. The theoretical developments associated with the Cartographic and Chicago Schools provide a useful foundation for the meso-level examination of crime patterns. In the more modern era, the focus of the GIS School has been toward micro-level geographic explanations of crime.

The Cartographic School

Some of the earliest maps of crime originate from France and were published in 1833 by Andre-Michel Guerry (1833), who showed, among other features, the distribution of violent and property crime across the jurisdictional divisions of France. These maps indicated that not only was there spatial variation in crime but that the risk of property crime and violent crime was often different in the same areas (Brantingham and Brantingham 1981). Analyzing French data at about the same time was Adolphe Quetelet (1842), who supplemented his maps with statistics showing spatial variations across France, and between social groups. These early pioneers are credited with founding what is termed the Cartographic School (Chainey and Ratcliffe 2005), and they inspired other studies into patterns of crime, such as Mayhew's (1862) study of offending patterns in London. These early studies set the foundation for illustrating that spatial patterns of crime were not random and prompted discussion on the factors that influenced these spatial variations in crime.

The Chicago School

The sociocultural triggers of crime and models for urban development

In the 20th century, more innovation followed with the research conducted by the Chicago School. This group of researchers included Clifford Shaw and Henry McKay, who drew on the spatial and temporal ideas of social ecology, forged by their predecessors, notably Ernest Burgess (1916). Shaw and McKay (1942) mapped, by hand, the residences of juvenile delinquents across Chicago. This pioneering mapping examined the sociocultural triggers of crime as Chicago expanded during a period of great economic growth. Shaw and McKay drew on Burgess's work by comparing socioeconomic and physical factors in different zones across the city.

Burgess introduced the zonal (or concentric) model of urban development in 1925. Burgess's idealized model, as shown in figure 1.1, had concentric zones radiating outward in bands from the city center, with each band representing a different stage of the city's development. The innermost zone (zone I), termed *the loop*, contained the central business district and had little residential development. Adjoining this was the *zone in transition* (zone II), an area taken over by business and light manufacturing industries, and which included the *factory zone*. The third zone (*zone of workingmen's homes*) was occupied by factory workers who had managed to escape the zone in transition but who were still tied to the city because of the need to work in the factories. Travel cost and time was a factor for these workers, so they lived in this third zone. The *residential zone* (IV) comprised high-class apartments or single-family suburban dwellings where the occupants accepted the higher costs of travel as the price of a better quality of life. Beyond the city limits was the *commuters zone* (zone V), where people lived in suburban areas or satellite towns, with a commute of up to an hour (Burgess 1925). To demonstrate the model, Burgess charted 1920s Chicago and overlaid his model onto this city's expanding and vibrant landscape.

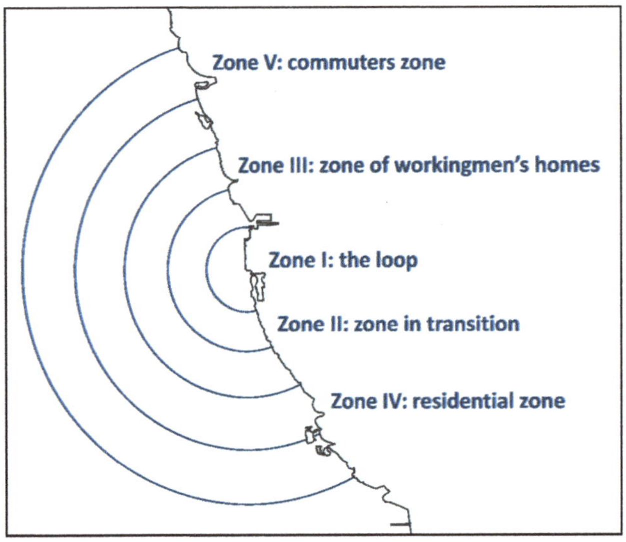

Figure 1.1. Burgess's concentric (zonal) circle model.

From a criminological perspective, the zone in transition was the area of most initial interest. Here, mobility was greatest, the availability of stimulus peaked, and there was a concentration of "juvenile delinquency, boys' gangs, crime, poverty, wife desertion, divorce, abandoned infants, [and] vice" (Burgess 1925, 59). Shaw and McKay (1942) went on to map the concentric zones using different bandwidths for different cities (e.g., in Chicago, the bandwidths were two miles, whereas in Philadelphia, they were one and a half miles). Their work spanned decades and formed the basis for much of American criminological thought into the latter half of the 20th century, especially in terms of explaining how delinquent areas become established (Brantingham and Brantingham 1981). In a meso sense, these studies led to the suggestion that as well as crime not being random in its spatial distribution, crime concentrates in certain areas. The studies also suggested that these high-crime areas persisted for some time, influenced by demographic and social mobility, and the availability and concentration of certain stimuli in these locations.

Why is crime likely to persist in certain areas?

If the concentration of the factors that influence why crime takes place in some locations and not in others remains unchanged, crime is likely to continue in these areas. In addition, if the factors that create favorable conditions for crime to occur can be identified, these may act as useful variables for determining where crime is most likely to occur.

The Burgess model worked well for North American cities in the 20th century but was less applicable outside this area. The development of North American cities took place over a fairly short time, whereas European urban development occurred over a considerably longer time. In more recent decades, cities in Latin America, Africa, the Middle East, and Asian countries have seen even more rapid development than experienced in North American cities during the 20th century. Urban geographers since have developed models that better explain the mosaic pattern of development in urban areas outside North America.

Knox (1994) describes an urban geography model of city expansion more fitting for a wider range of urban developments. This model includes stages of infilling after initial urbanization (in areas of the city where vacant land remains), downgrading when the housing stock ages, a thinning-out phase (characterized by rapid population turnover), and a final stage of rehabilitation or gentrification. From a crime perspective, the initial urbanization brings in many young families, which in turn may result in children in these areas reaching their teenage (and highest crime risk) years at the same time. The downgrading and thinning-out stages may also result in increases in crime because of social and structural neglect accelerating opportunities for crime, and the breakdown of social cohesion in the area because of population turnover. These points further strengthen the argument that if the specific theoretical conditions that give rise to crime can be identified, determining where these conditions exist may offer value in determining where crime is more likely to occur. The rehabilitation or gentrification stages may result in a reduction in crime, caused by the reintroduction of a more affluent population that can afford upgraded security features on their homes and other property (Knox 1994).

Social disorganization and collective efficacy

A theory that grew from the Chicago School was social disorganization. Social disorganization theory posits the idea that increased levels of delinquency, especially juvenile delinquency, exist because of the lack of a local social fabric in which the structure and culture of the community are not strong enough to provide a concerted influence over local residents (Shaw and McKay 1942). For example, social disorganization theory suggests that if there is a high degree of cultural heterogeneity and a high turnover of residents, the community is unlikely to agree to a common standard for behavior on the street, and that few residents are likely to know the young people on the street who are causing trouble, or know their families. With no clear rules of acceptable behavior and few sanctions available to curb adolescent exuberance (i.e., you cannot tell a child to stop misbehaving or you'll tell his parents, if you and the child both know that you do not know his parents), juvenile delinquency increases.

The practical testing of social organization theory has had its challenges because it can be difficult to construct variables that directly measure social disorganization (Chainey and Ratcliffe 2005). For example, it is unlikely that a household survey would be successful if it asked local residents to rate from 1 to 10 the level of social disorganization in their neighborhood, because few people would have a clear notion of what social disorganization is. Additionally, even if social disorganization could directly or indirectly be measured (using proxy variables), organizing a police response activity to affect this issue and improve the crime situation would be a challenge. Policing strategies are often limited when faced with these more systemic causes of crime and instead require contributions from other agencies involved in supporting social and community development.

In response to the difficulties in measuring social disorganization, some researchers have tried to measure its reverse, collective efficacy (sometimes referred to as *social efficacy*). Collective efficacy can be defined as the "social cohesion among neighbors combined with their willingness to intervene on behalf of the common good" (Sampson, Raudenbush, and Earls 1997, 918). Collective efficacy can be found in areas in which neighbors cooperate on issues of mutual interest, share some areas of agreement with the people who live around them, and are prepared to intervene if local youths behave in a manner unacceptable to local norms. Such levels of cooperation require enough implicit or explicit communication between neighbors to define and agree on the standard for local normative behavior. It is, therefore, argued that areas high in collective efficacy are well suited to resisting crime because some control can be exercised when confronted with the threat of crime or delinquent behavior.

Collective efficacy has been measured directly using community-based surveys that attempt to measure social and institutional neighborhood processes. Collective efficacy is related to the notion of *social capital*, a feature that some researchers have operationalized as the number of interactions that take place between neighbors. Social capital is a measure of the skills and social position that a person possesses that provide the power to effect a positive social change on the local environment. Sampson and his colleagues have led the research in this area, actively seeking to measure collective efficacy. Their survey of more than 8,000 Chicagoans included asking the respondents if they believed it was likely that neighbors could be counted on to intervene if children were spray-painting a local building or if a fight broke out (Sampson, Raudenbush, and Earls 1997). Their study also included census measures of race, poverty, and immigration, home ownership, and residential stability (among others) and concluded that collective efficacy could be reliably measured and could help residents control the level of violent crime. Sampson (2012) has built on these studies by further illustrating the relationships between crime and social conditions in Chicago.

Meso-level explanations of crime

Meso-level (neighborhood) explanations for crime from the Cartographic and Chicago Schools suggest that crime patterns can be interpreted if the factors that create favorable conditions for crime to occur are identified. Although some difficulties may exist in measuring some of the variables that have emerged from the thinking in the Chicago School (and more latterly, social disorganization, collective efficacy, and social capital), the identification of these factors may help inform the options for directing strategic policy that has a long-term, sustainable impact on reducing crime in the areas where crime endures. In the chapters that follow, spatial analytical techniques are presented that can help determine the factors that create favorable conditions for crime to occur.

The GIS School

Between the 1970s and 2000s, interest increased in environmental criminology, spatial crime analysis, and the investigation of offender patterns using geographic tools. The catalyst for this enthusiasm has been attributed to the development of the theory around defensible space, and the related principles associated with Crime Prevention through Environmental Design (CPTED) (Brantingham and Brantingham 1981, Jeffery 1971, Newman 1972). Building on the framework of CPTED and defensible space, several important advances were made during the early stages of this period, most notably:

- The theoretical developments of routine activity from Cohen and Felson (1979)
- The advances in crime pattern theory made by Brantingham and Brantingham (1981; 1984)
- The early application of spatiotemporal analysis techniques to crime by LeBeau (1987; 1992)
- The examination of crime across different spatial scales by Harries (1980)

These advancements helped fortify the theoretical principles of environmental criminology. However, it was the development of affordable geographic information systems (GIS) and the increasing technological developments in policing (such as the digitization and geocoding of crime records) that have allowed crime researchers and analysts to use the wealth of data recorded by police agencies and examine spatial patterns in this data. The next section examines the theoretical developments from environmental criminologists, a group referred to as the GIS School (Chainey and Ratcliffe 2005). These developments include the routine-activity approach, the rational-choice perspective, crime pattern theory, the least-effort principle, and the concepts of crime generators and attractors.

The theoretical approaches developed by environmental criminologists was a break from much of the criminology that came before, in which the theoretical focus had been toward explaining why people became involved in crime. Instead, environmental criminology was more oriented to the crime event, considering the nature of the crime offense, the offender's behavior, the risk of victimization, and the place where the crime event occurred. An environmental criminology approach encouraged the need for empirical examination of crime events, following the notion that the behavior of offenders was not random, their choice of victims was not random, and the risk of being a victim was unlikely to be random. These nonrandom features would then suggest that certain individuals or groups of people act in a manner that makes them more vulnerable to crime. In turn, this vulnerability means that the patterns observed in crime data can be used in some way to help better understand offending behavior, victimization, and why crimes take place at certain locations rather than others.

The routine-activity perspective

The routine-activity perspective originally started as a macro-level explanation of predatory crime (Cohen and Felson 1979) but has progressed over the years to provide a worthwhile mechanism to consider criminal opportunity more widely. The original work examined changing patterns of employment and the criminal opportunities that were created when fewer people stayed home during the day because they were out at their places of work.

The routine-activity perspective is based on the simple idea that for a crime to occur, three components are necessary: the presence of a likely offender, the presence of a suitable target, and the absence of a capable guardian. The target does not necessarily have to be a person, but instead could be buildings, vehicles, or other property and objects. Similarly, a guardian may not be just a person, such as a police officer, security guard, or even a neighbor, but could include closed-circuit television surveillance systems or a car alarm. The three components—offender, target, and lack of a guardian—must meet in time and space to provide the necessary chemistry for crime (Felson 1998). This meeting in time and space is not random but is dictated by the natural rhythm of daily life, involving people going about their routine activities.

The routine-activity perspective does not discuss only offenders, targets, and guardians, but adds important qualifiers to each of these components. That is, not all offenders are *likely* offenders, as some will lack the technical knowledge and skill to attack certain types of targets. Similarly, not all targets are *suitable* targets, as they may be inaccessible (such as a car parked in a locked garage) or too well defended. And although many objects and people can be guardians, at different times they may not be *capable* guardians. The routine-activity approach can be summarized by the following simple equation, (1.1):

$$\text{Likely offender} + \text{suitable target} - \text{capable guardian} = \text{a crime opportunity} \qquad (1.1)$$

The combination of these three components and their qualifiers then dictates that the risk of crime changes over time with the movement of people throughout the daily activities of their lives. That is, as these meetings in space and time are not random, the patterns observed in data could be considered in relation to these routine-activity principles.

Since Cohen and Felson's original work, Eck (1995) recast the concepts of the routine-activity approach into the crime problem analysis triangle (figure 1.2). The crime triangle introduced place as the third side of the triangle, with offenders and targets/victims forming the other two sides. This recasting of the routine-activity approach also led to the expansion of the original concept of guardianship by introducing the general term of *controller*. The concept of controllers is illustrated in figure 1.2 by the positioning of guardians in relation to victims and targets. Controllers were then added to the offender and place sides of the crime problem analysis triangle. For offenders, Felson (1995) introduced the concept of *handlers*. A handler is a person, a third party, who can influence the behavior of the offender. For example, a parent may be a handler, as may a teacher or any other person who knows or who could determine the name and identity of the person, and whose respect the offender might not want to lose. For the place side of the crime triangle, the concept of *place managers* was introduced (Eck 1995). Place managers are those who can control a place, even if they are not formally in charge of the area. In figure 1.2, the concept of place manager is redefined instead as *place management*. This redefinition is suggested because controlling a place need not be determined by the presence of a person (a place manager) but instead can be controlled by the better management of that place. For example, changing the physical design of an area (such as implementing security barriers in a parking lot or parking garage that make it more difficult for cars to be stolen) can make it less conducive to crime by improving the controls that make the place easier to manage.

Figure 1.2. The crime problem analysis triangle and associated controllers.

Crime opportunity and controllers

In practice, the crime problem analysis triangle helps focus analysis on the causes of crime, from a routine-activity and crime opportunity perspective, and the mechanisms that can influence those causes. The inclusion of controllers around the three sides of the triangle then helps illustrate that if criminal opportunity can be explained by drawing on theoretical perspectives, activities that involve one or more types of controllers can counter and minimize the crime opportunity.

The routine-activity perspective, therefore, indicates that the crime opportunities that emerge from the coming together of offenders, targets, and guardians, dictated by the rhythm of daily life, are unlikely to be random and instead form patterns in space and time. Following this consideration of the patterning of behavior (based on routine-activity principles), the occurrence of crime also could be prevented, as Felson (1995, 55) states, "Crime opportunity is the least when targets are directly supervised by guardians; offenders by handlers; and places by managers." Therefore, if geographic analysis techniques can observe spatial patterns in crime that can be explained in routine-activity terms, the interpretation of these patterns can help identify effective means for removing these criminal opportunities.

Target suitability, however, can change over space and time. Although the routine-activity perspective provides a general model for considering the likelihood of crime occurrence, available targets can differ in their attractiveness to a criminal. Clarke's (1999) CRAVED summary of the basic characteristics of a suitable, or hot, target provides a useful framework for determining the types of property that are at greater risk of being stolen.

CRAVED—the characteristics of targets that are prone to theft

- Concealable—things that easily can be concealed after being stolen are at more risk of being stolen.
- Removable—the easier a thing is to remove, the greater the risk of it being stolen.
- Available—objects that are plentiful are most at risk.
- Valuable—the more valuable an item is deemed to be and the higher its possible resale value on the stolen-goods market, the higher the risk of it being a target for theft.
- Enjoyable—the degree to which a product can be enjoyed determines its potential risk of theft.
- Disposable—many items are stolen so that they can be sold or traded to others; therefore, disposability is an important characteristic of an item's risk of theft.

Ratcliffe (2002) showed how the ideas of CRAVED products and the routine-activity approach could help explain vehicle crime in Sydney, Australia. Ratcliffe's analysis identified that vehicle crime in the affluent suburbs near the beaches was highest during the overnight periods, and that expensive cars were the main targets. The targeting of these cars fit the CRAVED criteria, not only because the vehicles were valuable but also because few residents in these areas had private garages where their cars could be parked (making them easier to remove). Vehicle crime was concentrated in the overnight period because the cars were suitable targets, were more available to thieves at night (i.e., the cars were being used by their owners during the day but were available on driveways or parked on the street at night), and the nighttime hours provided for few capable guardians.

The opportunity for crime and the spatial arrangement of targets under the principles of routine activities and CRAVED suggest that if patterns from recorded crime data identify targets that are prone to theft, these targets likely will continue to be high-risk items for theft unless one or more features of their CRAVED profile are addressed. The distribution of these attractive targets is not random, so an analysis of the spatial patterns of where items are stolen could explain why thefts take place in certain locations and not others. The initial challenge, therefore, is to apply spatial analysis techniques that help reveal where thefts take place. By doing so, the areas where these attractive targets are most prevalent is where prevention initiatives could be of most benefit. Thinking in terms of these theoretical principles can help interpret the patterns observed in data and help indicate potential ways to counter the crime activity.

Using CRAVED to consider how to address a property crime problem

Although CRAVED provides a useful framework for considering why certain items may be prone to theft, it also can help evaluate what can be done to address a property crime problem. For example, consider the theft of cell phones. What can the police do to reduce this crime by making it a less appealing target for theft? In practice, the police can do little to make the phone less concealable, more difficult to remove, less available, less valuable, and less enjoyable, but they could do something to address the disposability of the phones by focusing on the stolen-goods market. That is, make the stolen phones more difficult to sell by focusing investigative and enforcement efforts on the stolen-goods market. Although many of these selling markets may be on online, several are physically located in space. This factor provides another opportunity for spatial analysis to identify where these markets are so that crime reduction efforts can target these locations.

A review of the components of CRAVED also helps determine why there are changes over time in items that are prone to theft, and why certain items are rarely stolen. In the early 1990s, cell phones were not prone to theft because very few people owned them—they were not *available.* Since the 1990s, vehicle thefts have been reduced significantly because of improvements in the security of vehicles, making them more difficult to *remove.* Microwave ovens tend not to be stolen because there isn't really a stolen-goods market for them (*disposable*), and because they are inexpensive to replace (*valuable*).

When examining the spatial distribution of crimes that involve theft, CRAVED provides a useful framework for considering why certain areas are prone to theft more than others. For crimes against the person that do not involve theft, such as violent crime, different theoretical considerations are required. However, the notion of targets being *available* is important when considering crimes of violence and the geographic distribution of these targets. For example, an investigation into a series of sexual assault in Bath, England, identified that the offender would commit his crimes on Friday or Saturday nights and would focus his hunt for potential victims in one particular area of the city center because this was the time and place where potential targets were most *available*.

Rational choice

The rational-choice perspective (Clarke and Felson 1993; Cornish and Clarke 1986) provides a framework to consider offender decision-making when a crime opportunity is presented. It also can be used to consider likely strategies that will influence the decision-making of the offender, and hence counter their criminal behavior. Most offenders are known to make some sort of decision to commit a crime by weighing some of the pros and cons (i.e., the rewards against the chance of being caught) (Bernasco 2010; Clarke and Felson 1993; Cornish and Clarke 1986). This decision-making suggests that committing a crime is a (fairly) rational decision, and that an offender will commit an offense while trying to achieve some sort of desire or goal (Kennedy 2009). The goal may be to derive some sort of gain, as in the theft of a valuable item, a violent act against a rival group, or to achieve dominance over a partner as in the act of domestic abuse. Alternatively, the goal may be to seek personal pleasure or satisfaction as in sexual assault, harassment, or inflicting harm on a person in some other way. If the legitimate means of obtaining the offender's goal are not available, a decision may be made when a criminal opportunity becomes available. The decision may not be one that is fully calculated, as offenders may not consider all the consequences.

The criminal decision-making process generally can be considered to involve two parts. The first involves a long-term, multistage decision to become generally involved in criminal activity (*the criminal involvement decision*). The second part is a shorter-term, more immediate decision (*the criminal event decision*) to grasp an opportunity that is presented (Cornish and Clarke 1986). Factors such as alcohol, drugs, peer pressure, or limited education mean, of course, that not all offenders' decisions are purely rational, resulting in what has been termed *limited rationality*, also known as *bounded rationality* (Newman 1997). Although offenses are not necessarily committed with a high degree of planning, it is still useful to consider what factors influence decision-making in the commission of crime.

A useful framework for evaluating the decision-making that an offender is required to make is to think of choices that relate to effort, risks, rewards, ease of excuse, and provocations. Those with a keen eye who are familiar with situational crime prevention will recognize these five terms. Here, they are included and considered from an offender decision-making perspective to help explain components associated with rational choice theoretical principles. Table 1.1 lists several examples under each of these five terms, such as whether the location is easy for the offender to get to (*effort*), whether it is easy for the offender to be anonymous and avoid suspicion (*risk*), whether there are places to sell stolen items (*reward*), whether the rules for acceptable, legal behavior are not clear (*excuses*), and whether the offender is encouraged or pressured into committing a crime by the people with whom he or she associates (*provocations*). The effects of drugs and alcohol can limit the rationality of offenders (as a *driver* that influences why they decide to commit an offense), but what remains is the immediate decision-making an offender still is required to make when presented with an opportunity to commit a crime. For example, a person intent on committing a street robbery, while also being high on drugs at the time, is unlikely to behave in a completely random manner when at the point of crime commission. Instead, based on rational-choice perspective principles, the offender would consider whom best to target when in an area in which opportunities exist, and the chances of success in the commission of the crime (e.g., no presence of passersby to intervene, avoiding apprehension, the target not fighting back and injuring the offender, the target has an item that is worth stealing). In a spatial sense, the environment within which the offender operates varies from place to place, and so too, therefore, will the patterns of crime.

Table 1.1. Five terms associated with rational choice that influence offender decision-making

Effort	Rewards	Risks	Provocations	Excuses
• Location is easy to get to • Facilities are easy to access • Targets are easy to find • Targets are easy to remove • Tools/weapons are easy to get hold of	• Good financial benefit • Good demand for products • Places to sell stolen items • Imposition of power and dominance	• No one to intervene • No or little natural surveillance • No or little formal surveillance • Easy to be anonymous • No one appears to "control" the area	• Frustrations and stresses • Dispute • Retaliation • High emotional arousal • Peer pressure • Imitation of others	• Rules are not clear • No instructions on unacceptable/illegal behavior • Easy access to drugs and alcohol (as a "driver" to influence behavior) • Influence of other "drivers" (e.g., "I am poor, which is why I offend")

Crime pattern theory

Whereas the routine-activity approach provides a model to determine whether a crime has all the chemistry to occur, and the rational-choice perspective enables us to determine some of the thinking behind an offender's ultimate decision to commit a crime, crime pattern theory can help explain where and when crime is likely to occur. Crime pattern theory helps bring together the two areas of offender spatial distribution and offense spatial distribution by examining the "relationship of the offense to the offender's habitual use of space" (Bottoms and Wiles 2002, 638).

Crime pattern theory suggests that offenders are influenced by the daily activities and routines of their lives, so that even if they are searching for a criminal opportunity, they will tend to steer toward areas that are known or most familiar to them (Brantingham and Brantingham 1984). Two important theoretical concepts associated with crime pattern theory are the least-effort principle and the distinctions between how crime opportunities concentrate. These two theoretical concepts are considered in this section after an explanation of the other main theoretical principles associated with crime pattern theory.

Like offenders, we all have various routine activities in our lives. Most of us go to work, college, or school, and we usually go there from home. We also may go to shops and restaurants, bars, and cinemas. These places from which we carry out most of our activities are termed *nodes*, and connecting these nodes are *pathways* (the routes we take between our nodes). These frequent activities around our nodes, and the journeys we take along the pathways, create within us a "cognitive map" (Brantingham and Brantingham 1984, 358) of places, routes, and associations. Over time, these cognitive maps become a general list of well-known areas in which we feel comfortable. This environment consists not only of physical things, such as buildings, streets, and transport centers, but also the social and economic infrastructure through which we pass. Cities become an urban mosaic to us, with places in which we have no knowledge interspersed with well-known places (Chainey and Ratcliffe 2005). We also become familiar with the routes between these known areas. These islands of knowledge, and the routes that link them, become our *awareness space* (Brantingham and Brantingham 1981, 35).

Like us, offenders also have awareness spaces. They also move between places such as work, school, shops, and home, and for many offenders, the search for criminal opportunities takes place around these areas. Opportunities are not spaced evenly throughout the landscape, and some offenders will be able to take advantage of only some offense opportunities. Also, some of the awareness spaces will not be conducive to crime because of the presence of guardians or effective place management. Therefore, for each offender, we can formulate a model of awareness space and criminal opportunity space (with the implicit absence of guardianship). Where awareness space and criminal opportunity space intersect for each offender, we will find the areas of his or her crime occurrence. Crime pattern theory is, therefore, strongly connected with the interactions of criminals, and their physical and social environments.

The theoretical principles of crime pattern theory were modeled by Brantingham and Brantingham (1984), and an adaptation of the principles by Chainey and Ratcliffe (2005) can be seen in figure 1.3. Awareness space consists mainly of the places that are routinely frequented, and the routes between those places. We may be more familiar with distant places that we frequent regularly than local places just around the corner that we never visit; therefore, proximity does not always mean the same thing as familiarity.

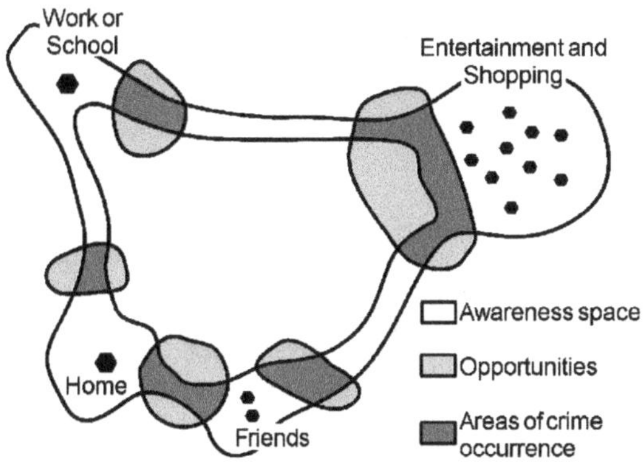

Figure 1.3. Hypothetical model of the creation of criminal occurrence space in which offender awareness space and opportunities coincide. Source: Chainey and Ratcliffe 2005.

The adaptation by Chainey and Ratcliffe (2005) of the original Brantinghams' model shown in figure 1.3 includes the location of friends as an influential component on offender awareness space. Although, for most people, work or school play a significant part in daily life, this can be less so for offenders. Wiles and Costello (2000) found, for example, that many offenders in Sheffield, England, who could have been in the workforce had never had full-time employment, whereas Rengert and Wasilchick (2000) reported interviews with many offenders who had quit legitimate employment to pursue a professional criminal career. These examples focused on residential burglars only; however, it is likely that the attractiveness of both lifestyle and the income of a life of crime are greater than poorly paid legitimate employment for many burglars, drug dealers, and robbers (Chainey and Ratcliffe 2005). Having a place of work is not a necessary requisite for forming an awareness space. The Sheffield study found that many offenders never had been employed, were often transient, and were influenced heavily by the location of friends and criminal peers (Wiles and Costello 2000).

Figure 1.3 also shows that some opportunities are outside the awareness space of the offender and are essentially unavailable. It is also important to consider that opportunities within the awareness space of an offender may have variable attractiveness. For example, for an offender who is intent on stealing a cell phone from a person, all people within their awareness space are theoretically potential targets, but some are more attractive targets than others. The areas with a greater proportion of targets that are potentially easier and more profitable are located within the opportunities space shown in figure 1.3. Within this opportunities space, certain targets will be more attractive than others, with the search for a specific target occurring within this area. This then means, according to crime pattern theory, that where an offender is most likely to commit his or her crimes will be in the areas in which the offender's awareness space and opportunity spaces overlap. These areas are indicated in figure 1.3 as the *areas of crime occurrence.*

Crime pattern theory is a general explanation of the crime patterns of offenders, and exceptions will always exist. Studies have found that some offending does not take place within an offender's own awareness space because of the influence that other people may have in determining where there are better opportunities to commit crime (such as when an offender co-offends with a person who is more experienced in committing crime) or by peers who introduce the offender to new areas and new opportunities (Rengert and Wasilchick 2000; Wiles and Costello 2000). When this happens, the spatial pattern of offenses can be different. However, for most offenders, the tendency is to offend in their awareness spaces.

There are several other reasons why offenders might commit offenses in familiar areas. It is helpful to know the layout of an area so if a quick getaway is required, the offender can avoid running up a dead-end street. Second, it has been suggested that offenders value feeling comfortable in an area and not feeling as if they stand out. This feeling of comfort has been suggested as a reason why offenders who live in poor neighborhoods do not often commit offenses in affluent areas (Rengert 1989). However, if a more affluent area immediately neighbors a poor area where offenders reside, the familiarity between the borders of the two may result in higher crime rates (Hirschfield, Bowers, and Brown 1995). In a similar vein to the poor-rich distinction are the white-black neighborhood distinctions found in the burglary patterns of offenders in the United States. Several studies have noted that white offenders avoid black neighborhoods and black offenders avoid white neighborhoods, each group feeling that the avoided areas were unsafe (Rengert and Wasilchick 2000; Wright and Decker 1994).

An important concept central to crime pattern theory is the general theoretical principle of least effort (Zipf 1949). In least-effort terms, physical space can be likened to a friction surface, in that it requires effort to cross it. The farther the distance to travel, the greater the cost in time and possibly money, with people usually exerting the minimum effort possible to complete their tasks. The least-effort principle, therefore, has an influence on the spatial behavior of offenders by explaining why there are spatial limits to the geographic coverage of an offender's awareness space, and why an offender chooses a certain location to commit crime. For example, the least-effort principle explains why, in figure 1.3, there are boundaries to the extent of the offender's awareness space and his or her knowledge of criminal opportunities.

While increased distance to commit crime increases the effort, it also increases the risk of crime commission and increases the possibility that the offender will stray into an unknown area. This helps explain why an offender's desire for spatial exploration to extend criminal opportunities is rare (Rossmo 2000). Offenders often are constrained by time and financial resources and lack the freedom to explore other opportunities or search farther afield for fresh prospects. The least-effort principle also can be used to explain that if opportunities for an offender to commit crime are equally available in two areas yet one area is much farther away, the offender is more likely to choose the nearest offense opportunity to minimize his or her effort in crime commission (Chainey and Ratcliffe 2005). This theoretical concept of minimizing the distance traveled to commit crime is supported by a wealth of empirical research that has shown that most offender journeys to commit crime tend to be short. For example, Wiles and Costello (2000) showed that offenders of burglaries to residential properties in Sheffield, England, on average traveled 1.9 miles from their home to their crimes. The least-effort principle is, therefore, a useful mechanism for thinking about the geographic extent of an offender's spatial behavior and the crime patterns that may result. In spatial crime analysis terms, it causes us to hypothesize that the person who committed an offense is likely to be local and familiar with the area, or in aggregate form, that the many offenses committed by individuals in the same area are most likely to involve offenders who live close to, or have a good degree of familiarity with, the area. In another sense, our understanding of the least-effort principle in relation to crime patterns can encourage us to ask if there is anything about the crimes being committed that suggest the offender is *unlikely* to be local and possess no prior familiarity of where the offenses have taken place. For example, would offenders active in oil tapping from pipelines operate in a completely random manner in choosing their locations to commit these offenses, or is the location of these offenses influenced by the effort it takes to travel to where these pipelines are located and any necessary familiarity an offender requires to be successful in this crime commission?

An offender's choice of location to commit a crime, and the patterns that result, additionally can be explained in terms of how opportunities to commit crime concentrate (i.e., how the opportunity space, shown in figure 1.3, forms). The concentration of crime opportunities can be explained in two main ways: crime generators and crime attractors. A crime generator is an area in which large numbers of people or other targets are drawn (Brantingham and Brantingham 1995). These places are crime generators because they provide times and places of many opportunities for offenders. For example, busy train stations and shopping areas may experience high levels of crime principally because of the large number of place users and targets at these locations (Clarke and Eck 2003). By comparison, crime attractors are places that create specific criminal opportunities and, in doing so, attract motivated offenders seeking these specific opportunities (Brantingham and Brantingham 1995). The lure of a known criminal opportunity draws offenders to the area, enticing them with the knowledge that the area has a reputation for a specific type, or types, of illicit opportunity. For example, areas close to schools provide an attractive location for street robbers looking to target lone schoolchildren on their walk home and robbing them of their cell phone.

Crime generators and crime attractors, and geographic scale

Crime generators and crime attractors offer definitions to distinguish between areas in which crime is likely to concentrate. They are typically used in a meso- and micro-geographic sense to explain that the high levels of crime are the result of many targets being available, or specific opportunities exist for criminal commission.

Crime pattern theory provides a model to help explain why an offender chooses a certain location to commit a crime. It illustrates that this decision-making is informed by the offender's awareness of the area and the familiarity of opportunities, and by his or her previous experience of crime commission. Crime pattern theory, therefore, provides a further theoretical explanation for why spatial patterns of crime are not random. Although the theoretical focus of crime pattern theory is on explaining the geographic patterning of crime committed by individual offenders, the aggregation of many offenders being attracted to similar targets means the theory also can help explain aggregate geographic patterns of crime that may be observed in data. The application of geographic analysis techniques then may help us identify these aggregate spatial patterns from which the principles of crime pattern theory can be drawn to help explain why crime occurs in certain locations, and determine from this synopsis offender behavior patterns.

Key takeaways on the geographic theories of crime

Crime pattern theory and the least-effort principle provide a framework for understanding the criminal spatial landscape. The awareness spaces referred to in crime pattern theory are made up of different structures that provide a changing pattern of opportunities to commit offenses depending on the environmental backdrop—the social, psychological, economic, physical, and temporal landscape the offender passes through.

These structures change across space and affect the type of criminal opportunities that are available to motivated offenders. The motivations of offenders, though, are bounded by the rationality in their decision-making. Offenders prefer to commit offenses in areas in which they are comfortable and, by their preference, to minimize effort. The theoretical explanations for the criminal spatial landscape, therefore, provide the source for interpreting the spatial patterns observed in crime data and determine why crime may occur in certain locations and not others.

Chapter summary

Concepts such as the routine-activity approach, rational choice, crime pattern theory, the least-effort principle, crime attractors, and crime generators are valuable in conceptualizing the underlying criminal landscape and the spatial decision-making processes of offenders. The theoretical principles developed from the Cartographic and Chicago Schools, namely social disorganization and collective efficacy, further illustrate that the geographic distribution of crime is not random and that neighborhood-level explanations can contribute to helping understand why crime takes place at certain locations rather than others.

The inherent spatial quality to crime, and the knowledge that crime tends to occur at certain locations rather than being randomly distributed, places us in a better position to consider how to use spatial analysis techniques to identify and examine patterns of crime. Theory is vital for the analysis of crime because it helps us move from just providing interesting descriptions about the crime patterns we observe to interpreting and understanding why these patterns occur.

References

Bernasco, W. 2010. *Offenders on Offending: Learning about Crime from Criminals*. Cullompton: Willan.

Bottoms, A. E., and P. Wiles. 2002. "Environmental Criminology." In *The Oxford Handbook of Criminology*, third edition, edited by M. Maguire, R. Morgan, and R. Reiner. Oxford: Oxford University Press.

Brantingham, P. J., and P. L. Brantingham. 1981. *Environmental Criminology*. London: Sage.

Brantingham, P. J., and P. L. Brantingham. 1984. *Patterns in Crime*. New York: Macmillan.

Brantingham, P. J., and P. L. Brantingham. 1995. "Criminality of Place: Crime Generators and Crime Attractors." *European Journal of Criminal Policy and Research* 3:5–26.

Burgess, E. W. 1916. "Juvenile Delinquency in a Small City." *Journal of the American Institute of Criminal Law and Criminology* 6:724–28.

Burgess, E. W. 1925. "The Growth of the City: An Introduction to a Research Project." In *The City*, edited by R. E. Park, E. W. Burgess, and R. D. McKenzie, 47–62. Chicago: University of Chicago Press.

Chainey, S. P., and J. H. Ratcliffe. 2005. *GIS and Crime Mapping*. London: Wiley.

Clarke, R. V. 1999. "Hot Products: Understanding, Anticipating, and Reducing Demand for Stolen Goods." *Police Research Group: Police Research Series*, paper 112.

Clarke, R. V., and J. Eck. 2003. *Become a Problem-Solving Crime Analyst in 55 Small Steps*. London: University College London.

Clarke, R. V., and M. Felson. 1993. "Introduction: Criminology, Routine Activity, and Rational Choice." In *Routine Activity and Rational Choice*, edited by R. V. Clarke and M. Felson, 259–94. New Brunswick, NJ: Transaction Publishers.

Cohen, L. E., and M. Felson. 1979. "Social Change and Crime Rate Trends: A Routine Activity Approach." *American Sociological Review* 44:588–608.

Cornish, D., and R. Clarke. 1986. *The Reasoning Criminal: Rational Choice Perspectives on Offending*. New York, NY: Springer-Verlag.

Eck, J. 1995. "A General Model of the Geography of Illicit Retail Marketplaces." In *Crime and Place*, edited by D. Weisburd and J. E. Eck, 67–93. Monsey, NY: Criminal Justice Press.

Felson, M. 1995. "Those Who Discourage Crime." In *Crime and Place*, edited by D. Weisburd and J. E. Eck, 53–66. Monsey, NY: Criminal Justice Press.

Felson, M. 1998. *Crime and Everyday Life: Impact and Implications for Society*. Thousand Oaks, CA: Pine Forge Press.

Guerry, A.-M. 1833. "Essai sur la statistique morale de la France." Précédé d'un rapport à l'Académie des Sciences. Paris: Chez Crochard.

Harries, K. 1980. *Crime and the Environment*. Springfield, IL: Charles C. Thomas.

Hirschfield, A., K. J. Bowers, and P. J. B. Brown. 1995. "Exploring Relations between Crime and Disadvantage on Merseyside." *European Journal on Criminal Policy and Research* 3:93–112.

Jeffery, C. R. 1971. *Crime Prevention through Environmental Design*. Beverly Hills, CA: Sage Publications.

Kennedy, D. M. 2009. *Deterrence and Crime Prevention: Reconsidering the Prospect of Sanction*. New York, NY: Routledge.

Knox, P. L. 1994. *Urbanization: An Introduction to Urban Geography*. Englewood Cliffs, NJ: Prentice-Hall.

LeBeau, J. L. 1987. "The Methods and Measures of Centrography and the Spatial Dynamics of Rape." *Journal of Quantitative Criminology* 3:125–41.

LeBeau, J. L. 1992. "Four Case Studies Illustrating the Spatial-Temporal Analysis of Serial Rapists." *Police Studies* 15:124–45.

Mayhew, H. 1862. *London Labor and the London Poor*. London: Griffin Bohn.

Newman, G. 1997. "Introduction: Towards a Theory of Situational Crime Prevention." In *Rational Choice and Situational Crime Prevention: Theoretical Foundations*, edited by G. Newman, R. Clarke, and S. G. Shoham, 1–23. Dartmouth, England: Ashgate.

Newman, O. 1972. *Defensible Space: Crime Prevention through Urban Design*. New York: Macmillan.

Quetelet, A. 1842. *A Treatise in Man*. Edinburgh: Chambers.

Ratcliffe, J. H. 2002. "Aoristic Signatures and the Temporal Analysis of High Volume Crime Patterns." *Quantitative Criminology* 18:23–43.

Rengert, G. 1989. "Behavioral Geography and Criminal Behavior." In *The Geography of Crime*, edited by D. J. Evans and D. T. Herbert, 161–75. London: Routledge.

Rengert, G. F., and J. Wasilchick. 1985. *Suburban Burglary: A Time and Place for Everything*. Springfield, IL: C. C. Thomas Publishing.

Rengert, G. F., and J. Wasilchick. 2000. *Suburban Burglary: A Tale of Two Suburbs*. Springfield, IL: C.C. Thomas Publishing.

Rossmo, K. 2000. *Geographic Profiling*. Boca Raton, FL: CRC Press.

Sampson, R. J. 2012. *Great American City: Chicago and the Enduring Neighborhood Effect*. Chicago: University of Chicago Press.

Sampson, R. J., S. W. Raudenbush, and F. Earls. 1997. "Neighborhoods and Violent Crime: A Multilevel Study of Collective Efficacy." *Science* 277:918–24.

Shaw, C. R., and H. D. McKay. 1942. *Juvenile Delinquency and Urban Areas*. Chicago: Chicago University Press.

Wiles, P., and A. Costello. 2000. *The "Road to Nowhere": The Evidence for Traveling Criminals*. London: Home Office.

Wright, R., and S. Decker. 1994. *Burglars on the Job*. Boston, MA: Northeastern University Press.

Zipf, G. K. 1949. *Human Behavior and the Principle of Least Effort*. Oxford: Addison-Wesley Press.

CHAPTER 2

Statistical measures for spatial patterns, and the journey to crime

Objectives

In this chapter, we introduce statistical measures that help examine general spatial patterns in crime. These statistical measures include centrographic statistics, spatial autocorrelation techniques, and tests for clustering. We also discuss techniques for measuring the journey to crime. This group of spatial statistics techniques can help exhibit how the individual behavior of offenders tends to operate toward a common aggregate tendency, and how these patterns can be observed. Simple measures are used to track spatial movements in offending behavior, observe that offenders are more likely to travel short distances in their commission of crime, measure how crime levels in one area may be more similar to crime levels in areas nearby, and for testing if hot spots are present. Although there are always exceptions to these principles (which are discussed in the chapter), the examples provide an important initial illustration of the value in examining the spatial characteristics of offender behavior and geographic patterns of crime.

Key learning points

- Examine and recognize that the spatial behavior of offenders most often aggregates toward certain common tendencies, such as how most distances that offenders travel to commit crime tend to be short and that crime patterns tend to cluster.
- Consider Tobler's First Law of Geography and its application to crime, that things that are closer together tend to be more similar than things that are farther apart.

- Introduce and provide guidance on the use of centrographic statistical techniques (such as the spatial mean, the weighted spatial mean, and the standard deviational ellipse), spatial autocorrelation techniques (i.e., Moran's I and Geary's C), tests for clustering (e.g., the nearest neighbor index), and techniques for measuring the journey to crime.

Introduction

Crime data can exhibit patterns in offending behavior, victimization, and the areas in which most crime tends to take place. If crime were purely random, there would be little use for this book. A useful place to start in examining spatial patterns of crime is to consider techniques that reveal patterns of central and common tendency in how offenders behave and how crime is spatially distributed. Although exceptions to these patterns are always likely, being aware of central and common tendency and how to measure whether they are present can reveal exceptions to these patterns. To recall a couple of theoretical principles from chapter 1 relating to crime pattern theory: it is more likely that offenders will commit crime in the places with which they are familiar rather than in places they have never been, and that places in which there are opportunities to commit crime are not available everywhere. Thus, there are restrictions on what offenders decide to do and where offenders can commit crime, and these restrictions then lead to patterns in the distribution of crime.

To begin, statistical techniques for measuring spatial patterns of central tendency are introduced, and their practical application is discussed. Next, what often is referred to as the First Law of Geography is considered in terms of its relevance to crime, including techniques for examining whether this theoretical principle is observed in patterns of crime (spatial autocorrelation techniques) and technical guidance on how to apply these techniques. Spatial autocorrelation techniques are useful in many circumstances for examining spatial patterns in data when data exists in certain forms. Spatial autocorrelation measures require data to be aggregated to a geographic unit, such as a census block. When crime data exists in the form of geographic points, other techniques can be used to examine the spatial distribution of crime. The nearest neighbor index is a technique that can be applied to point data and is useful in initially determining whether hot spots of crime are present in the crime data. Guidance is provided in this chapter on how to use the nearest neighbor index.

Another important type of aggregate offending behavior observed in crime data is the *journey to crime*—the distance offenders travel to a location in which they commit a crime. In this chapter, the key findings from this observation are presented, guidance on how to measure the journey to crime is provided, and limitations on the complexity of its use are discussed. The application of the principles of the journey to crime is also discussed, including how this observation is used for geographic profiling—a technique that helps determine where an offender is likely to live (or is spatially anchored on the basis of some other location relating to his or her life, and their crime commission).

Throughout the chapter, examples are provided to illustrate the application of each spatial technique.

Centrographic statistics: The spatial mean, weighted spatial mean, and spatial standard deviational ellipse

Centrographic statistics provide the means for examining general spatial patterns of central tendency in crime data. Centrographic techniques are simple, so they should not be expected to reveal anything sophisticated or particularly insightful that leads directly to how an offender is caught or how a crime problem is solved. However, because of their simplicity, these statistics provide a useful starting point for examining spatial patterns of crime.

Centrographic statistics can show the following general patterns of central tendency:

- The *center of gravity* of a set of crimes
- How crimes of the same type have moved in their general geographic distribution over time
- How a crime series associated with an offender has moved in its general geographic distribution over time

The spatial mean and the weighted mean center

The spatial mean (sometimes referred to as the *centroid* or *mean center*) is a simple measure of central tendency in the spatial distribution of data. The spatial mean is usually applied to point data rather than line or polygon data. The spatial mean provides an indication of the center of gravity in the distribution of the data. Figure 2.1 illustrates an example of the spatial mean applied to the distribution of 15 crime points. The spatial mean is calculated as the mean of the geographic coordinates of these 15 points and requires data to be projected for it to be measured accurately.

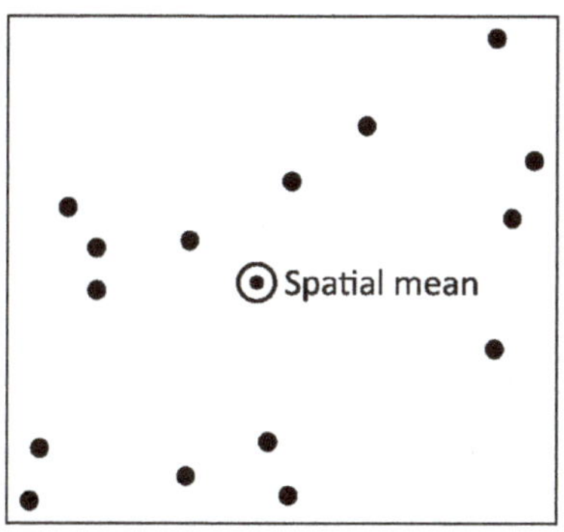

Figure 2.1. The spatial mean of 15 points.

The spatial mean was used in some of the earliest attempts at examining the spatial profiling of serial offenders. For example, LeBeau (1987) used the spatial mean to indicate the geographic movement over time of a serial rapist's offending activities in San Diego, California. The spatial mean also was used by the Yorkshire Ripper investigative review team in England to indicate that Peter Sutcliffe (the Yorkshire Ripper) was more likely to live in Bradford than Sunderland, from where the police had received a hoax phone call from a person who claimed to be the Ripper (Kind 1987). Figure 2.2 shows the application of the spatial mean over the duration of a series of eight sexual assaults in the city of Bath, England. Each black square represents the spatial mean as the series developed, from the first two offenses to the inclusion of the last of the eight offenses. Although the spatial mean cannot be relied on as an accurate investigative tool for determining the anchor points of serial offenders (Chainey and Ratcliffe 2005), it can be used to provide some indication of how a series of offenses has changed over time, and show whether a consistent center of gravity to the offenses is apparent. We discuss the application of centrographic measures for geographic profiling in chapter 7.

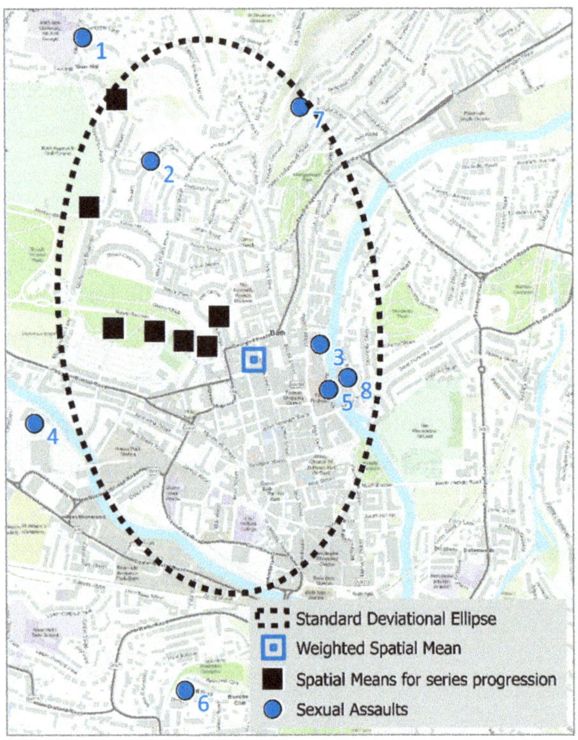

Figure 2.2. Centrographic statistics for a series of sexual assaults in Bath, England, including the movement of the spatial mean over the progression of the series, the weighted spatial mean, and the standard deviational ellipse.

The weighted spatial mean is based on the same principles as the spatial mean but gives more weight to certain points. For example, it can give more weight to recent incidents rather than to older ones, or to certain types of crime. Figure 2.2 also shows the weighted spatial mean, based on giving greater weight to more recent offenses (i.e., a weight of eight for the eighth offense and a weight of one for the first offense).

Spatial standard deviational ellipse

The standard deviational ellipse can indicate the dispersal and direction of a group of offenses around the spatial mean. Along the longer axis of the ellipse, if a line was drawn from end to end, it would pass through the spatial mean. Similarly, if a line was drawn for the shorter axis of the ellipse that joins the closest edges, this line also would run through the spatial mean. The greater the dispersal of the crime data points, the larger the ellipse. If the ellipse is close to circular in shape, the crime points are evenly distributed in all directions around the spatial mean. If the ellipse is narrow, the crime data points follow a linear pattern. Figure 2.2 shows an example of the standard deviational ellipse applied to the Bath sexual assault series. The ellipse is narrow from north to south, reflecting the underlying distribution and orientation of the eight assaults. Most tools for calculating the spatial standard deviational ellipse require the user to specify the number of standard deviations. It is advisable to start by running the tool to only one standard deviation because this is the most common parameter used in practice. Then, you can experiment with different standard deviations once you are familiar with the technique.

Spatial autocorrelation techniques

Observations close together tend to be more similar than those farther apart (Tobler 1970). The pattern described in Tobler's First Law of Geography is present in many forms of nature, economic and social behavior, and crime. This relationship between geographic observations being similar when they are close together is referred to as *positive spatial autocorrelation.*

The Moran's I statistic measures spatial autocorrelation. Moran's I compares the similarity in values between each location and its near neighbors (Eck et al. 2005; Levine 2010; O'Sullivan and Unwin 2003). Where values in a geographic unit are high and are surrounded by other geographic units with similarly high values, positive spatial autocorrelation exists. When areas with low levels of crime are surrounded by areas with high levels of crime and high-crime areas are surrounded by spatial units with few crimes, this pattern displays negative spatial autocorrelation.

The calculation of Moran's I requires an intensity value and the position (i.e., geographic coordinates) of each event to be examined. In practice, crime data must be aggregated to some form of geographic unit in which its position is represented by the centroid of the geographic unit (e.g., the centroid of a grid cell or census area). This centroid point then is assigned an intensity value that represents the number of observations (i.e., crimes) in the geographic unit (Eck et al. 2005). The Moran's I result varies between −1.0 and +1.0. When values in a geographic unit are greater than zero and surrounded by other geographic units with similarly high values, positive spatial autocorrelation exists. When areas with low values are surrounded by high-crime areas and high-crime areas are surrounded by geographic units with low levels of crime, the Moran's I result would be less than 0, indicating negative spatial autocorrelation. For these reasons, Moran's I has been used to identify whether hot spots are present in spatially distributed crime data (Eck et al. 2005; Levine 2010)—i.e., where geographic units with high levels of crime are surrounded by other geographic units with similarly high values. The statistical significance of the Moran's I result can be tested against a theoretical distribution (one that is normally distributed) by dividing by its theoretical standard deviation (Levine 2010). This determines whether the spatial autocorrelation pattern observed in the data is statistically significant.

Figure 2.3 shows the distribution of burglaries across Newcastle, England, for a one-year period. The pattern of the distribution of crime appears to show that areas in which burglary levels are high tend to be surrounded by areas that also have high levels of burglary. The Moran's I value for the distribution of these burglaries was 0.252 ($p < .01$—i.e., significant to a confidence threshold of greater than 99 percent), indicating that positive spatial autocorrelation was observed in the data. That is, areas are surrounded by areas with similar values.

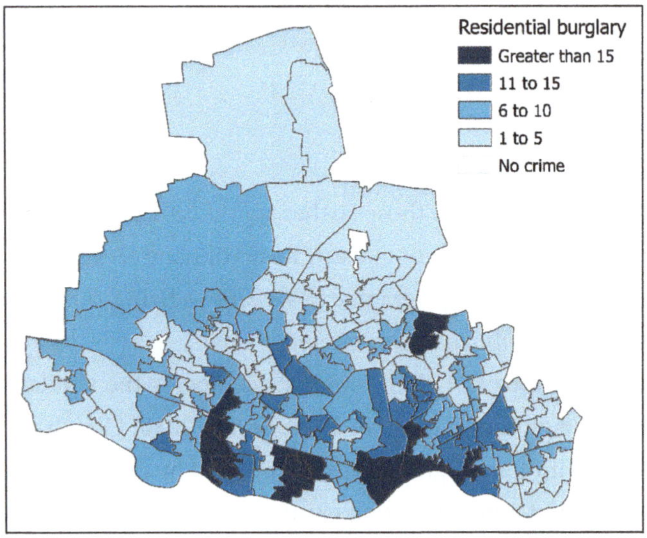

Figure 2.3. House burglaries by census units in Newcastle, England. The Moran's I value for the distribution of these burglaries was 0.252 ($p < .01$).

Positive spatial autocorrelation may indicate not only that areas with high values are surrounded by other areas that also have high values, but that areas with low values are surrounded by other areas with low values. That is, evidence of positive spatial autocorrelation using Moran's I relates to areas surrounded by similar values rather than exclusively relating to areas in which high levels of crime cluster.

Geary's C statistic is an alternative measure of spatial autocorrelation that involves measuring the deviations in intensity values of each point with one another. Like Moran's I, the Geary's C statistic requires data to be aggregated to some form of geographic unit (e.g., a grid cell or census area) to which an intensity value is assigned (e.g., the number of crimes). The values of Geary's C typically vary between zero and two, where values less than one indicate evidence of positive spatial autocorrelation, and values greater than one indicate evidence of negative spatial autocorrelation. The Geary's C statistic can be tested for significance against a theoretical distribution (one that is normally distributed) by dividing by its theoretical standard deviation (Levine 2010). Like Moran's I, this statistic can be used to determine the level of statistical significance in the Geary's C result. Also, like Moran's I, evidence of positive spatial autocorrelation in data that is examined using Geary's C relates to areas that are surrounded by similar values rather than exclusively referring to areas in which high (crime) levels cluster.

Using Moran's I and Geary's C for detecting hot spots

Although Moran's I and Geary's C are useful statistics for examining whether spatial autocorrelation is present in the data, this is not quite the same as determining whether hot spots are present in crime data. Positive spatial autocorrelation under Moran's I and Geary's C requires further investigation to identify whether the pattern refers to areas of high crime surrounded by other areas of high crime or areas of low crime surrounded by other areas of low crime. Also, if you have access to point data of crime incidents, rather than losing some spatial detail in the original data by aggregating the point data to geographic units to perform Moran's I or Geary's C, a test for spatial clustering that uses point data is likely to be preferred.

Nearest neighbor index

The nearest neighbor index (NNI) test (sometimes referred to as the *nearest neighbor ratio* or the *average nearest neighbor test*) compares the spatial distribution of point data against random variation (Clark and Evans 1954). If the result generated from the NNI test is one, the data is randomly distributed. If the NNI result is less than one, the data shows evidence of clustering. An NNI result greater than one indicates evidence of a uniform pattern in the data. The statistical significance of the result can be derived (at least approximately) by comparing the theoretical distribution of the nearest neighbor distances between points under complete spatial randomness, with the observed nearest neighbor distances (Bailey and Gatrell 1995).

Figure 2.4 shows the distribution of house burglaries for a one-year period in Newcastle, England, as points. The same data was used to create the choropleth map in figure 2.3 and the corresponding Moran's I statistic. The points certainly appear to indicate that burglaries are spatially clustered. The NNI result for this data was 0.506 ($p < .01$), indicating that the data is indeed significantly clustered. The NNI provides a useful preliminary measure for hot spot analysis because it indicates whether clustering is present and whether this clustering is significant.

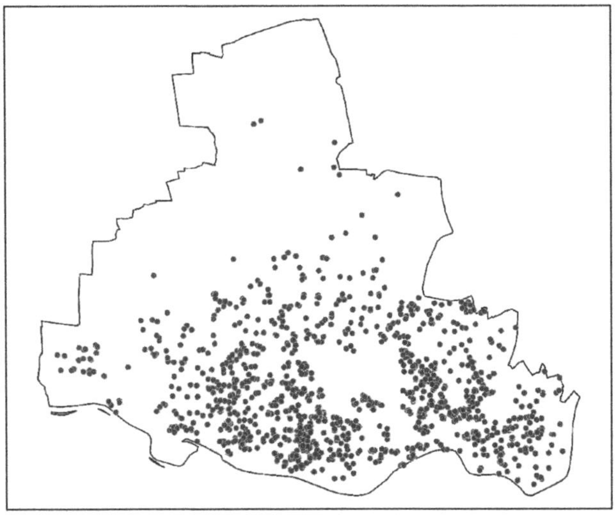

Figure 2.4. The distribution of house burglaries in Newcastle, England, shown as points.

A challenge that many analysts and researchers often have is determining whether they have enough data to produce a hot spot map. For this purpose, the NNI is a useful test to determine whether clustering is present in a data sample, and if present, a hot spot map can be produced. The results from the following experiment illustrate this. Crime data for two study areas (area A and area B) and for a number of crime types (residential burglary, theft from the person, theft from a vehicle, theft of a vehicle, and assaults) was organized into the temporal periods shown in table 2.1. For example, one day of residential burglary data was selected and the NNI test was applied, followed by applying the NNI to two days of residential burglary data, and so on. This experiment was used to determine the typical temporal period of input data required to generate hot spot maps for these two study areas. The result of each test was noted, including whether it was statistically significant to $p < .05$. If there were no crime events or only one crime event for a temporal input period, the nearest neighbor statistic could not be applied.

Table 2.1. Temporal periods of input data used for NNI tests

Temporal input data periods	1 day	2 days	3 days	4 days	5 days	6 days	1 week
	2 weeks	3 weeks	4 weeks	5 weeks	6 weeks	7 weeks	8 weeks

NNI tests also were performed on input data for each of the study areas and for each crime type for a set of uniform categories for the number of events. Table 2.2 lists these categories of input data—for example, the 10 burglaries committed immediately before a study area's measurement dates were selected, and the NNI test was applied. This approach would determine the typical number of events required before clustering was detected. The result of the test was noted, including whether it was statistically significant to $p < .05$. Using these results, an additional set of NNI tests was carried out to determine the exact number of events in each dataset in which clustering was statistically significant ($p < .05$).

Table 2.2. Uniform categories of input data used for NNI tests

Uniform input data periods (number of crime records)			
5	35	65	110
10	40	70	120
15	45	75	130
20	50	80	140
25	55	90	150
30	60	100	

Table 2.3 shows that for area A, only three days of data on thefts from vehicles was required before clustering was evident. Residential burglary and theft from the person data showed evidence of clustering using only four and five days of data, respectively, whereas theft of vehicles data required one week and three days of data before clustering was observed. Before these respective thresholds were reached, the crime data either was randomly distributed (in a spatial statistical sense) or was uniformly distributed ($p < .05$). That is, one week and three days of theft from vehicle data for study area A typically would be required to generate a meaningful map identifying where hot spots were located. If crime data was used for less than one week and three days, it is likely that no hot spots yet would be apparent. Table 2.3 shows the exact number of crime events that were required in area A before clustering was significant. This number ranged from 45 offenses for residential burglary to 66 offenses for theft from vehicles.

Table 2.3. Amount of crime data, by volume and retrospective temporal period, required in area A and area B for clustering to be statistically significant

	Area A		Area B	
	Days/weeks of data for evidence of clustering	*n* of cases for evidence of clustering	Days/weeks of data for evidence of clustering	*n* of cases for evidence of clustering
Residential burglary	4 days	45	2 weeks, 4 days	65
Theft from the person	5 days	52	2 weeks, 5 days	34
Theft of vehicle	1 week, 3 days	64	16 weeks, 5 days	63
Theft from vehicle	3 days	66	1 week, 1 day	37
Assaults	4 days	42	1 week, 5 days	51

$p < .05$

In area B, eight days to 16 weeks and five days of data were required before clustering was evident across all crime types (see table 2.3). Like area A, theft from vehicles required the fewest number of days of recorded crime data before clustering was significant (eight days), followed by residential burglary (two weeks, four days) and theft from the person (two weeks, five days). One week and five days of assault data was required for clustering to be evident. Over 16 weeks of theft of vehicle data was required before clustering was observed.

The results from these NNI tests show the differences between study areas and between crime types in the amount of crime data required before clusters are statistically evident, and in turn the point at which this data can be turned into hot spot maps of crime. For example, although one week of residential burglary, theft from the person, and theft from vehicle data for area A likely would be sufficient for performing some level of hot spot mapping, mapping only one week of crime data for any of the crime types examined from area B would show only random spatial patterns of these crime types. In all cases for area B, over one week of crime data was required for each crime type for clustering to be evident. In the case of theft of vehicles in area B, over 16 weeks of crime data was required for clustering to be observed. Analysis of the number of crime events required showed that clustering became evident when 34 theft from the person offenses in area B were tested, whereas 66 theft from vehicle offenses in area A were required before clustering was evident.

These results show that, although in some cases short retrospective periods of crime data (of less than a week) or small volumes of crime (e.g., 34 crimes) may be enough for clusters to be observed, the specific retrospective point when clusters are evident varies. In practitioner terms, simply choosing a retrospective period, whether it be a retrospective number of days or a retrospective volume of crime, and expecting clusters to be present is not sufficient if you then expect hot spots to appear on a map. The results from these tests illustrate the value in performing the NNI test as a preliminary stage to hot spot mapping to ensure that hot spots are present in the data examined, particularly when only small volumes of data are available.

Centrographic statistics provide a useful, albeit simple means of identifying central tendency in patterns of crime data. Moran's I and Geary's C statistics indicate whether observations close together tend to be similar, and hence whether positive or negative spatial autocorrelation exists. The NNI is a useful measure for determining whether clustering is present in point data. Each of these techniques can be used to help better understand the distribution of crime and aggregate patterns of offending. Another pattern of offender behavior that is important to consider is the journey that offenders make to the places where they commit crime.

The journey to crime

Offenders are restricted in how they move around space, tending to follow basic rules regarding their movement. In chapter 1, we introduced the behavioral concept of the least-effort principle, which suggests that people usually will exert the minimum effort possible to complete their tasks. Tasks include visiting shops or supermarkets, performing recreational activities, visiting friends, traveling to work, or journeying to crime. This theory implies a decay in the frequency of activity as distance increases. That is, in the routine of performing daily activities, more tasks likely are to be completed with short trips rather than more tasks being completed that require traveling long distances. Tasks requiring longer distances tend to be completed less often. In a crime sense, offenders tend to travel short distances (on average) to commit their crimes rather than traveling long journeys to crime. This is not to say that offenders do not travel long distances to commit crime—for example, take the case of offenders traveling across international borders to commit crime. But it does say that most offender journeys to crime likely will be short rather than long. If long distances are traveled, a base is likely to be present at their destination location from which they commit crime (with the offending likely to be close to this base location rather than a long distance away).

The principle of least effort and distance decay in the frequency of offending activities has been observed in numerous studies of the journey to crime (see Rossmo 2000 for an extensive review). Frisbie et al. (1977) performed one of the first analyses of journey to crime distances, showing that in Minneapolis more than 50 percent of residential burglary offenders traveled less than 0.5 miles from their homes to their crimes, commercial burglars tended to travel slightly farther (committing 50 percent of incidents within 0.8 miles), and perpetrators of assaults and commercial robbers tended to travel the longest distances, committing 50 percent of their crimes within 1.2 miles of their home. One of the other most comprehensive studies into the journey to crime was performed by Wiles and Costello (2000) in Sheffield, England. They showed that offenders of burglaries to residential properties on average traveled 1.9 miles from their home to their crimes, theft from vehicle offenders traveled on average 2 miles, and those who committed theft of vehicles on average traveled 2.4 miles to their crimes. They also showed that younger offenders tended to travel shorter distances than older offenders (most likely because younger offenders have smaller awareness spaces than more experienced offenders), and that over a 30-year period, distances traveled to crime as car ownership increased had increased only slightly (with most journeys to crime remaining short).

Although the principles of least effort and distance decay work as basic theoretical concepts, the reality is more complex. Real distance decay functions can vary in shape and linear orientation in response to different types of movement behavior. For example, convenience shopping trips tend to be short and high in their frequency, to purchase items such as milk or bread from the closest possible source. Comparison shopping, which may involve buyers seeking more expensive items such as furniture or a new car, is more likely to involve longer trips to search for these more expensive goods, to find where the best deals are, and to review a greater range of products. Although convenience and comparison shopping on the internet have reduced the number of physical journeys we each now may take, the principles relating to making these types of journeys remain. The differences in distance decay functions for activities such as shopping also apply to crime. Opportunity-based criminal behavior may be more frequent over shorter distances compared with organized and planned criminal behavior, which may involve farther distances to complete criminal acts.

Although there are always exceptions to the distances offenders travel, the likelihood that the journey to crime is short rather than long is an important aggregate behavioral principle. For instance, if crime in an area recently has increased, it is more likely that those involved in these crimes live locally rather than a long distance away, or if not living locally, have a local base from which they are operating.

Measuring the journey to crime requires data with two locations recorded: the offender's home (or some other anchor point relevant to each offender) and the location in which each offender has committed a crime. For obvious reasons, the offender for each offense must be known to make this calculation. Although the offenders for all crimes are unlikely to be known, the data relating to known offenders typically provides a good sample to work with. Three methods are used most for measuring the journey to crime (figure 2.5):

- Euclidean distance: the straight-line distance between two points—i.e., from the offender's home address to the crime location
- Manhattan distance: the shortest distance between two points, with distance constrained to horizontal or vertical directions
- Street route distance: the shortest-path distance following the street network

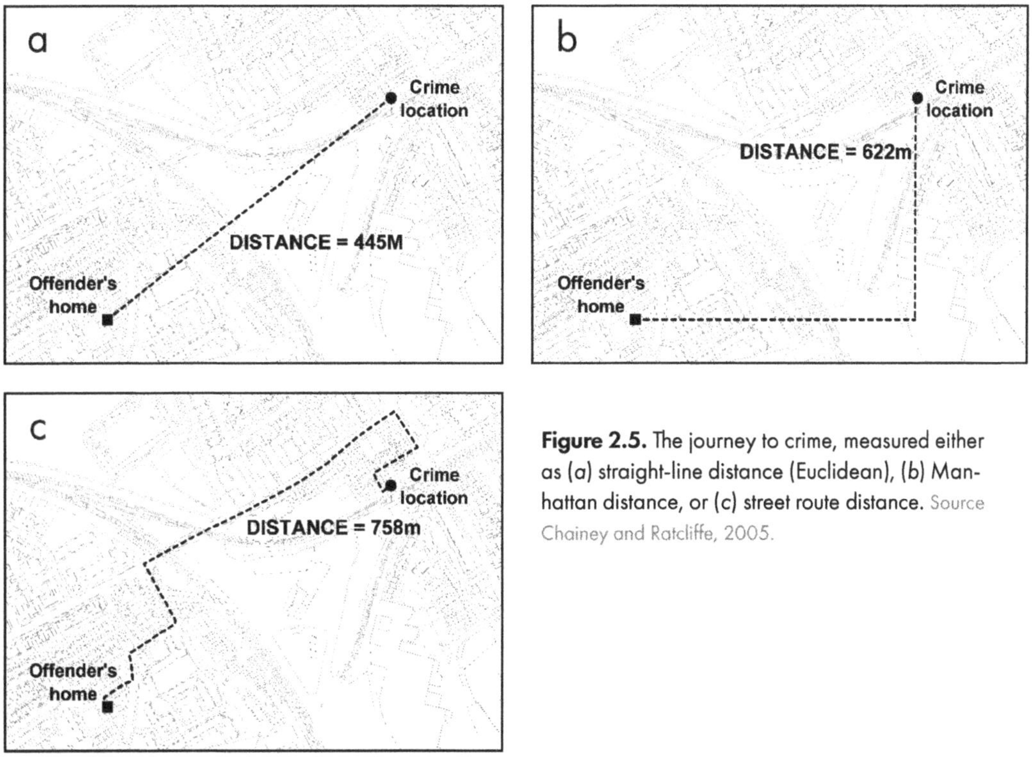

Figure 2.5. The journey to crime, measured either as (*a*) straight-line distance (Euclidean), (*b*) Manhattan distance, or (*c*) street route distance. Source Chainey and Ratcliffe, 2005.

The Euclidean and Manhattan distance methods are often the simplest to calculate, but they do not take into account any of the physical barriers that may obstruct an offender's route to committing a crime, such as railway lines, rivers, buildings, or open spaces that are not accessible. This has led to the assumption that the street route tends to be the more accurate of the methods for measuring the journey to crime, but it is generally the longer of the three methods (as shown in the single example in figure 2.5). However, several studies that have examined differences in the average journey to crime for multiple offenses and offenders have shown that Manhattan and street route distance methods generate similar results. Table 2.4 provides an example of these types of results from an analysis of offenses for residential burglary, street robbery, and vehicle thefts in south London. In each case, the average distance that offenders traveled to the crimes they committed was short (even in a well-connected city in which there is a high degree of easy mobility using either private or public transportation). Additionally, the average journey to crime was almost the same distance for Manhattan and street route distance results for each of the crime types examined. In each case, the results using the Euclidean method were 20 percent less than street route and Manhattan distances. These results suggest that Manhattan and street route measures are the best methods to use when measuring the journey to crime, or when applying the principles of journey to crime in the spatial analysis of criminal behavior, such as for geographic profiling purposes. Rossmo and Davies (2006) found similar findings to those shown in table 2.4 in a study of the journey to crime of stranger rape offenders: the average journey to crime calculated using the Manhattan and street route distance methods was almost identical.

Table 2.4. Journey to crime distances based on Euclidean, Manhattan, and street route methods for one year of residential burglary, street robbery, and vehicle theft (where the offender was known) in south London, England

Type of crime	*n*	Euclidean distance (km)	Manhattan distance (km)	Street route distance (km)	Ratio between Euclidean and Manhattan distances	Ratio between Euclidean and street route distances	Ratio between Manhattan and street route distances
Residential burglary	169	2.892	3.616	3.621	0.80	0.80	1.00
Robbery from the person	1085	3.101	3.843	3.811	0.81	0.81	1.01
Theft of a vehicle	203	3.405	4.291	4.264	0.79	0.80	1.01

Most journey to crime studies, however, make the same assumption: that the journey to crime begins from home. This assumption mainly has been because the data available for these journey to crime studies includes only the location of each offense and home address of the offender who is known to have committed the offense. The journey to crime may start at numerous other places, such as a place of work, a school location (for a young offender who still attends school), or other frequented premises such as bars, nightclubs, or the homes of family or friends they visit. Although these other spatial anchors are important to consider, especially for techniques such as geographic profiling, the principles of journey to crime still remain: offender journeys are likely to be short, rather than long, from any anchor point.

Several studies also have examined the journey offenders make to the first, second, third, and other subsequent crimes in a series (Buchler and Leineweber 1991; Canter 1994; Davies and Dale 1995; LeBeau 1985; Rossmo 1995; Warren, Reboussin, and Hazelwood 1995). These studies have produced mixed results, and in practice are often of limited use because the analysis excludes unknown crimes the offender may have committed as part of the series; the studies also do not consider other crimes of different types that are not part of the series (but committed by the same person), any offense that may have originated from a location other than the offender's home, nor the fact that the offender may have moved during the series of offenses.

Although difficulties exist with measuring the journey an offender takes to commit a crime as well as interpreting patterns of offender mobility, immense value still can be drawn from the findings of any journey to crime analysis. For example, in the case of nondomestic child murder, in 91 percent of cases, the child's body was found within 800 meters of where the child was last seen, and in two out of three cases, the killer was in the area in which he or she encountered his or her victim because the offender "belonged" there (i.e., lived or worked in the area). In one out of five cases, the offender who murdered the child lived within 75 meters of the encounter site (Rossmo

2000). Furthermore, offenders (for many types of crimes) are known to return to the site where they committed a crime. For example, in the case of the Son of Sam, a murderer who operated in New York City, "on the nights when he couldn't find a victim to kill, he would go back to the scene of an old crime to relive the crime and to fantasize about it. Now that's a heck of a piece of information to store somewhere to see whether other offenders do the same thing" (Porter 1983).

Chapter summary

Centrographic statistics, spatial autocorrelation techniques, and tests for clustering provide a useful way to examine spatial patterns of crime and observe how the behavior of offenders tends to operate toward a common aggregate tendency. Centrographic statistics provide a simple means of observing central tendency in criminal behavior, in terms of both individual offending behavior and the behavior of multiple offenders. Spatial autocorrelation techniques help determine whether crime levels in one area are surrounded by other areas of similar crime levels, and the NNI determines whether clustering is present in point data. The findings from numerous studies of the journey to crime show that the distances that offenders travel to commit their crimes tend to be short. Being familiar with the ways to apply these techniques helps you interpret spatial patterns of crime and understand offending behavior.

References

Bailey, T. C., and A. C. Gatrell. 1995. *Interactive Spatial Data Analysis.* Reading, MA: Addison-Wesley.

Buchler, H., and H. Leineweber. 1991. "The Escape Behavior of Bank Robbers and Circular Blockade Operations by the Police." In *Police Research in the Federal Republic of Germany*, edited by E. Kube and H. Storzer. Berlin: Springer-Verlag.

Canter, D. 1994. *Criminal Shadows.* London: HarperCollins.

Chainey, S. P. and J. H. Ratcliffe. 2005. *GIS and Crime Mapping.* London: Wiley.

Clark, P. J., and F. C. Evans. 1954. "Distance to Nearest Neighbor as a Measure of Spatial Relationships in Populations." *Ecology* 35 (4): 445–53.

Davies, A., and A. Dale. 1995. *Locating the Stranger Rapist.* Police Research Group special interest series issue 3. London: Home Office Police Department.

Eck, J. E., S. P. Chainey, J. G. Cameron, M. Leitner, and R. E. Wilson. 2005. *Mapping Crime: Understanding Hot Spots.* Washington, DC: National Institute of Justice.

Frisbie, D. W., G. Fishbine, R. Hintz, M. Joelson, and J. B. Nutter. 1977. *Crime in Minneapolis: Proposals for Prevention.* St. Paul, MN: Community Crime Prevention Project, Governor's Commission on Crime Prevention and Control.

Kind, S. 1987. "Navigational Ideas and the Yorkshire Ripper Investigation." *Journal of Navigation* 40 (3): 385–93.

LeBeau, J. L. 1985. "Some Problems with Measuring and Describing Rape Presented by the Serial Offender." *Justice Quarterly* 2 (3): 385-98.

LeBeau, J. L. 1987. "The Methods and Measures of Centrography and the Spatial Dynamics of Rape." *Journal of Quantitative Criminology* 3:125–41.

Levine, N. 2010. *CrimeStat III: A Spatial Statistics Program for the Analysis of Crime Incident Locations*. Houston, TX: Ned Levine and Associates, Washington, DC: National Institute of Justice.

O'Sullivan, D., and D. J. Unwin. 2003. *Geographic Information Analysis*. Hoboken, NJ: Wiley.

Porter, B. 1983. "Mind Hunters." *Psychology Today* (April).

Rossmo, K. 1995. "Place, Space, and Police Investigations: Hunting Serial Violent Criminals." In *Crime and Place*, edited by J. E. Eck and D. Weisburd. Washington, DC: Police Executive Research Forum.

Rossmo, K. 2000. *Geographic Profiling*. Boca Raton, FL: CRC Press.

Rossmo, K., and A. Davies. 2006. *Investigating the Use of Geodemographics for Identifying Stranger Rapists*. Home Office Research Study, Research, Development and Statistics Directorate, Home Office.

Tobler, W. 1970. "A Computer Movie Simulating Urban Growth in the Detroit Region." *Economic Geography* 46 (2): 234–40.

Warren, J., R. Reboussin, and R. Hazelwood. 1995. *The Geographic and Temporal Sequencing of Serial Rape*. US Department of Justice Report. Washington, DC: National Institute of Justice, Office of Justice Programs.

Wiles, P., and A. Costello. 2000. *The "Road to Nowhere": The Evidence for Traveling Criminals*. Research Study 207. London: Home Office.

CHAPTER 3

Hot spot analysis: Examining spatial concentrations and clusters of crime

Objectives

In this chapter, we examine the use of two categories of techniques for analyzing and mapping hot spots of crime: measuring and mapping crime concentration at the micro-place geographic scale and spatial clustering techniques. Examining concentrations of crime at the micro-place level (such as street segments) helps identify the specific locations of high levels of crime. Spatial analysis techniques that identify clusters of crime indicate how the level of crime in each location is spatially associated with crime in immediate surrounding locations. There are many types of spatial clustering techniques. In this chapter, we focus on kernel density estimation, local Moran's I, and the Getis-Ord Gi* statistic. Both categories of techniques—micro-place crime concentration analysis and spatial clustering techniques—help determine where resources can be targeted to reduce crime. Technical details on the hot spot analysis techniques are provided, with examples. The chapter begins by defining what a hot spot is, and then draws from the theoretical principles described in chapter 1 to explain why hot spots occur.

Key learning points

- Define what a hot spot is, describe the theoretical principles on why hot spots exist, and explain why hot spot analysis is important for prioritizing the targeting of police and public safety resources for reducing crime.
- Introduce and provide guidance on measuring and mapping crime concentration at the micro-place level. This includes using Weisburd's (2015) bandwidths for measuring crime concentration, and visualizing patterns of micro-place crime concentration.

- Introduce and provide guidance on spatial clustering techniques, including advice on suitable technical parameters that must be entered for these techniques to work (such as the bandwidth distance for kernel density estimation) and identifying areas in which clusters are statistically significant.

Introduction

A hot spot is an area of high crime concentration relative to the distribution of crime across the entire study area (Chainey and Ratcliffe 2005, Sherman 2009). Hot spots can exist at different geographic scales of interest, whether it is at the city level for identifying neighborhoods in which crime is highest or at the neighborhood level for identifying streets, or clusters of buildings, in which crime is observed to highly concentrate. Hot spot analysis involves using a technique or a combination of techniques to identify concentrations and clusters of crime.

The mapping of hot spots of crime has become common practice in police and public safety agencies across the world and has been applied to many forms of crime: from residential burglary to street robbery, from vehicle crime to violent crime, and from antisocial behavior to homicides. The primary purpose of hot spot analysis is to identify visually the locations in which crime levels appear to be highest, and from this, determine where to target and deploy resources to tackle the crime problem. Examples include the following:

- Using hot spot analysis to support the operational briefing of police patrols (Goldsmith et al. 2000; Harries 1999; Osborne and Wernicke 2003). For example, hot spot maps were used by the Metropolitan Police in London to show recent incidents (i.e., in the last four days) against a backdrop (hot spot map) of the distribution of crime over the previous month to help in the operational briefing of neighborhood patrol teams (Home Office 2005).
- A more focused use of hot spot analysis for informing the deployment of police patrols is hot spot policing. This involves the targeting of police patrols to high-crime locations (Sherman, Gartin, and Buerger 1989; Braga, Papachristos, and Hureau 2014), where the patrols typically remain for a period. In two recent studies of hot spot policing, targeted patrols reduced violent crime by 23 percent in Philadelphia (Ratcliffe et al. 2011), and reduced street robberies in Montevideo, Uruguay, by 23 percent (Chainey, Serrano-Berthet, and Veneri 2020).
- The use of intelligence products has become a key component in modern policing. Intelligence products include problem profiles that aim to better understand crime problems and strategic intelligence assessments that identify the key issues and threats that require prioritized attention in a jurisdictional area. Hot spot analysis is a common feature within

these products, informing operational tactics, identifying persistent problem areas, and supporting a targeted approach for crime reduction (Chainey and Chapman 2013). For example, in the West Midlands Police Force, hot spot mapping was used to identify priority neighborhoods across the city of Birmingham to help determine the persistent problem areas that required dedicated focus (Home Office 2005).

- Hot spot analysis also has been a common feature in CompStat-style performance meetings. CompStat is a policing management meeting process, with meetings held regularly (from every week to every month, depending on adoption by police agency) to inform on recent events and trends, determine future activity, and build in a process of accountability. The use of hot spot analysis is a common feature in these meetings for reviewing performance and determining future actions (for examples, see Chainey and Ratcliffe 2005).

Key theoretical principles behind hot spots

It is the characteristics of the places in which hot spots form that explain why crime concentrates in some areas and not others. On one level, these characteristics relate to the neighborhood. For example, crime activity is more likely in places where the social ties within a community are too weak to influence how people behave, or where social ties may promote illegal behavior. On another level, these characteristics of places where hot spots exist relate to the opportunities, interactions, and behaviors of people in the specific places where crimes occur. For example, if a person has the motivation to commit a crime such as a cell phone robbery, the opportunities to do so may be most present in certain locations. These neighborhood and situational factors are next considered in more detail.

Neighborhood characteristics

Social disorganization theory posits the idea that higher levels of crime are present in certain neighborhoods because the social fabric in these neighborhoods is not strong enough to positively influence the people who live in the neighborhood. Collective efficacy refers to the ability that people who live (or work) in the neighborhood possess to control the behavior of individuals. If there are no clear rules for acceptable behavior, and there are few sanctions available to counter illegal activity, crime (as the theory suggests) is more likely to occur. In certain neighborhoods, the collective efficacy of an area indeed may strongly promote illicit behavior (e.g., in areas where there is high distrust of the police and a lack of respect for rule setting).

Situational characteristics

Regardless of our background, we all are rational actors who seek to maximize benefits while reducing costs and risks. We consider information from our environment when making decisions and act accordingly by weighing the gain we would expect against the risks and costs associated with our actions. These principles apply not only to law-abiding people but also to those who commit crime. Offenders decide to commit a crime when and where they think the possible benefits are greater than the effort involved and are greater than the risks of being caught.

We all also have daily routines, which in turn can help predict where and when we will be and what it is we are likely to be doing. In a crime sense, this is based on the simple idea that the behavior of offenders and their potential victims helps explain the occurrence of crime. Recall from chapter 1, for a crime to occur, three components are necessary: the presence of a likely offender, the presence of a suitable target, and the absence of a capable guardian, with these three components meeting in time and space to provide the necessary chemistry for crime (Felson 1998). This meeting in time and space is not random but is dictated by the natural rhythm of daily life—people going about their routine activities, such as potential targets going to and from work. The combination of the three components (offender, target, and guardian) and their qualifiers (likely, suitable, and capable) then dictates that the risk of crime changes over time with the movement of people throughout the daily routine activities of their lives. These meetings in time and space are not random and will, therefore, tend to take place where and when the crime opportunities are most present. This theoretical perspective helps explain why hot spots of crime are hot spots during certain hours of the day.

Crime pattern theory suggests that offenders will tend to steer toward areas that are known to them to commit crime—the offender's awareness space. Opportunities for crime are not evenly spaced across the landscape. Therefore, within an offender's awareness space is the need to select certain areas in which opportunities are more present. Although there are always exceptions, the majority of offenders will tend to commit most of their crimes in the places where they know the opportunities are present, where they know the best ways to escape, and where they feel comfortable, do not feel as if they stand out, or can easily appear to be anonymous. In hot spot analysis terms, these principles can help us consider that the person who committed an offense is likely to have been lured by their awareness of the rich opportunities for crime, albeit directed to certain places because opportunities to commit crime are not equally distributed across space.

Therefore, crime hot spots exist at specific places because of a combination of factors that make these places particularly favorable for crime to occur. These favorable factors relate to the normal, everyday activities of people that place them in these environments, the unequal geographic distribution of opportunities to commit crime, and an offender's assessment of the risks, efforts, and rewards for committing a crime.

Techniques for analyzing and mapping hot spots of crime

Many techniques can be used for analyzing and mapping hot spots of crime. These techniques can be as straightforward as representing each crime event as a point and observing the geographic distribution of these points, using functions within a GIS for creating spatial ellipses representing the location and orientation of clusters of crime, thematically shading administrative areas (e.g., census zones or police beats), or representing the distribution of crime as a continuous surface that relates to the density of the geographic distribution of crime and its association with neighboring areas. Identifying hot spots using point mapping has become outdated since the proliferation of GIS software and the increasing sophistication of mapping techniques. Using spatial ellipses (see Block and Block 2000 for an example) has become less popular as newer techniques have become more widely available, and thematically shading administrative areas is subject to the modifiable areal unit problem, or MAUP (Openshaw 1984), which can make it problematic for hot spot analysis (Chainey, Tompson, and Uhlig 2008).

In this chapter, we focus on two types of hot spot analysis techniques: measuring and mapping crime concentration at the micro-place level and spatial clustering techniques that consider the spatial association of crime levels between areas. We remind you also of the nearest neighbor index measure described in chapter 2, which is a useful preliminary test for determining whether hot spots are present in the data you want to map.

Measuring and mapping crime concentration at the micro-place level

The analysis of geographic patterns of crime increasingly has become concerned with examining crime at the micro-place level. Micro-places are street junctions, street segments (the part of a street between two junctions), or other fine geographic units such as small grid cells. Part of the reason for the geographic unit of study becoming ever finer is because of continual improvements in the spatial level at which crime data is recorded reliably. Furthermore, the increasing move to a micro-place focus is also based on the recognition that city, subcity, and neighborhood-level analysis hides much of the spatial heterogeneity apparent in criminal activity. Failing to recognize the spatial heterogeneity of crime can result in identifying unsafe neighborhoods that are generally safe (Sherman, Gartin, and Buerger 1989) (i.e., when only a small area within the neighborhood is unsafe) or identifying neighborhoods as being safe when some micro-places within them are unsafe.

Spatial heterogeneity

Spatial heterogeneity describes the uneven distribution of a phenomenon over space. Regarding crime, spatial heterogeneity is shown by areas displaying concentrations of crime, in that these crimes are unevenly distributed across space.

Since the 1980s, numerous studies have found that crime significantly concentrates at small geographic scales (for a full review, see Lee et al. 2017 and Chainey et al. 2019). This has led Weisburd (2015) to propose a law of crime concentration, oriented to micro-places, positing that crime concentrates within certain spatial bandwidths: for a cumulative proportion of 25 percent of crime, the bandwidth for the proportion of micro-places is between 0.4 and 1.6 percent; and for a cumulative proportion of 50 percent of crime, the bandwidth for the proportion of micro-places is between 2.1 and 6 percent. These bandwidths were defined from a study of crime concentration for a selection of cities in the United States. Since then, numerous other studies have illustrated the applicability of these bandwidths to crime patterns in other parts of North America (Gill, Wooditch, and Weisburd 2017), in Europe (Bernasco and Steenbeek 2017), in Latin America (Chainey et al. 2019), and in Asia (Amemiya and Ohyama 2019). Crime concentration also has been observed to be highly stable over time (Weisburd et al. 2004), and when crime increases, the largest increases take place in hot spots (Chainey and Monteiro 2019). Many other studies also have shown that interventions targeted at specific places in which crime previously has concentrated can have a significant impact on reducing crime (Braga, Papachristos, and Hureau 2014).

In the four steps that follow, the technical process for measuring and mapping crime concentration at the micro-place level is described and illustrated with examples.

Step 1: Preparing micro-place geographic unit data

First, the data for the geographic unit at which crime concentration will be examined must be prepared. Micro-place geographic units can be either street segments, street junctions, a combination of street segments and street junctions, or micro-grid cells. Street segments exist as linear geographic objects, street junctions as point objects, and micro-grid cells as polygon objects. Each micro-place unit must be referenced by a unique identifying number. The geographic objects represent the unit to which crime data is to be aggregated (described in step 2). If street segments are available, this geographic unit is the preferred unit of choice. Street segments are considered to offer an ideal micro-place unit of study for crime concentration because they act as behavioral settings along which social activities are organized (Felson 1987; Smith, Frazee, and Davison 2000; Weisburd, Groff, and Yang 2012). In some situations, crime data is not referenced to street segments but street junctions. In these cases, the street junction can be used as the micro-place unit of study. Quite often, crime data in a single dataset is referenced to either street segments or street junctions.

In this case, both micro-place units can be used to examine concentrations of crime. In some settings, street segment data is not available. Then, small grid cells can be used as an alternative micro-place geographic unit (see Chainey and Monteiro 2019 for an example of the use of grid cells for examining concentrations of crime).

The first part of step 1 involves selecting the micro-place unit of study. Next, the content and quality of the micro-place unit of study should be checked. For street segments, this process should include reviewing how the street segments are represented, such as ensuring these geographic units are indeed split between street junctions (rather than represented as a continual line across many street junctions), and checking that only one street segment exists between each junction. When a street has a central divide between two sides of a roadway, the street segment data may represent this street as two separate segments—one for each side of the roadway. When such situations occur, only one street segment should be used in the analysis. If it is not possible to remove each duplicate street segment, the impact of these street segment duplicates should be considered when interpreting the results. Checks for street junction data and grid cells primarily should review the completeness of this data for the study area and ensure that each unit has a unique identifying reference number.

Street segments vary in length. It is, therefore, useful to measure the mean, minimum, and maximum lengths of the street segments used. This will aid comparison against other studies of crime concentration. In Weisburd's (2015) study of crime concentration, the length of street segments in seven cities in the United States ranged from 56 meters to 207 meters. The mean across all cities was 144 meters. In a study of crime concentration for 37 cities in Latin America, the mean street length for each city varied between 100 meters and 231 meters, and was 137 meters across all the cities (Chainey et al. 2019). Information on mean street lengths has been used to define the size of grid cells when street segments have not been available for examining crime concentration, with most studies using grid cells of 100 meters or 150 meters.

The final part of step 1 involves calculating the number of micro-place geographic units in the study area. Table 3.1 provides examples of this measure for three cities: New York City (USA), Montevideo (Uruguay), and Rio de Janeiro (Brazil).

Table 3.1. Metrics on micro-place units

City	Micro-place geographic unit	Number of geographic units	Size metrics
New York City, USA	Street segments	87,279	Average length = 120 m
Montevideo, Uruguay	Street segments	33,975	Average length = 117 m
Rio de Janeiro, Brazil	Micro-grid cells	41,944	150 × 150 m

Step 2: Joining of geographically referenced crime data to micro-place units

The first part of this second step involves selecting the geographically referenced crime data to be analyzed. The geographic referencing of crime data usually follows one of two approaches. Crime data either is referenced geographically by using street files (as is often the case in the United States) or is referenced geographically to address points and land parcels, using land and property gazetteers (as is common in England). In this step, we assume crime data already is geographically referenced using either of these approaches (see Chainey and Ratcliffe 2005 for more details on geographic referencing processes for crime data). Then, we need to choose the type of crime to examine—e.g., street robberies, assaults, or residential burglaries.

The second part of step 2 involves joining the crime data to its respective micro-place unit. If the crime data was referenced geographically using street files, this process is already done. However, it is important that for each crime record, the unique identifier for the street segment (e.g., its unique ID number) to which the crime record is referenced geographically is also recorded in the crime data.

If the crime data is referenced geographically using address points and land parcels, it must be joined to its relevant micro-place unit. For street segment (or street junction) micro-place units, using a function within a GIS joins each crime record to its corresponding micro-place unit. The most straightforward method is to join the crime record to its nearest street segment. This results in each crime record being assigned the unique identifying details of the street segment (e.g., its unique ID number). If the micro-place unit of study is grid cells, the aggregation process involves joining each crime record to the grid cell within which the crime record is located. This joining process results in each crime record being assigned the unique identifying reference for the grid cell within which it is located.

Step 3: Calculation of crime concentration levels across the micro-place units

The first part of step 3 involves calculating how many crime records there are in the data to be analyzed. Next, the crime data must be aggregated using the micro-place identity number to calculate a count of the number of crimes in each micro-place unit (e.g., the number of crimes per street segment). This function can be performed in a GIS or a spreadsheet data software tool, such as Microsoft Excel, with the use of pivot tables. Once the data is arranged in a table, with the table listing each micro-place unit and the count of crime in each unit, the data is then rank ordered from highest to lowest, based on the number of crimes in each unit. Then, the percentage of crimes in each micro-place unit relative to all crimes is calculated, followed by the calculation of the cumulative percentage for each micro-place unit, across all the micro-place units.

Table 3.2 provides a sample of this type of output, listing street segment ID numbers, the number of crimes in each street segment, the proportion of crime in each street segment, and a cumulative percentage total. It shows that the top-ranked street segment had 281 crimes, representing

0.55 percent of all crimes in the study area. The seven street segments that had the most crimes accounted for 2.52 percent of all crime in the study area. This study area included more than 40,000 street segments (table 3.2 shows only the first, rank ordered, seven street segments from this dataset). The full table (not shown) lists all 40,000 street segments, the number of crimes in each segment, the proportion of crime in each street segment, and the cumulative total percentage across all street segments.

Table 3.2. An example of micro-place unit crime concentration calculation for a sample of street segments

Street segment ID number	Number of crimes	Proportion of all crime	Cumulative total
429654	281	0.5512%	0.5512%
374397	237	0.4649%	1.0161%
372887	171	0.3354%	1.3516%
377381	160	0.3139%	1.6654%
399773	155	0.3041%	1.9695%
374398	146	0.2864%	2.2559%
410257	137	0.2687%	2.5246%

Note: This example shows only the first seven highest-ranked street segments, for a data sample that consisted of over 40,000 street segments.

Once the table is created listing all micro-place units, with the count of crime for each unit, the proportion of all crime, and cumulative total, a measure of crime concentration can be generated that compares with Weisburd's (2015) bandwidths for crime concentration. This comparison involves using the cumulative total column in the table to determine how many micro-place units accounted for 25 percent of crime and how many micro-place units accounted for 50 percent of crime. Table 3.3 provides an example, showing that in a city consisting of 33,975 street segments, 25 percent of all robberies were concentrated into 0.8 percent of all street segments, and 50 percent of robberies were concentrated into 2.8 percent of all street segments. As these crime concentration levels are toward the lower range of the bandwidths proposed by Weisburd, these results suggest that robberies were highly concentrated in this city.

Table 3.3. Crime concentration in a sample city

Number of street segments	Number of robberies	% of street segments accounting for 25% of robberies (*n* streets)	% of street segments accounting for 50% of robberies (*n* streets)
33,975	8,971	0.8% (287)	2.8% (938)

Step 4: Mapping of high crime concentration micro-places

Step 4 involves mapping the locations of the micro-places in which crime is observed to highly concentrate. If street segments or street junctions are the micro-place unit of analysis, a useful way to visualize the results is to use graduated symbols of the geographic objects. Figure 3.1a provides an example, showing the street segments in an area of a city that accounted for 25 percent of crime and 50 percent of crime (across the entire city). Figure 3.1b shows, for the same area, an example of crime concentration using micro-grid cells, showing those grid cells that accounted for 25 percent and 50 percent of all crime.

Figure 3.1. (a) Street segments accounting for 25 percent and 50 percent of crime, and (b) micro-grid cells accounting for 25 percent and 50 percent of crime.

Once the micro-places are mapped and statistics relating to crime concentration are generated, these results then can be used to determine where to target police and public safety resources to counter these high spatial concentrations of crime (e.g., hot spot policing patrols). In chapter 4, we examine these spatial concentration patterns further by considering the temporal patterns within these areas of high-crime concentration.

Checklist: Measuring and mapping crime concentration at the micro-place level

Step 1: Preparing micro-place geographic unit data

- ☐ Choose geographic micro-place unit of study (e.g., street segment).
- ☐ Check geographic unit data for content and quality.
- ☐ Measure mean length of street segments or define size of micro-grids.

Step 2: Joining of geographically referenced crime data

- ☐ Join each crime record to its relevant micro-place unit (if necessary).
- ☐ Check that each crime record contains the unique identifying details of the micro-place unit in which the crime record is located.

Step 3: Calculation of crime concentration levels across the micro-place units

- ☐ Create a table listing each micro-place unit and its count of crime, proportion of crime, and cumulative total.
- ☐ Calculate crime concentration statistics.

Step 4: Mapping of high crime concentration micro-places

- ☐ Map the micro-place units that account for 25 percent of crime and 50 percent of crime.

Examining crime concentration across all micro-places—Lorenz curves and Gini coefficients

Although measuring the proportion of micro-places that account for 25 percent and 50 percent of crime in an area is useful and straightforward to calculate, Lorenz curves and Gini coefficients can help examine the concentration of crime across all micro-places in a study area. Figure 3.2a is an example of a Lorenz curve, showing the proportion of crime against the proportion of micro-place units across all micro-place units. The Lorenz curve is steep, indicating that for this study area, crime is highly concentrated. The Gini coefficient measures the area under the curve. The closer the Gini coefficient is to 1, the more concentrated the crime pattern. In this example, the Gini coefficient is 0.98.

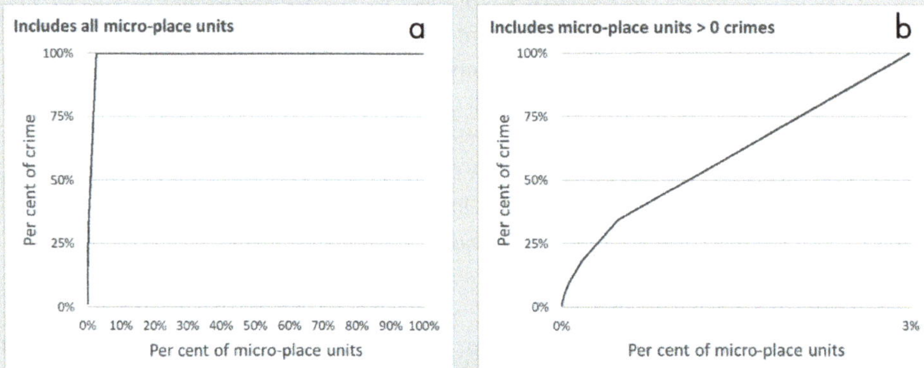

Figure 3.2. (a) A Lorenz curve showing the concentration of crime for all places, and (b) a generalized Lorenz curve of crime concentration for only places with crime.

Often, when examining crime concentration across micro-places, there are more micro-place units than crimes. This disparity causes a bias in the results. To address it, generalized versions of the Lorenz curve and the Gini coefficient can be calculated that include only those micro-place units in which at least one crime took place. Figure 3.2b shows an example of a generalized Lorenz curve. This approach is most useful when comparing different study areas in which the crime concentration is high. For example, an analysis of crime concentration between two cities found that both cities had Gini coefficients of 0.98. The generalized Gini coefficients varied between 0.27 for city A and 0.40 for city B, suggesting that the concentration of crime was greater in city B than in city A. (See Bernasco and Steenbeek 2017 for more details on generalized Lorenz curves and generalized Gini coefficients.)

Hot spot mapping using spatial analysis techniques

Several spatial analysis techniques can be used for mapping hot spots of crime. We focus in this chapter on three spatial analysis techniques identified as being the most useful for mapping crime hot spots: kernel density estimation, local Moran's I, and the Gi* statistic. You already may be familiar with some or all of these techniques but less familiar with how to determine which parameters to use (e.g., kernel density estimation bandwidth size) and the differences in the output that each technique produces. We address these points in this chapter.

Kernel density estimation

Kernel density estimation (KDE) is one of the most common mapping techniques for visualizing hot spots of crime. KDE has become popular as a hot spot mapping technique because of its wide availability in GIS software, its accuracy in hot spot identification, and the aesthetic look of the resulting map in comparison with other techniques (Chainey and Ratcliffe 2005; Chainey, Tompson, and Uhlig 2008; Eck et al. 2005).

To apply KDE to hot spot mapping, the crime data must be geographically referenced as point data. The KDE function is then applied to these points to produce a continuous surface estimate representing the density of the point distribution. The continuous surface is represented as grid cells covering the entire study area, with each cell having a density value. In mathematical terms, KDE is expressed as:

$$f(x,y) = \frac{1}{nh^2} \sum_{i=1}^{n} k\left(\frac{d_i}{h}\right) \qquad (3.1)$$

where $f(x,y)$ is the density value at location (x,y), n is the number of points, h is the bandwidth, d_i is the geographic distance between incident i and location (x,y), and k is a density function, known as the *kernel*. Each x,y location refers to the centroid of the grid cell. The kernel k can take many forms, although the results between different functions produce similar density values (Bailey and Gatrell 1995). A common choice for k is the quartic function (Chainey and Ratcliffe 2005; Levine 2010). Figure 3.3 shows an example of a crime hot spot map using KDE, showing those areas in which the density of crime is highest—the hot spots.

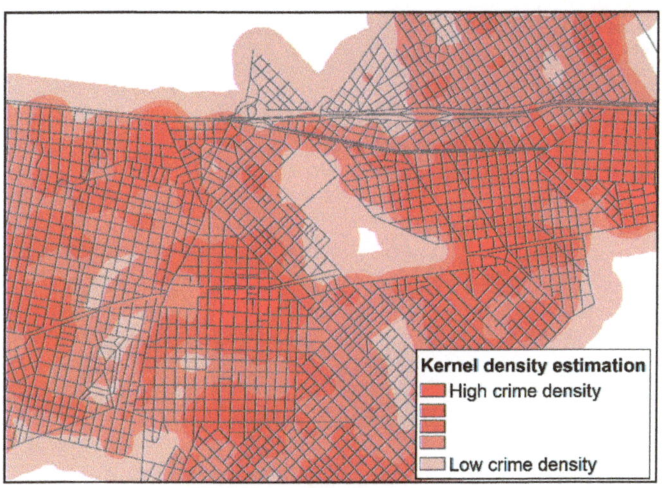

Figure 3.3. A KDE hot spot map of the distribution of crime.

Typically, when applying the KDE technique, you are required to enter two parameters: the cell size and the bandwidth. As shown in equation (3.1), the density value calculated for each location is affected by the total number of points, their spatial distribution, and the bandwidth. The cell size does not affect the density value but affects the visual spatial resolution of the hot spot map. Figure 3.4 shows three examples of KDE hot spot maps, based on the same crime data and same bandwidth but for different cell sizes. The patterns of crime hot spots are the same, with the maps differing only in terms of how the hot spots are represented visually. The KDE map shown in figure 3.4a used a cell size of 50 meters and appears quite blocky. The KDE map shown in figure 3.4c used a cell size of 10 meters, and hence is much finer in its appearance. Figure 3.4b used a cell size of 30 meters.

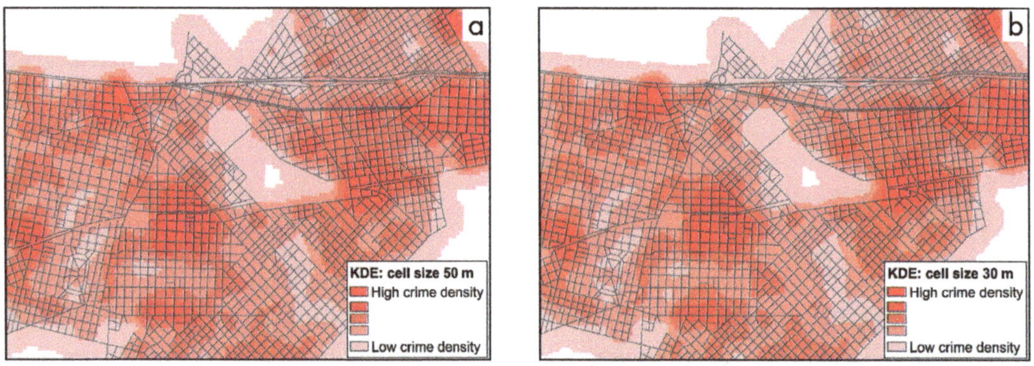

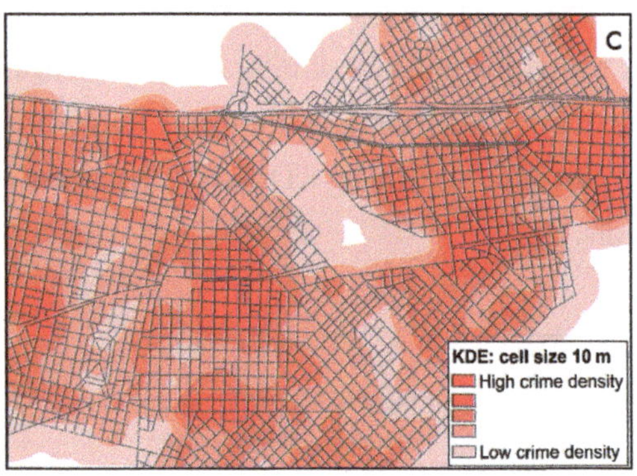

Figure 3.4. KDE hot spot maps of crime using the same bandwidth but a cell size of (a) 50 meters, (b) 30 meters, and (c) 10 meters.

KDE cell size and its effect on the results of a hot spot map—an analogy using HDTV

An analogy that may be useful when explaining the influence of cell size on KDE hot spot mapping output is the comparison between watching a TV program in standard resolution versus high definition. Let's say you are watching your favorite sports team on TV in standard (640 × 480 pixel) resolution, and then you remember you also can watch the same game in full HD (1920 × 1080 pixel resolution) on another TV channel. So you turn over to the full HD channel, and the picture is much better, but it does not change the fact that your sports team is still losing! What you still are seeing is (depressingly) the same, but at least you now are enjoying it (or maybe not if your team is losing badly) in crystal-clear, vivid high resolution. Similarly, with a KDE hot spot map, if you change the cell size to much smaller, the map still tells the same story (i.e., the number and location of the hot spots has not changed), but now your hot spot map looks a lot nicer.

The cell size of 30 meters was chosen for producing the KDE hot spot map in figure 3.4b following the advice from Chainey and Ratcliffe (2005), who suggest that a good starting point for the KDE cell size is to divide the shorter side of the study area's minimum bounding rectangle (MBR) by 150. Although simple to calculate and not particularly scientifically informed, experience suggests it works well, as shown in figure 3.4b. From this starting cell size, additional KDE maps can be produced easily using smaller cell sizes to improve the map's visual appearance. When selecting the cell size, the main point is that many calculations of the density value (f) must be performed if the cell size is small, compared with many fewer calculations of f if the cell size is large. Nowadays, this consideration tends not to be a problem with the memory-processing capacity of most computers.

As referred to above, the number of points, their spatial distribution, and the bandwidth are the three components that affect kernel density values. When generating a KDE hot spot map of crime using data for a study area for a certain duration, the number of crime incidents across the study area remains the same. Therefore, the total number of incidents would not influence changes in the density estimate. The spatial distribution of the crime incidents across the study area is static, and hence also would not influence changes in the density estimate. Therefore, it is the bandwidth that influences different values of the density value (f) at each x,y location.

Several bandwidth size optimization routines such as the mean integrated squared error (Bowman and Azzalini 1997; Fotheringham, Brunsdon, and Charlton 2002), Akaike correlation coefficient, and the cross validation method (Brunsdon 1995; Fotheringham, Brunsdon, and Charlton 2002; Silverman 1986) are available, but these approaches tend to produce large bandwidths, and are considered unsuitable for exploring local spatial patterns of the density distribution of crime (Chainey 2013). Others suggest an approach of experimenting with different sizes of bandwidth (Bailey and Gatrell 1995; Eck et al. 2005; Chainey and Ratcliffe 2005). A good starting bandwidth is to measure the shorter side of the study area's MBR, divide by 150, and multiply this value by five

(Chainey 2013). Although simple to calculate, and not particularly scientific, it produces good initial output. From this, other KDE maps can be produced, experimenting with different bandwidths, leaving it to the user to choose the map they consider the most useful.

Figure 3.5 shows three KDE crime hot spot maps using different bandwidths. Figure 3.5a was produced using a bandwidth of 150 meters, following the procedure described by Chainey (2013). The KDE hot spot map in figure 3.5b was produced using a bandwidth of 500 meters, and the KDE hot spot map in figure 3.5c was produced using a bandwidth of 50 meters. In all cases, a grid cell size of 10 meters was used. The first thing to notice is the difference in the number of hot spots shown in the three maps. All KDE maps were produced using a natural breaks (Jenks) classification process (see Eck et al. 2005 for details on classification methods for KDE hot spot mapping). In figure 3.5b, only three hot spots (shown by the high crime density thematic class) are apparent, compared with more than 10 shown in figure 3.5c. In figure 3.5c, it is difficult to discern the general spatial patterns of crime. The choice of bandwidth for producing a KDE map comes down to user preference, considering what the map is used for. Based on experience in seeing many KDE hot spot maps, it is best not to produce something that looks too spiky, as in figure 3.5c, and not something as general and unspecific as the hot spot map in figure 3.5b. To identify where hot spots of crime are in a study area, my preference would be to use figure 3.5a.

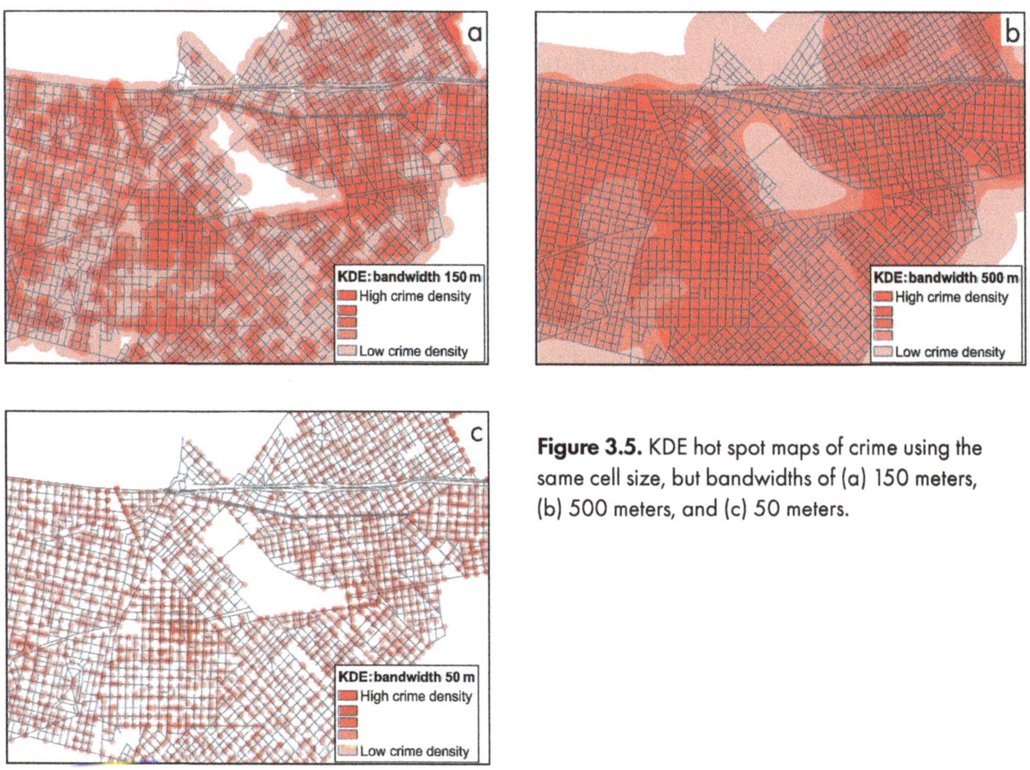

Figure 3.5. KDE hot spot maps of crime using the same cell size, but bandwidths of (a) 150 meters, (b) 500 meters, and (c) 50 meters.

Harm hot spots

Density mapping techniques such as KDE apply an equal weight to each crime incident. Mapping harm is an alternative approach to mapping crime density. *Harm* refers to the consequences of criminal activities (Von Hirsch and Jareborg 1991; Ratcliffe 2015), with each crime incident weighted according to the consequential impact it has (i.e., each crime point is weighted by its harmful effect). Most harm studies use sentencing guidelines to calculate the harm weights, which involves multiplying each crime incident by the number of days an offender is imprisoned (Curtis-Ham and Walton 2018; Sherman, Neyroud, and Neyroud 2016).

Most studies of *crime hot spots* involve the examination of a specific type of crime rather than many types of crime, and hence in most cases, a harm hot spot mapping approach is not necessary—the harm hot spots will show the same crime hot spots, albeit with a wider range of values for the harm map because of the multiplication of each incident by its harm weight. However, in situations in which several types of crime are combined to produce a hot spot map, a harm hot spot map can be useful. Harm hot spot mapping can be performed using KDE by assigning the harm weight to each point.

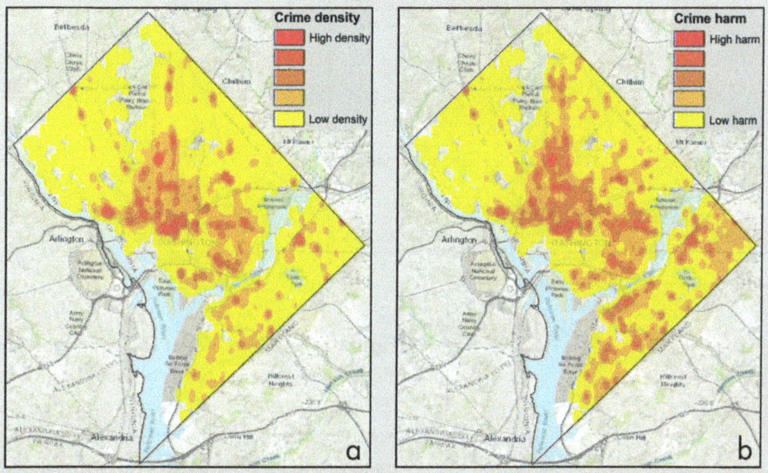

Figure 3.6. (a) A crime density hot spot map and (b) a crime harm hot spot map of robberies and thefts for the District of Columbia.

Figure 3.6a shows a crime density hot spot map (using KDE) of robberies and other thefts for the District of Columbia. Figure 3.6b shows a crime harm hot spot map, using the same data but with each crime incident weighted by its harm value. The harm weights were determined using the Cambridge Crime Harm Index (Sherman, Neyroud, and Neyroud 2016), one of the most universally used measures of harm. Robberies have a more harmful effect than thefts. For instance, a robbery in which a weapon was used generates a Cambridge Crime Harm Index value of 2,641, whereas a minor theft has a Cambridge Crime Harm Index value of 272. The two maps highlight similar areas but, as can be seen, also show subtle differences—the main hot spot in the central, northern area of the District is shown in both maps, but in the crime harm hot spot map, the eastern and southeastern areas of the District are emphasized more.

Figure 3.5a showed the locations of hot spots across an entire study area. Then, if you want to examine hot spots in more specific geographic detail, each hot spot can be examined by producing additional KDE maps for each of these hot spots. Figure 3.7 shows an example of this approach—figure 3.7a uses a bandwidth of 150 meters and cell size of 10 meters to identify the main hot spots in a study area. The main hot spot then becomes the area of attention, with a second KDE hot spot map being produced for this area to identify more precisely the specific locations of the hot spots of crime. Figure 3.7b was generated using a bandwidth of 100 meters and a cell size of 2 meters.

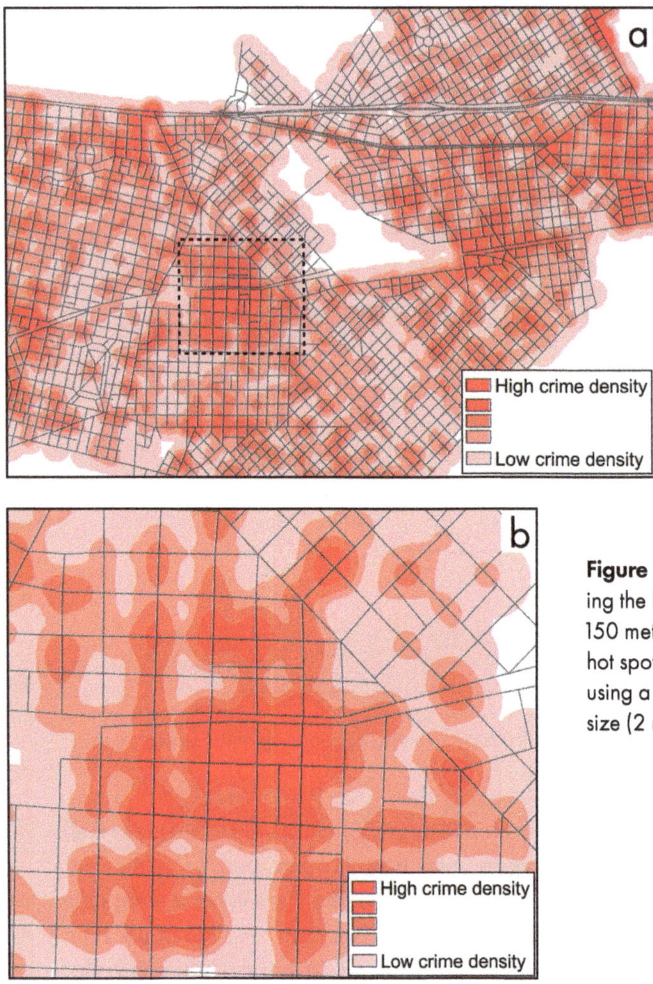

Figure 3.7. (a) A KDE hot spot map for identifying the key strategic areas of focus (bandwidth 150 meters, cell size 10 meters) and (b) a KDE hot spot map for a specific focus area generated using a smaller bandwidth (100 meters) and cell size (2 meters).

KDE is not without its weaknesses. Hot spot maps produced using KDE can fail to identify areas in which there is a high and compact concentration of crime because the bandwidth used (especially if large) may smooth out the density values in these areas. Additionally, when using KDE, it is up to you to choose what is defined as *hot* by deciding the value for the top thematic class. In the examples shown in this chapter, we standardized this procedure by using the natural breaks thematic classification method for all the maps produced. However, most GIS software offers several options to determine a thematic classification method, leading to further subjectivity in defining what is *hot* in the hot spot mapping output generated.

KDE hot spot mapping across a network

Most applications of hot spot mapping with KDE use crime data that is geographically referenced as points. KDE also can be applied to line objects, such as street segments. Figure 3.8 shows an example, comparing a KDE map produced across a planar surface (as shown in the other KDE examples in this chapter) and a KDE map produced for a network. Xie and Yan (2008) provide a good example of the application of linear KDE, and Partridge (2013) provides details of an alternative method for mapping crimes along routes. Performing these techniques on linear objects, however, is limited because there are not many software tools to do so.

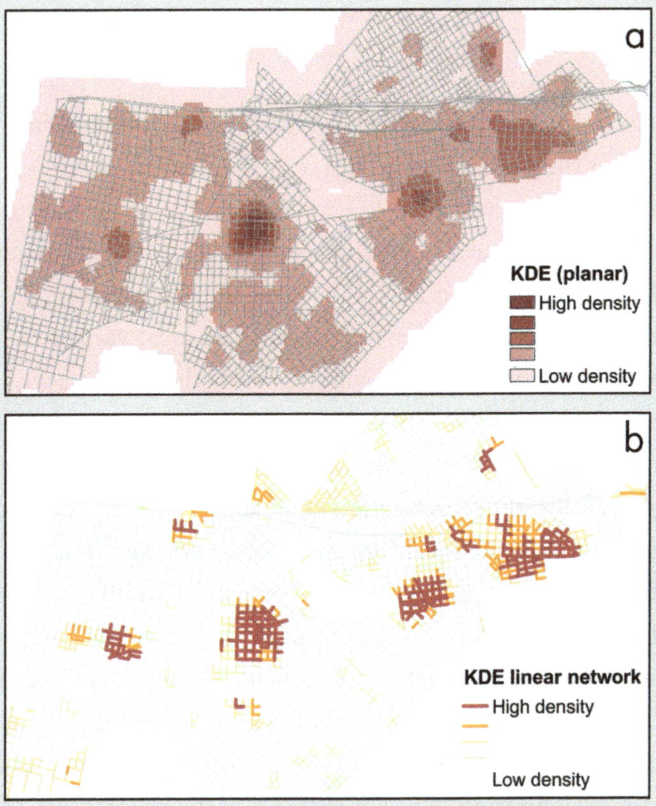

Figure 3.8. A comparison of (a) KDE applied across a planar surface and (b) KDE applied across a network of streets.

Determining what is hot

Hot spot maps generated using KDE are useful for showing where crime concentrates but may fail to determine unambiguously what is hot (in hot spot analysis terms) from what is not hot. That is, a KDE hot spot map may fail to separate the significant spatial concentrations of crime from those spatial distributions of less interest, or from random variation. Spatial statistical techniques that determine whether the patterns of crime concentration are statistically significant include local Moran's I and the Gi* statistic. These techniques identify the association between a single point and its neighbors, up to a specified distance from the point (Anselin 1995). Each of these techniques requires data to be aggregated to some form of geographic unit (e.g., census output areas or grids) and a spatial neighborhood to be defined for each of these geographic units. This could be units that are adjacent to or units within a specified radius. Importantly, both the local Moran's I and Gi* statistic techniques can determine what is hot by determining whether a cluster of crime is statistically significant.

Local Moran's I

Local Moran's I is a variant of its global spatial autocorrelation equivalent (described in chapter 2). The global Moran's I statistic compares the similarity in values (e.g., the number of crimes) between each location and its nearest neighbors. Where values in a geographic unit are high and surrounded by other geographic units with similarly high values, positive spatial autocorrelation exists. If areas with low values are surrounded by high-crime areas and high-crime areas are surrounded by areas with little crime, negative spatial autocorrelation can be observed. The local variant of Moran's I essentially applies Moran's I to each local spatial unit in comparison with its neighbors. The local Moran's I can be used to determine for each observation the extent to which there is spatial association (i.e., clustering of similar values around that observation). Typically, to apply local Moran's I, you are required to enter one parameter: the search radius that defines the neighbors to include around an observation in the calculation of the local Moran's I (often referred to as the *distance band* or *threshold distance*). In many ways, the search radius is like the bandwidth measure used for kernel density estimation, except that a suitable value is slightly easier to determine. Most often, our interest is in the local association between a geographic unit and its nearest neighbors, and most usually the surrounding neighbors that share a boundary with the geographic unit. The size of the search radius, therefore, should be at least the distance from the centroid of the observation to the centroids of all its eight surrounding neighboring cells. For example, if the cell size was 125 meters, a search radius to apply would be 178 meters, as this is the farthest distance to include the eight neighbors that surround the cell [the cells on the diagonal—i.e., $\sqrt{(125^2 + 125^2)}$].

An exact test of significance is not possible for local Moran's I because the distribution of each statistic is not known, and the expected value of I and the variance of I are complicated mathematically. Instead, high positive or high negative standardized scores of I [termed $Z(I_i)$] can be taken as indicators of similarity or dissimilarity (Levine 2010). A high positive value for the $Z(I_i)$s indicates the spatial clustering of similar values (either high or low), whereas a high negative score indicates clustering of dissimilar values (a high area surrounded by low areas or a low area surrounded by areas with high values). The higher the value of $Z(I_i)$, the more the observation is similar (positive) or dissimilar (negative) to its neighbors. Thus, the local Moran's I statistic can be used to identify crime hot spots that, in a spatial sense, are statistically significant.

Figure 3.9 shows an example of a local Moran's I hot spot map. The map was generated using grid cells, with a count of crime for each grid cell. The search radius was set so that it includes only the closest cells around each observation. Figure 3.9a shows several areas in which the observations were determined to be statistically significant. Recall that local Moran's I is a test of spatial association. Therefore, the areas shown as significant are those in which there is a clustering of similar values around that observation. Figure 3.9a shows both hot spots (grid cells with high values surrounded by cells also with high values) and cold spots (grid cells with low values surrounded by cells with low values). Therefore, for each area of significance, you are usually required to review the results to determine whether they represent a statistically significant hot spot or a statistically significant cold spot. Figure 3.9b shows the distinction between hot spots and cold spots from the local Moran's I results.

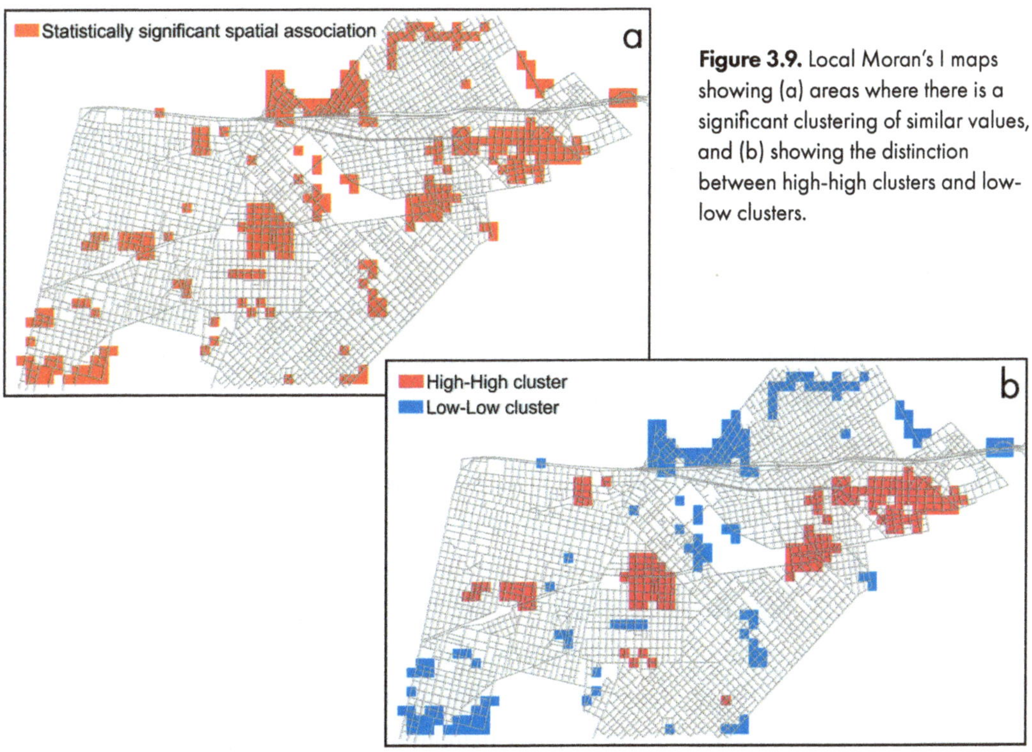

Figure 3.9. Local Moran's I maps showing (a) areas where there is a significant clustering of similar values, and (b) showing the distinction between high-high clusters and low-low clusters.

Ratcliffe (2010) also shows a useful application of local Moran's I for identifying emerging patterns of crime—where areas that were mainly responsible for an overall increase in crime in a city were surrounded by other areas that also were largely responsible for the citywide increase. We examine techniques for analyzing emerging crime areas in chapter 6.

The Gi* statistic

The Gi* (pronounced G-i-star) statistic compares local averages with global averages to identify areas that are significantly different in comparison with what is generally observed across the whole study area (Ord and Getis 1995). Using data aggregated into geographic units (e.g., grid cells or census areas), the sum of values for the unit of interest and its neighbors within a user-defined radius is compared with all the values in the geographic units for the entire study area. When there is a clustering of high values, the Gi* values in these areas will be positive. When there is a clustering of low values, the Gi* values will be negative (Anselin 1995; Chainey and Ratcliffe 2005; Getis and Ord 1996; Ord and Getis 1995).

The output generated from the Gi* statistic comprises standardized z-scores that can be used to determine whether the Gi* value for each cell of interest is statistically significant. The z-scores indicate the place of a value in a dataset relative to the mean, standardized with respect to the standard deviation. When z equals zero, the observation is equivalent to the mean of the entire sample. When z is less than zero, the observation is a value less than the mean. When z is greater than zero, the observation is a value greater than the mean. When a normal distribution of the observations is assumed, z-score values of 1.96, 2.576, and 3.291 are used to determine $p <= .05$, (95% confidence), $p <= .01$ (99% confidence), and $p <= .001$ (99.9% confidence) statistical significance levels, respectively. That is, if a Gi* result for an observation is greater than or equal to 1.96, the observation is determined to be statistically significant to $p <= .05$. The Gi* statistic, therefore, can distinguish between those areas that, in a statistical sense, are hot from those that are not hot (i.e., where there is significant spatial concentration in the distribution of crime data across a study area in comparison with the spatial distribution of crime that is not significant).

The Gi* statistic typically requires you to determine the value of one parameter—the lag distance. Like the search radius described for local Moran's I, the lag distance is the radius around each cell that determines which neighboring cells to include in the calculation of Gi*. Again, like local Moran's I, our interest is in the local association between a geographic unit and its nearest neighbors, and most usually the immediate surrounding neighbors. The lag distance, therefore, should be at least the distance of the radius from each cell's centroid that has a coverage that will consider all its immediate surrounding neighbors.

An issue that confronts the application of many spatial analysis techniques (relevant to Gi* and local Moran's I) is the multiple use of observations in analysis calculations. A statistical assumption that must be upheld is that the determination of each z-score value is independent. With many spatial analysis techniques (especially cluster analysis techniques), treatment of statistical tests usually is applied not only to the observation but also to a defined set of neighbors. Each observation is used not just once, but multiple times because it is included as a neighbor for many other observations. This issue of testing multiple times results in the increased probability of incorrectly rejecting the null hypothesis—referred to as a Type I error. Failure to account for these effects results in the overidentification of clusters using Gi* and local Moran's I (Anselin 1995; Getis and Aldstadt 2004; Ord and Getis 1995).

Several approaches exist to correct for multiple testing. False discovery rate (FDR) controlling procedures seek to reduce the expected proportion of Type I errors (i.e., false discoveries) when the issue of multiple testing is present (for an example of FDR applied to geographic data, see Caldas de Castro and Singer 2006). However, FDR approaches tend not to be that conservative, resulting in a higher chance of Type I errors remaining. Family-wise error rate (FWER) approaches seek to reduce the probability of even one false discovery and, therefore, tend to be more conservative than FDR approaches (Brunsdon and Charlton 2011).

Ord and Getis (1995) suggest a Bonferroni correction (a FWER procedure) to address the problem of multiple testing associated with the application of the Gi* statistic. The Bonferroni procedure has the advantage of being straightforward to calculate but can be conservative if there are many tests. Also, the Bonferroni correction procedure comes at the cost of increasing the probability of producing Type II errors (i.e., the failure of detecting whether an effect is present), and can result in a lack of statistical power (i.e., the probability that the null hypothesis is rejected when clustering occurs) (Brunsdon and Charlton 2011). As previously advised, the number of surrounding areas used in the computation of each Gi* statistic is small (e.g., when grid cells are used as the geographic unit, only the eight surrounding neighbors for each cell are used). Hence, although conservative, the Bonferroni correction is considered accurate in these circumstances (Rogerson 2004). In addition, a conservative determination of statistical significance thresholds is considered a positive feature for hot spot analysis because it may be useful in identifying small specific areas for focused policing and crime prevention targeting.

Figure 3.10 shows Gi* statistic results for a study area. Figure 3.10a shows statistically significant crime hot spots using an uncorrected procedure. The crime hot spots are displayed according to four thematic thresholds: not significant (i.e., Gi* values were less than 1.96); significant to $p <= .05$ (i.e., Gi* values were greater than 1.96); significant to $p <= .01$ (i.e., Gi* values were greater than 2.576); and significant to $p <= .001$ (i.e., Gi* values were greater than 3.291). Figure 3.10b shows statistically significant crime hot spots using the FDR correction procedure, and figure 3.10c shows statistically significant hot spots using the Bonferroni correction procedure. The Bonferroni method is the more conservative of the two correction procedures, shown by displaying a smaller number of hot spots in comparison with the two other maps.

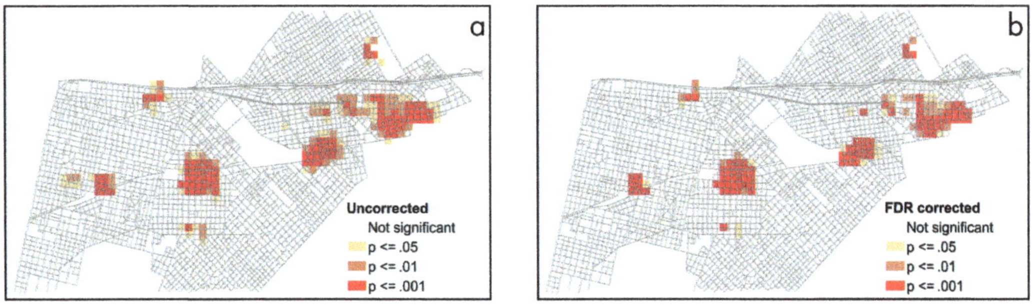

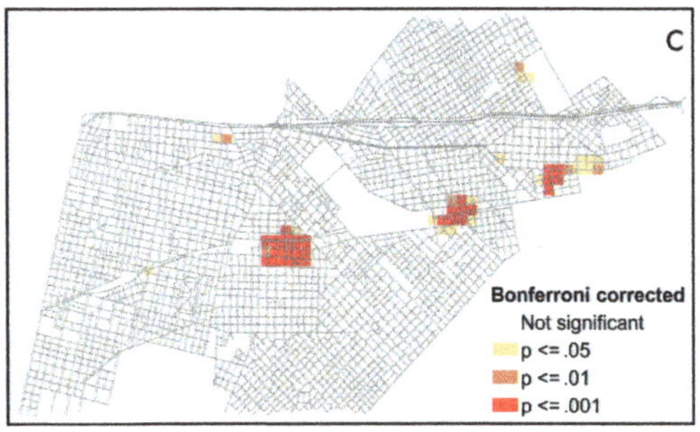

Figure 3.10. Gi* crime hot spot maps: (a) uncorrected, (b) FDR corrected, and (c) Bonferroni corrected.

Applying an FDR and Bonferroni correction to local Moran's I and Gi* statistic results in ArcGIS

In ArcGIS® software, a function exists in the local Moran's I and Gi* tools for applying the FDR correction to determine significance levels. Simply enable the FDR function when calculating the local Moran's I or Gi* statistic to use this correction procedure.

A Bonferroni correction procedure for determining significance levels involves a two-step process:

First, for a significance level of *p* <= *.05*, you must calculate the number of geographic units (e.g., grid cells) representing 5% of the study area: 0.05/number of cells covering the study area.

Second, you must determine the z-score that represents this corrected percentile. The most practical way is to use a percentile to z-score calculator to determine the z-score value that represents the *p* <= *.05* statistical significance threshold. Many tools exist online to calculate this value, with one such online tool available from MeasuringU at http://www.measuringusability.com/zcalcp.php.

The two-step process then can be repeated for determining Bonferroni corrected *p* <= *.01* and *p* <= *.001* statistical significance threshold values, substituting the initial calculation using 0.01/number of cells covering the study area and 0.001/number of cells, respectively.

For example, for an area comprising 256 cells, a Bonferroni corrected *p* <= *.05* statistical significance threshold value was calculated as 3.725 (i.e., the result of 0.05/256, which then was converted to a standardized z-score). That is, under a Bonferroni correction, any cell for this study area that has a standardized local Moran's I or Gi* value that is greater than or equal to 3.725 is statistically significant to $p <= .05$ (compared to an uncorrected standardized z-score of 1.96 for $p <= .05$).

The use of the Gi* statistic for mapping hot spots of crime helps remove the ambiguity of defining areas that are hot (as in the case of using KDE). Additionally, a problem sometimes associated with KDE is that, in some locations, it may smooth over or into areas in which there is no recorded crime because it is not constrained to the high detail of the underlying geography of the crime point distribution. An advantage of the Gi* statistic is that it compares local averages against global averages and can exclude in an analysis areas in which it is impossible for certain crimes to occur (e.g., for residential burglary, this would include parkland and rivers, or any other areas where there is no residential housing). Areas that can be excluded can be identified in a GIS, and the grid cells that cover these areas could be extracted from the full grid coverage so they do not influence the global average. Another problem associated with KDE is that it may smooth away the peaks in areas in which the specified bandwidth aggregates high values with neighboring low values, and as a result overlooks areas in which the concentration of crime is tightly compact. The Gi* statistic is better at avoiding this problem. As shown in figure 3.11, the KDE map and Gi* map are similar, as they are both effective in identifying the locations where hot spots are located. However, the map showing the Gi* statistic results identifies several small areas in which crime was observed to significantly concentrate, with these areas not appearing as hot spots in the KDE map.

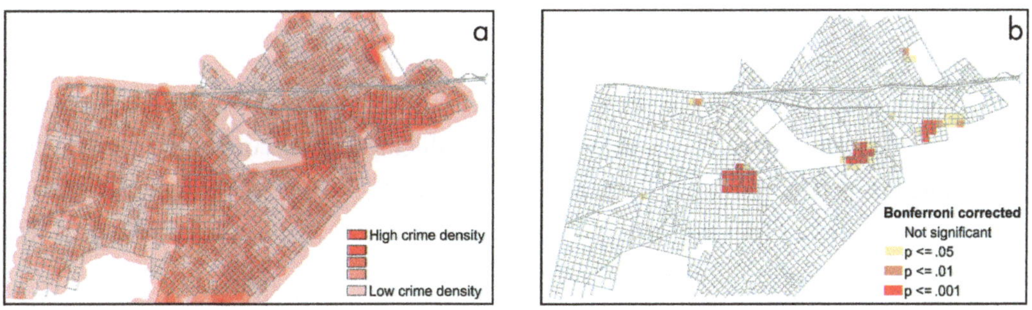

Figure 3.11. A comparison between (a) KDE hot spot mapping output, and (b) the hot spot mapping output produced using the Gi* statistic.

In this chapter, we have looked at two categories of hot spot analysis techniques: measuring and mapping crime concentration at the micro-place geographic scale and spatial clustering techniques. Each category of techniques has advantages and disadvantages, so in practice it is often useful to use a combination of techniques when determining which areas to target and deploy resources. For instance, crime concentration measures initially would help indicate the extent of the concentrated pattern of crime, such as the proportion of micro-places that account for 25 percent of crime. Then, mapping the micro-place units that account for 25 percent of crime would reveal the specific locations in which these crimes are most concentrated, supported with a Gi* statistic hot spot map identifying the most significant clusters of crime (see figure 3.12 for an example).

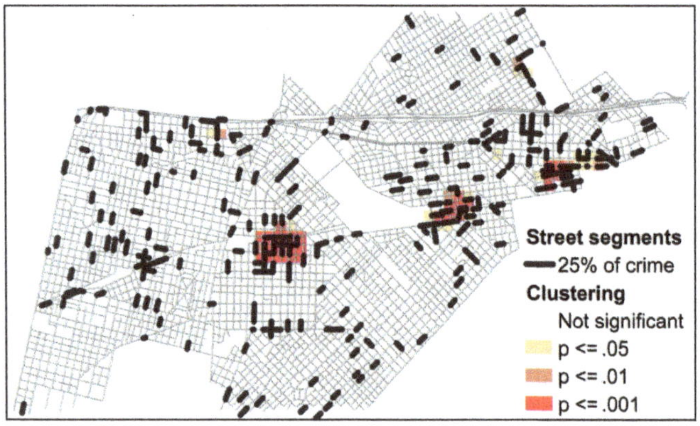

Figure 3.12. Street segments that experienced high levels of crime concentration with Gi* hot spots.

Chapter summary

Hot spot analysis is one of the most used techniques for examining the geographic distribution of crime. In this chapter, we have examined the use of two categories of techniques for analyzing and mapping crime hot spots.

Examining crime concentration at the micro-place scale provides the means to identify the specific locations in which crime is observed to concentrate. Measures, such as comparing with Weisburd's bandwidths of crime concentration and producing Lorenz curves and Gini coefficients for examining crime concentration more fully, can be used to indicate the extent of crime concentration. When the areas of high concentration are mapped, this then shows where interventions (such as hot spot policing) specifically can be targeted.

Spatial clustering techniques, such as KDE, local Moran's I, and the Gi* statistic, indicate how the level of crime in each location is spatially associated with crime in neighboring locations. In particular, the Gi* statistic not only identifies where clusters of crime are located but also can be used to determine what is hot, in statistical terms.

Together, when considered alongside the reasons why hot spots form, both categories of techniques can help determine where resources can be targeted and help indicate the type of activity to perform in these areas.

References

Amemiya, M., and T. Ohyama. 2019. "Toward a Test of the 'Law of Crime Concentration' in Japanese Cities: A Geographical Crime Analysis in Tokyo and Osaka." *Crime Science* 8:11.

Anselin, L. 1995. "Local Indicators of Spatial Association—LISA." *Geographical Analysis* 27:93–115.

Bailey, T. C., and A. C. Gatrell. 1995. *Interactive Spatial Data Analysis*. Reading, MA: Addison-Wesley.

Bernasco, W., and W. Steenbeek. 2017. "More Places than Crimes: Implications for Evaluating the Law of Crime Concentration at Place." *Journal of Quantitative Criminology* 33 (3): 451–67.

Block, R., and R. B. Block. 2000. "The Bronx and Chicago—Street Robbery and the Environs of Rapid Transit Stations." In *Analyzing Crime Patterns: Frontiers and Practice*, edited by V. Goldsmith, P. G. McGuire, J. H. Mollenkopf, and T. A. Ross. Thousand Oaks, CA: Sage.

Bowman, A. W., and A. Azzalini. 1997. *Applied Smoothing Techniques for Data Analysis: The Kernel Approach with S-Plus Illustrations*. Oxford: Oxford University Press.

Braga, A. A., A. V. Papachristos, and D. Hureau. 2014. "The Effects of Hot Spots Policing on Crime: An Updated Systematic Review and Meta-Analysis." *Justice Quarterly* 31 (4): 633–63.

Brunsdon, C. 1995. "Estimating Probability Surfaces for Geographical Point Data: An Adaptive Kernel Algorithm." *Computers and Geosciences* 21:877–94.

Brunsdon, C., and M. Charlton. 2011. "An Assessment of the Effectiveness of Multiple Hypothesis Testing for Geographical Anomaly Detection." *Environment and Planning B: Planning and Design* 38 (2): 216–30.

Caldas de Castro, M., and B. H. Singer. 2006. "Controlling the False Discovery Rate: A New Application to Account for Multiple and Dependent Tests in Local Statistics of Spatial Association." *Geographical Analysis* 38:180–208.

Chainey, S. P. 2013. "Examining the Influence of Cell Size and Bandwidth Size on Kernel Density Estimation Crime Hotspot Maps for Predicting Spatial Patterns of Crime." *Bulletin of the Geographical Society of Liege* 60:7-19.

Chainey, S. P., and J. Chapman. 2013. "A Problem-Oriented Approach to the Production of Strategic Intelligence Assessments." *Policing: An International Journal of Strategies and Management* 36 (3): 474–90.

Chainey, S. P., and J. H. Ratcliffe. 2005. *GIS and Crime Mapping*. London: Wiley.

Chainey, S. P., and J. Monteiro. 2019. "The Dispersion of Crime Concentration during a Period of Crime Increase." *Security Journal* 32 (3): 324–41.

Chainey, S. P., G. Pezzuchi, N. O. Guerrero Rojas, J. L. Hernandez Ramirez, J. Monteiro, and E. Rosas Valdez. 2019. "Crime Concentration at Micro-Places in Latin America." *Crime Science* 8:5.

Chainey, S. P., L. Tompson, and S. Uhlig. 2008. "The Utility of Hotspot Mapping for Predicting Spatial Patterns of Crime." *Security Journal* 21:1-2.

Chainey, S. P., R. Serrano-Berthet, and F. Veneri. 2020. "The Impact of a Hot Spot Policing Program in Montevideo, Uruguay: An Evaluation Using a Quasi-experimental Difference-in-Difference Negative Binomial Approach." *Police Practice and Research*. DOI:10.1080/15614263.2020.1749619.

Curtis-Ham, S., and D. Walton. 2018. "The New Zealand Crime Harm Index: Quantifying Harm Using Sentencing Data." *Policing: A Journal of Policy and Practice* 12 (4): 455–67.

Eck, J. E., S. P. Chainey, J. G. Cameron, M. Leitner, and R. E. Wilson. 2005. *Mapping Crime: Understanding Hot Spots*. Washington, DC: National Institute of Justice.

Felson, M. 1987. "Routine Activities and Crime Prevention in the Developing Metropolis." *Criminology* 25:911–32.

Felson, M. 1998. *Crime and Everyday Life: Impact and Implications for Society*. Thousand Oaks, CA: Pine Forge Press.

Fotheringham, A. S., C. Brunsdon, and M. Charlton. 2002. *Geographically Weighted Regression: The Analysis of Spatially Varying Relationships*. Chichester: Wiley.

Getis, A., and J. Aldstadt. 2004. "Constructing the Spatial Weights Matrix Using a Local Statistic." *Geographical Analysis* 35 (2): 90–104.

Getis, A., and J. K. Ord. 1996. "Local Spatial Statistics: An Overview." In *Spatial Analysis: Modeling in a GIS Environment*, edited by P. Longley and M. Batty. Cambridge: GeoInformation International.

Gill, C., A. Wooditch, and D. Weisburd. 2017. "Testing the 'Law of Crime Concentration at Place' in a Suburban Setting: Implications for Research and Practice." *Journal of Quantitative Criminology* 33 (3): 519–45.

Goldsmith, V., P. G. McGuire, J. H. Mollenkopf, and T. A. Ross. 2000. *Analyzing Crime Patterns: Frontiers of Practice*. New York: Altamira Press.

Harries, K. 1999. *Mapping Crime: Principles and Practice*. Washington, DC: United States National Institute of Justice.

Home Office. 2005. *Crime Mapping: Improving Performance. A Good Practice Guide for Front Line Officers*. London: Home Office.

Lee, Y., J. E. Eck., S. O, and N. Martinez. 2017. "How Concentrated Is Crime at Places? A Systematic Review from 1970 to 2015." *Crime Science* 6 (6): 1-16.

Levine, N. 2010. *CrimeStat III: A Spatial Statistics Program for the Analysis of Crime Incident Locations*. Houston, TX: Ned Levine and Associates and Washington, DC: National Institute of Justice.

Openshaw, S. 1984. "The Modifiable Areal Unit Problem." *Concepts and Techniques in Modern Geography 38*. Norwich, England: Geobooks.

Ord, J. K., and A. Getis. 1995. "Local Spatial Autocorrelation Statistics: Distributional Issues and an Application." *Geographical Analysis* 27 (4): 286–306.

Osborne, D. A., and S. C. Wernicke. 2003. *Introduction to Crime Analysis: Basic Resources for Criminal Justice Practice*. New York: Haworth Press.

Partridge, H. 2013. *Hot Routes*. JDiBrief Series. London: UCL Jill Dando Institute of Security and Crime Science.

Ratcliffe, J. H. 2010. "The Spatial Dependency of Crime Increase Dispersion." *Security Journal* 23:18–36.

Ratcliffe, J. H. 2015. "Harm-Focused Policing." *Ideas in American Policing* 19:1-12.

Ratcliffe, J. H., T. Taniguchi, E. Groff, and J. Wood. 2011. "The Philadelphia Foot Patrol Experiment: A Randomized Controlled Trial of Police Patrol Effectiveness in Violent Crime Hotspots." *Criminology* 49 (3): 795–831.

Rogerson, P. A. 2004. "The Statistical Significance of the Maximum Local Statistic." In *Spatial Econometrics and Spatial Statistics*, edited by A. Getis, J. Mur, and H. G. Zoller. Basingstoke: Palgrave McMillan.

Sherman, L. 2009. "Hot Spots." In *The SAGE Dictionary of Policing*, by A. Wakefield and J. Fleming. London: Sage Publications.

Sherman, L., P. Gartin, and M. E. Buerger. 1989. "Hot Spots of Predatory Crime: Routine Activities and the Criminology of Place." *Criminology* 27:27–55.

Sherman, L., P. W. Neyroud, and E. Neyroud. 2016. "The Cambridge Crime Harm Index: Measuring Total Harm from Crime Based on Sentencing Guidelines." *Policing: A Journal of Policy and Practice* 10 (3): 171–83.

Silverman, B. W. 1986. *Density Estimation for Statistics and Data Analysis*. London: Chapman and Hall/CRC.

Smith, W. R., S. G. Frazee, and E. L. Davison. 2000. "Furthering the Integration of Routine Activity and Social Disorganization Theories: Small Units of Analysis and the Study of Street Robbery as a Diffusion Process." *Criminology* 38 (2): 489–524.

Von Hirsch, A., and N. Jareborg. 1991. "Gauging Criminal Harm: A Living-Standard Analysis." *Oxford Journal of Legal Studies* 11:1.

Weisburd, D. 2015. "The Law of Crime Concentration and the Criminology of Place." *Criminology* 53:133-57.

Weisburd, D., E. R. Groff, and S. Yang. 2012. *The Criminology of Place: Street Segments and Our Understanding of the Crime Problem*. Oxford: Oxford University Press.

Weisburd, D., S. Bushway, C. Lum, and S. Yang. 2004. "Trajectories of Crime at Places: A Longitudinal Study of Street Segments in the City of Seattle." *Criminology* 42:283–322.

Xie, Z., and J. Yan. 2008. "Kernel Density Estimation of Traffic Accidents in a Network Space." *Computers, Environment, and Urban Systems* 35 (5): 396–406.

CHAPTER 4

The temporal analysis of crime

Objectives

In this chapter, we examine ways to analyze temporal patterns of crime. The focus is on temporal analysis techniques that build on the spatial analysis techniques presented in previous chapters, and that provide a foundation for several techniques in the chapters that follow. We begin by examining time discretely. This involves examining techniques to analyze when crime concentrates by day of the week and time of day. We then illustrate the importance of considering temporal patterns and spatial patterns together. We illustrate the use of temporal concentration profiles for areas and describe a simple technique, the hot spot matrix, that highlights differences in crime patterns on the basis of their spatial and temporal characteristics. We then describe measures for analyzing the stability of crime. This includes a simple but useful measure of variability, and a more advanced technique, Andresen's S Index—a statistical technique for determining whether the differences between two patterns are stable or not stable.

In later chapters, we describe other ways to consider time and space together (e.g., for evaluating the impact of geographically targeted interventions) and several other more advanced spatial and temporal analysis techniques, such as the analysis of emerging problem areas and near repeat victimization. The main objectives of this chapter are to introduce you to why an analysis of temporal patterns is crucial to understanding crime and to provide you with the key starting points for temporal crime analysis.

Key learning points

- Understand the importance in examining the temporal patterns of crime, particularly the temporal concentration of crime by the day of the week and the time of day. This includes considering how to analyze data when a time and date range exists in crime records.
- Consider how temporal patterns and spatial patterns are related, and introduce a simple method (the hot spot matrix) that will help you consider time and space together rather than as discrete patterns.
- Understand how to apply measures that examine the stability of crime patterns over time: a simple coefficient of variability measure and Andresen's S Index.

Introduction

It is likely (I hope) that you have reached this chapter and recognize that crime has an inherent geographical component: it has to happen somewhere, be committed by someone whose activities are associated with physical locations, and include a victim or some other target that also is geographically located in some manner. Crime also has an inherent temporal component: it has to happen at some time. *When* a crime takes place, it is highly unlikely that this temporal component is random.

The commission of crime is linked to opportunities. Opportunities to commit crime are not everywhere, nor are they present all the time. Neither are offenders always present whenever a crime opportunity is available. As I write this, one of my teenage sons has just cycled home from school. He has come in through the front door and left his rather nice mountain bike on the driveway, unlocked, a perfect opportunity for any *motivated offender* to steal it. But right now, there is no offender lurking on our street. My son remembers he has left his bike out front and pops outside and brings it into the garage, so it's now safe and secure (and no longer a *suitable* target). If he had left his bike outside overnight, the chances of a person walking by and seeing a nice, unlocked mountain bike, and the opportunity for theft (under the cover of darkness), would be much higher. *Time* is important when analyzing crime because it can help us identify patterns and inform our understanding about why crime occurs (such as whether it's associated with absent-minded teenagers not taking care of their property).

There are several ways to consider the temporal aspects of crime. We can analyze how crime changes over time, we can examine when crime occurs, and we can examine how a series of crimes may be connected to the offending of offenders or a criminal group. In this chapter, we describe several starting points for analyzing the temporal component of crime. We begin by describing analysis techniques that examine patterns of crime by day of the week and time of day. This includes the use of data clocks—a technique for identifying when crime concentrates. We also discuss what to do when your data contains a time range—for instance, for crime records of burglary when a *from time* (e.g., when the person left their house) and *to time* (when the person returned home and found their house had been burgled) are recorded. This starting point to the techniques discussed in this chapter involves analyzing the temporal component of crime with little consideration of its association with spatial patterns of crime. We quickly address this by illustrating the value of data clocks for different areas, and by doing so help illustrate why a consideration of time and space together is important. We then extend the basic concept of using data clocks for different areas by illustrating the use of Ratcliffe's (2004) hot spot matrix to help consider what may be behind different combinations of spatial and temporal patterns of crime.

We then describe how to apply measures that examine the stability of crime patterns over time. We focus on measures of stability because a key question we often must answer is whether the patterns we are observing now are different in comparison with what has been observed before. We describe two techniques for this type of analysis: variability and Andresen's S Index.

Analyzing temporal concentrations of crime

Simple temporal analysis of when crime occurs involves aggregating the number of crimes by year, season, month, day of the week, or hour of the day to generate a count for each temporal unit of analysis. From this tally, levels of crime can be compared. For example, figure 4.1 shows counts of crime (on the basis of aggregating crime records for each temporal unit) for street robberies in a medium-sized city by months in the year, days of the week, and the time of day. Albeit simple to calculate, the charts usefully show variation during the course of the year (possibly because of seasonal factors rather than an increasing crime trend), with robberies highest on Fridays, and variation across the time of day with peaks in the late-afternoon/early evening period, and an acute peak between 22:00 and 22:59.

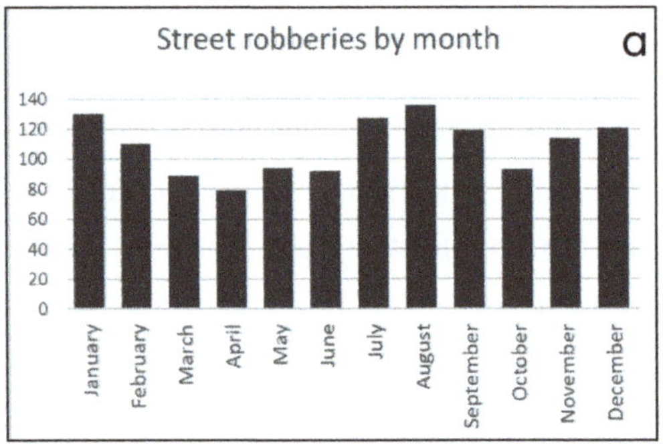

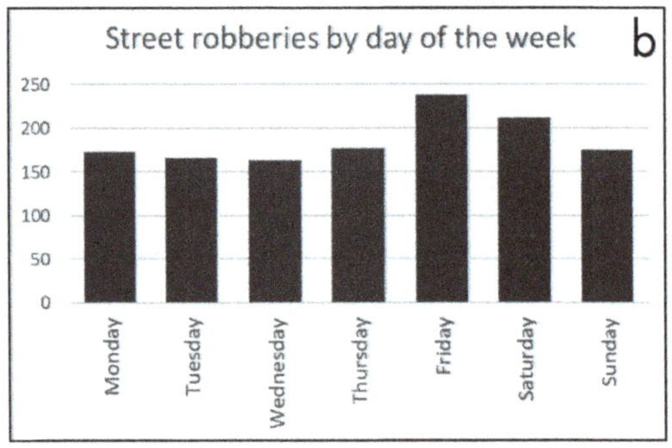

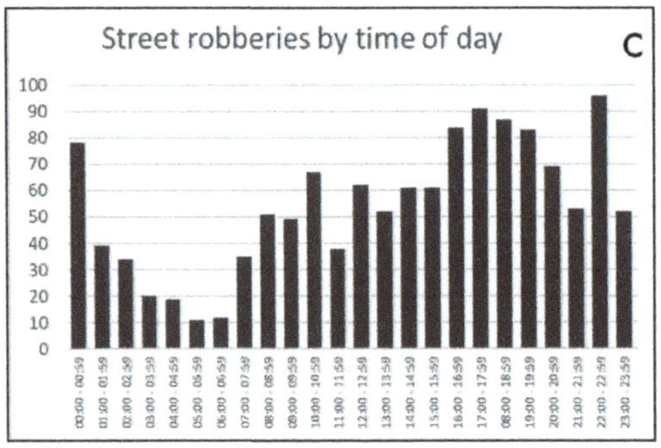

Figure 4.1. Charts showing patterns of street robbery for a medium-sized city, by (a) months, (b) days of the week, and (c) the time of day.

Data clocks

A slightly more advanced way to examine temporal patterns is with the use of a *data clock*. A data clock can be used to examine patterns of crime by the day of the week and the time of day together. To produce a data clock requires coding each crime record by the day on which it took place (e.g., 1 for Monday, 2 for Tuesday, and so on) and coding each crime record for the hour interval in which it took place (e.g., 1 for 00:00–00:59, 2 for 01:00–01:59, … and 24 for 23:00–23:59). With some skill (and patience), this can be done in Microsoft Excel, but several software tools are available to automate the production of data clocks.

Using the same data used to generate figures 4.1b and 4.1c, the data clock in figure 4.2 shows that it is only at certain times on certain days of the week that robberies concentrate. For example, figure 4.1b shows that most robberies occurred on Fridays. The data clock for Fridays shows that it was between 16:00 and 20:00 when robberies first peaked, and then peaked again between 21:00 and 1:00. Figure 4.1c indicates that the overall time of day peak for robberies was between 22:00 and 23:00 hours. The data clock shows that these peaks were present only on Tuesdays and Fridays. Figure 4.1b shows that Monday was one of the days in the week when the least number of robberies occurred, but the data clock in figure 4.2 shows that on Mondays, between 17:00 and 18:00, there is a high concentration of robberies.

Figure 4.2. A data clock of street robberies.

The data clock in figure 4.2 was generated using the function of the same name in ArcGIS Pro. All it required was for me to select the file for analysis and specify the column, or columns, containing the date and time of day information on when each robbery occurred. This tool, however, requires the date and time information to be in *date* and *time* format so that the software recognizes, for example, that February 10, 2020, was a Monday and that 00:00 is an actual time and not a zero value. It is advised that before you run your data through a data clock tool, you review your date and time data to ensure it is correctly formatted and complete. For instance, if you have some records that do not have a value for the time of day, you will need to consider how your process or the software you use for generating data clocks treats blank records. What you must avoid are blank records being treated as taking place at 00:00, because if you have many blank records, the tool will indicate incorrectly that many crimes took place at midnight.

Overnight concentrations of crime

Most data clocks provide output that allows you to visualize temporal patterns of crime only by using midnight as the start and end of the period of interest. Although presenting data for each day of the week chronologically is simpler to organize and present using a data clock, care must be taken when interpreting the results if the crime being studied exhibits patterns of interest that take place from the late evening hours and into the early morning hours.

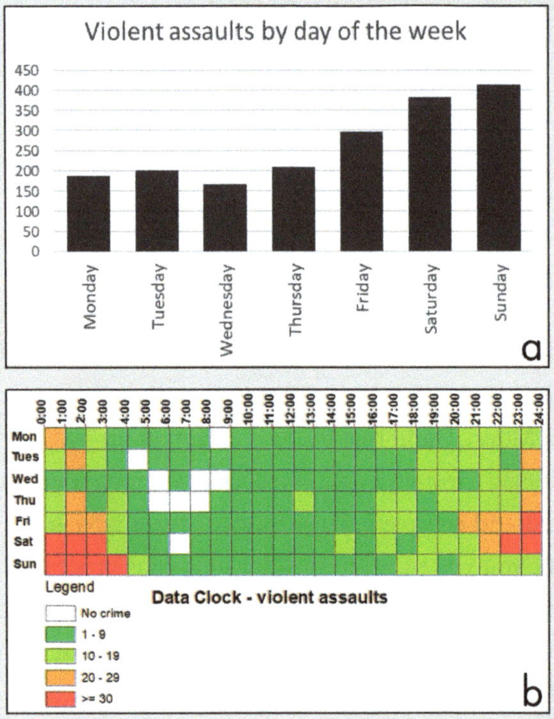

Figure 4.3. (a) A day of the week chart, and (b) a data clock for violent assaults.

In figure 4.3, two temporal charts are shown, created using the same data. Figure 4.3a shows the day of the week when violent assaults occurred, with the peak on Sundays. Figure 4.3b shows a data clock for these violent assaults. Rather than the peak in violent assaults being across the day and night on Sunday, the peak is in the early hours of Sunday morning, extending over from Saturday night. When analyzing day of the week and time of day profiles of crime, it is important to think about the temporal settings in which these crimes occurred. So, in interpreting the results shown in figure 4.3b, it would be better and more accurate to state that assaults concentrated on Friday nights and into the early hours of Saturday, and on Saturday nights and into the early hours of Sunday. This assessment ensures our audience recognizes that Friday and Saturday nights are a problem for assaults, and that Sunday nights are not a problem (it's the early hours of Sunday, continuing from Saturday night, that is the problem for assaults).

Aoristic data clocks

Crimes against the person, such as robberies and assaults, usually have only a single date and time entered into the crime record. For example, if you were assaulted, you most likely would be able to tell the police the date of the assault and the time it took place, closest at least to the nearest hour. So, the date and time information in the crime record may look something like this:

- Date: 01/24/2020
- Time: 23:30

For crimes against property, often there are two date-and-time entries in the crime record—the date and time when the victim left the property and the date and time the victim returned to find the property was broken into (e.g., their home), damaged (e.g., their car), or stolen (e.g., a bicycle left unlocked on the driveway to their house). What is entered into the crime record may look something like this:

- Committed from date: 01/24/2020
- Committed from time: 07:15
- Committed to date: 01/24/2020
- Committed to time: 18:45

In the example above, this could be a residential burglary, in which the owner of the home that was burgled left for work at 07:15, returned at 18:45 to find their house was broken into, and items had been stolen. For these types of crime, an aoristic analysis is necessary. An aoristic analysis involves considering the date and time range and assigning equal probability values to an appropriate temporal unit of analysis (Ratcliffe 2002). In most cases, single-hour intervals are the most appropriate units for temporal aoristic analysis. For example (and as illustrated in figure 4.4), if the crime record stated that the crime took place on a particular day (e.g., 01/24/2020), and between 11:15 and 13:30, for each of the three one-hour intervals during which the crime could have occurred, each interval is assigned a probability value of .333 (to three decimal places). For a crime that took place 01/24/2020, and between 07:15 and 18:45, there would be 12 one-hour intervals during which the crime could have occurred. For this record, each interval is assigned the probability value of .083. For a crime record that had a *committed from date* of 01/23/2020, *committed from time* of 19:30, a *committed to date* of 01/25/2020, and *a committed to time* of 10:30, there would be 40 one-hour intervals, spread over three days, during which the crime could have occurred. For this record, each interval is assigned the probability value of .025. If the aoristic analysis solely examines time of day, note that as the time interval of 10:00–10:59 is represented twice across the time range (and as there are many other one-hour time intervals represented more than once for this crime record), this one-hour time interval contributes twice to the total of all probabilities when all the one-hour time intervals are summed. In the example shown in figure 4.4, the distribution of when crime occurs starts to become clearer as more crimes are treated to this aoristic analysis. If the aoristic analysis examines both day of the week and time of day, the analysis would recognize that the date 01/23/2020 was a

Thursday and so would contribute the probability value of only .025 to the one-hour intervals covered by the time range for Thursdays, and likewise for the probability values for 01/24/2020 (for the one-hour intervals for Fridays), and 01/25/2020 (for the one-hour intervals for Saturdays).

	Day range	00:00	01:00	02:00	03:00	04:00	05:00	06:00	07:00	08:00	09:00	10:00	11:00	12:00	13:00	14:00	15:00	16:00	17:00	18:00	19:00	20:00	21:00	22:00	23:00
Crime 1: 24 Jan 11:15 to 24 Jan 13:30	*01/24/2020*												0.333	0.333	0.333										
Crime 2: 24 Jan 07:15 to 24 Jan 18:45	*01/24/2020*								0.083	0.083	0.083	0.083	0.083	0.083	0.083	0.083	0.083	0.083	0.083	0.083					
Crime 3: 23 Jan 19:30 to 25 Jan 10:30	*01/23/2020*																				0.025	0.025	0.025	0.025	0.025
	01/24/2020	0.025	0.025	0.025	0.025	0.025	0.025	0.025	0.025	0.025	0.025	0.025	0.025	0.025	0.025	0.025	0.025	0.025	0.025	0.025	0.025	0.025	0.025	0.025	0.025
	01/25/2020	0.025	0.025	0.025	0.025	0.025	0.025	0.025	0.025	0.025	0.025	0.025													
Sum of probabilities		0.050	0.050	0.050	0.050	0.050	0.050	0.050	0.133	0.133	0.133	0.133	0.441	0.441	0.441	0.108	0.108	0.108	0.108	0.108	0.050	0.050	0.050	0.050	0.050

Figure 4.4. Aoristic calculation of the time of day on which three crimes occurred.

Why not just use the midpoint between a date and time range for each crime record to produce a data clock rather than use the aoristic approach?

This question was answered by researcher Matt Ashby in a study he conducted. Matt showed that aoristic analysis is much more accurate for determining when crimes occur than either the midpoint in a date and time range, the committed from point, or the committed to point. Data clocks of crime, in which many of the records contain date and time ranges, tend to show only when a person leaves their property if based on the committed from data, when they return to their property if based on the committed to data, or some artificial point in time (e.g., the midpoint) that does not accurately reflect when most of the crimes take place. The findings from his research are available in an open-access journal article (Ashby and Bowers 2013).

Figure 4.5 shows the data clock for an aoristic analysis of residential burglaries for a medium-sized city. The values in each cell are the sum of the probability values for each one-hour interval, across the days of the week. The analysis is based on 1,304 records for a one-year period. Table 4.1 shows that the recorded data on when the burglaries took place indicates that for 68 percent of incidents, the burglary took place on a particular day (rather than the recorded data stating the date of the incident spanned more than one day). For those burglaries in which the crime date range was for less than one day, over half of these incidents had a time range of less than one hour. Less than 5 percent of the burglary records had a date range that was of five days or more.

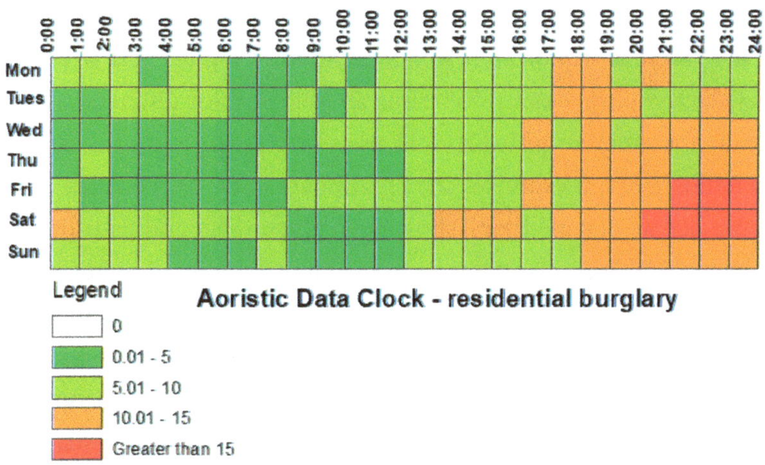

Figure 4.5. Aoristic data clock of residential burglaries for a medium-sized city.

Table 4.1. Data used for calculating the aoristic data clock in figure 4.5

Number of days in date range	*n* of crimes	% of crimes
Same day	888	68.1%
1 day	254	19.5%
2 days	49	3.8%
3 days	35	2.7%
4 days	16	1.2%
5 days or more	62	4.8%
Total	**1,304**	**100%**

a

Number of hours between crimes on the same day	*n* of crimes	% of crimes
<= 1 hour	458	51.6%
> 1 to <= 3 hours	158	17.8%
> 3 to <= 6 hours	118	13.3%
> 6 to <= 12 hours	135	15.2%
Greater than 12 hours	19	2.1%
Total	**888**	**100%**

b

Note: (a) Number of days between committed from date and committed to date, and (b) number of hours between crimes committed on the same day.

Figure 4.5 shows that the peak in residential burglaries was on Friday and Saturday evenings. This may be because offenders are taking advantage of when the residents of the properties that are burgled are out socializing. The city that is the subject of the analysis has a large university student population. The analysis that could follow could examine whether houses and apartments in which students lived were particularly vulnerable to residential burglary, and whether the concentrated criminal activity on Friday and Saturday evenings was because of students' homes being targeted by offenders when these university students were out socializing on these evenings. This example illustrates how a good temporal analysis can prompt hypotheses that can then be subject to further analysis to help understand the observed crime patterns.

Data clock profiles for different areas

The data clocks produced in the previous section were for single areas. A logical next step in using data clocks is to consider how the temporal patterns of crime are different for different areas. This can include creating data clocks for different jurisdictional areas or creating separate data clocks for crime hot spots. Figure 4.6 illustrates an example. The area shown in figure 4.6 is the area that the aoristic data clock in figure 4.5 was created for—showing Friday nights and Saturday nights between 20:00 and midnight as the peak periods. When examined more closely, each hot spot has a different temporal profile from the other, and different from the study area as a whole, to wit:

- In hot spot A (figure 4.6b), burglaries were most concentrated on Friday nights into the early hours of Saturday morning, in addition to midafternoon to early evenings on Sundays and other afternoon and evening peaks on Mondays, Tuesdays, and Wednesdays.
- In hot spot B (figure 4.6c), the main peak was on Sundays between early afternoon and late evening. Other peaks in burglary were on Friday evenings and the late afternoon and early evening on other weekdays.
- In hot spot C (figure 4.6d), the main peaks were on weekdays (except Tuesdays), from late afternoon to early evening, as well as early evening on Sundays.
- In hot spot D (figure 4.6e), the main peak was on Saturday evening until midnight.
- The data clock shown in figure 4.5 aggregates the temporal profiles for these four main hot spots. This example illustrates that to better understand crime, examining specific temporal patterns in each area is important because it then can direct further analysis that may help explain the differences in the patterns.

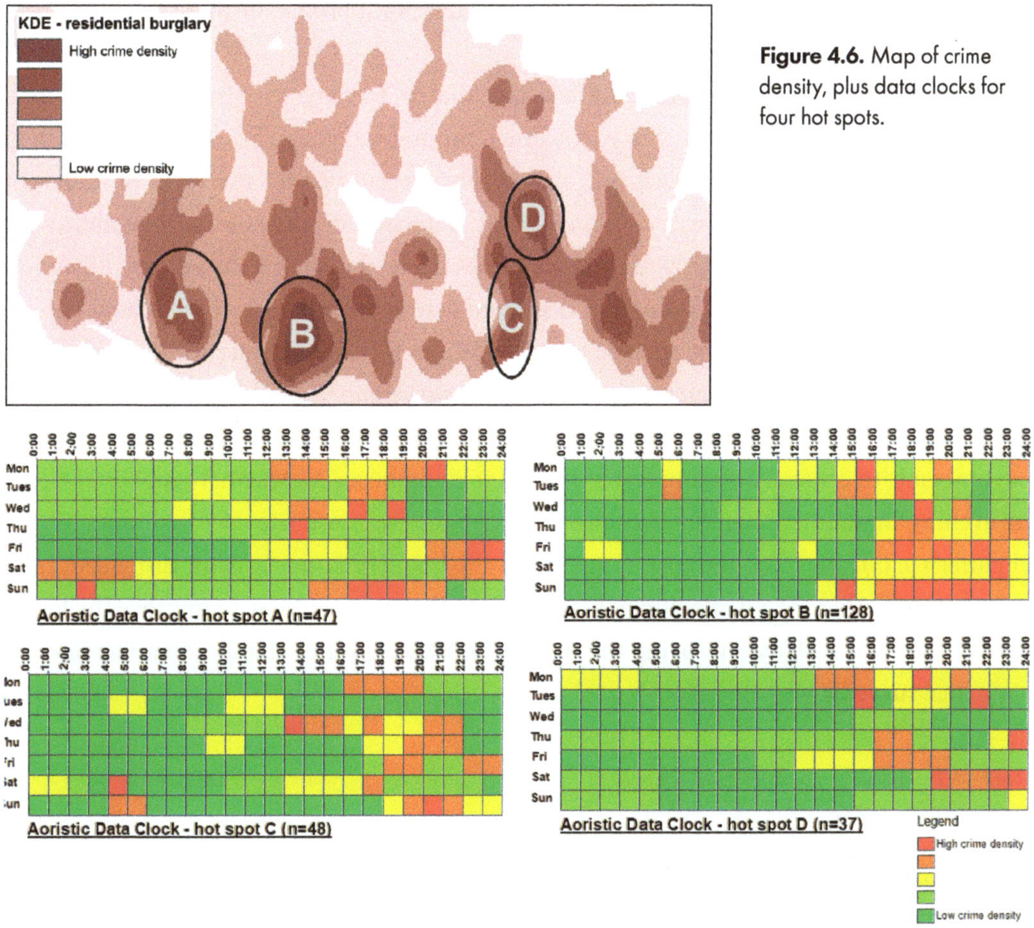

Figure 4.6. Map of crime density, plus data clocks for four hot spots.

The hot spot matrix

The hot spot matrix (Ratcliffe 2004) is a useful way to categorize temporal and spatial patterns, consider how these categories in combination can help interpret these patterns, and prompt ideas on how to counter these patterns. The hot spot matrix is organized into three types of temporal patterns and three types of spatial concentration.

Each category of *spatial concentration* refers to the *dispersal* of crime across a crime hot spot:

- Hot point—crimes are centered on a single location, caused by repeated victimization at this location.
- Clustered—crimes are clustered around a small number of locations within the hot spot.
- Dispersed—crimes are distributed around the whole hot spot.

The three *temporal categories* are as follows:

- Acute—crimes take place during a single short period of time.
- Focused—crimes take place during a short block of time.

- Diffused—crimes are equally spread across the temporal profile, or there is no clear pattern of the times when the crimes occur.

Table 4.2 illustrates an example of the hot spot matrix for street robbery. On the x-axis, the spatial categories are listed, and on the y-axis, the temporal categories are listed, each with graphics that illustrate the category. In each cell, suggestions are entered for the types of interventions that may work best to counter each combination of patterns. For example, high-visibility patrols, deployed to the area during the period most robberies occur (acute or focused), are likely to work best when the spatial pattern is clustered or dispersed across the hot spot. If the spatial pattern is a hot point, interventions should be specific to this location. If the temporal profile for the hot point is acute and the objective is to catch offenders, covert surveillance would be a good option. If the objective is more toward reducing victimization, attention would be toward what makes the hot point vulnerable to crime. For example, the hot point could be an ATM, where robberies occur late in the evening. A solution could be to close the ATM during these late evening hours. If the temporal pattern is diffused, and the spatial pattern in the hot spot is clustered or dispersed, the most effective option, rather than deploying police patrols to the area, could be to improve the landscape features of the area, such as improving the lighting or making other changes that increase natural and informal surveillance.

Table 4.2. A hot spot matrix for street robbery

Street robbery	**Hot point** 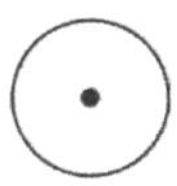	**Clustered** 	**Dispersed** 
Acute 	• Target protection (to reduce vulnerability) • Covert surveillance of offending	• Targeted high-visibility patrols to clusters • Stop, check, and frisk/search	• High-visibility patrols across hot spot
Focused 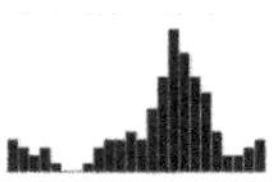	• Target protection (to reduce vulnerability)	• Targeted high-visibility patrols to clusters • Stop, check, and frisk/search • Community engagement (to reduce risk of victimization)	• High-visibility patrols across hot spot • Stop, check, and frisk/search • Community engagement (to reduce risk of victimization)
Diffused 	• Target protection (to reduce vulnerability)	• Targeted street layout/landscape improvements • Targeted improvements in natural surveillance	• Street layout/landscape improvements • Improvements in natural surveillance

When using the hot spot matrix, it is important not to become too concerned about each exact categorization of a spatial and temporal pattern and its definition. The hot spot matrix is designed to recognize the value of considering spatial and temporal patterns together, rather than separately, and then, prompt ideas on the most effective way to counter each combination of patterns.

Analyzing the stability of spatial patterns of crime

Policing and public safety interventions targeted to areas in which crime previously has occurred typically assume one thing—it is these same places in which crime also will take place in the future. At this point, we are not going to get into a long debate about spatial crime prediction. We consider aspects of spatial prediction in other chapters. Instead, at this point, it is important that we understand the stability of our spatial patterns. That is, how the crime patterns we observe change over time. By analyzing the stability of crime, we then can answer the question about whether the places where we have observed crime, and where we are suggesting resources should be targeted, are unlikely to change significantly in the near term.

A consistent finding across many research studies has shown the persistent longevity of crime at certain places (Brantingham and Brantingham 1981; Groff, Weisburd, and Morris 2008; Shaw and McKay 1942; Weisburd et al. 2004). That is, spatial patterns of crime, particularly hot spots of crime, tend to be stable. Other research, however, has contradicted this observation, and has stated that spatial patterns of crime tend to move around (Johnson and Bowers 2004; Johnson, Lab, and Bowers 2008). These findings into the fluid spatial patterns of crime are based mainly on the principles of foraging behavior of offenders (operating in short spates in their crime commission) (Johnson, Summers, and Pease 2009), and principally have been examined fully only for residential burglary rather than a wider range of crime types. This more short-term mobility of offending behavior means that if areas are identified as places for the police to target (on the basis of a recent spate of crime), any targeting of policing or crime prevention activity to these areas may arrive after the offender has moved on. The theoretical principles of offender foraging behavior are discussed in chapter 5 in relation to patterns of repeat and near repeat victimization. At this point, we suggest there is value in testing the stability of our spatial patterns of crime to determine the extent to which the patterns we observe have been stable, and hence are likely to continue to show the same level of stability until something is done to counter the crime problem we are examining. We also examine the distinction between the fluidity of spates of crime and the stability of hot spots of crime when we discuss crime risk in chapter 9.

In this section, two techniques for examining the spatial stability of crime are described: the temporal stability index and the S Index.

Temporal stability index

The stability of crime patterns can be measured using a homogeneity index, a measure commonly used in social science research to summarize the distribution of data across nominal categories (Blau 1977; Chainey and Ratcliffe 2005; Gibbs and Martin 1962). The homogeneity index (which, from this point, for the purposes of crime analysis will be referred to as the *temporal stability index*) takes crime data for an area that has been segmented into equal temporal periods and measures whether crime in the area is a result of crime occurring over a short period or whether the level of crime has been stable over a longer period. The temporal stability index (TSI) is computed using the equation (4.1):

$$TSI = 1 - \Sigma p_i^2 \qquad (4.1)$$

where p_i is the proportion of crime within each time category of the total number of crimes in the area of study for the temporal period being analyzed. The TSI is bounded by a minimum value of 0 and a maximum value of $1-1/n_i$, where n_i is the total number of observed categories. A value close to the maximum indicates the data is heterogeneous—that is, the level of crime is dispersed equally across the observed categories. A value of 0 indicates complete homogeneity—the crimes are concentrated entirely within only one category (Haberman and Ratcliffe 2012). Let us illustrate with an example.

Table 4.3 shows data for two neighborhoods over a 24-week period. The 24-week period is organized into six four-week periods. The number of crimes in each neighborhood for each four-week period is calculated. The volume of crime for each four-week period is expressed as a fraction of the total number of crimes for the entire period of interest. This value for each four-week period is then squared, and these values are summed. One minus the sum of this value represents the TSI value for crime in the area of study over the 24-week period. In this example, the maximum possible TSI is 0.8333 [(1 – (1/6)]. The example in table 4.3 shows 18 crimes in neighborhood A and 18 crimes in neighborhood B over the entire period. In neighborhood A, the 18 crimes were spread equally over the six four-week periods, with three crimes experienced in each period. Once the calculations described have been made, the TSI for this neighborhood was 0.8333, equal to the maximum TSI value. This result illustrates that crime levels in neighborhood A over the six four-week periods have been stable. In contrast, neighborhood B has a TSI of 0 (following the same calculations as above). This results because all 18 crimes occurred in one of the six four-week periods.

Table 4.3. An example of the temporal stability index for two neighborhoods

		Four-week periods							
		1	**2**	**3**	**4**	**5**	**6**	**Sum**	**TSI**
Neighborhood A	*n* crimes in n'hood = 18	3	3	3	3	3	3		.8333
	Fraction of crime	.167	.167	.167	.167	.167	.167		
	Fraction of crime^2	.0278	.0278	.0278	.0278	.0278	.0278	**.1667**	
Neighborhood B	*n* crimes in n'hood = 18	0	0	0	0	18	0		0
	Fraction of crime	0	0	0	0	1	0		
	Fraction of crime^2	0	0	0	0	1	0	**1**	

This example was based on two neighborhoods. The TSI can be applied to as many geographic units as necessary, and of any size, such as jurisdictional areas or hot spots. The following example illustrates the use of the TSI for examining the stability of residential burglary hot spots in the city of Newcastle, England (using data for a one-year period). With use of the Gi* statistic, 16 hot spots were identified ($p <= .05$). Several of the hot spots were small hot spots made up of single cells or a small number of cells. If a hot spot contained six or fewer crimes for the one-year period used in the analysis, these hot spots were not included in the TSI analysis, following the recommendation of Haberman and Ratcliffe (2012). This calculation reduced the number of hot spots for the TSI analysis to seven. Data for the one-year period was organized into 13 four-week periods. The maximum possible TSI for a 13 by four-week analysis period was 0.923. That is, if any of the seven hot spots in Newcastle had a TSI close to this maximum value, the more stable the hot spot, suggesting that the hot spot experienced a consistent high level of crime over the previous year. The minimum TSI was 0.

Table 4.4 lists the TSI results for the seven residential-burglary hot spots. Each hot spot had a TSI of at least 0.85, suggesting that crime levels in these hot spots had been stable over the previous year. The largest hot spot, experiencing 114 burglaries over this period, had the highest TSI value. These results are useful, as they suggest that the hot spots are stable—they are places in which the concentration of crime has endured consistently over the retrospective past. In turn, these hot spots are likely to continue unless something is done to reduce the concentration of crime in these areas (i.e., they are locations in which interventions for reducing crime are likely to be beneficial in this regard).

Table 4.4. TSI values for residential burglary hot spots in the city of Newcastle, England

Hot spot #	**1**	**2**	**3**	**4**	**5**	**6**	**7**
Hot spot TSI	0.920	0.879	0.880	0.861	0.862	0.874	0.857
***n* crimes in hot spot**	114	43	23	29	21	25	21

The S Index

The S Index, developed by Martin A. Andresen (2009), is a spatial point pattern test that applies a nonparametric Monte Carlo approach for measuring the degree of similarity between two sets of spatial data. This process can include taking crime data for one period and comparing its distribution to crime data for a different period. Free software for applying the S Index is available from Simon Fraser University, at http://www.sfu.ca/~andresen/spptest/spptest.html.

The S Index is a global measure and varies between zero and one. The closer the S Index value is to one, the more similar (stable) the pattern. Andresen (2009) suggests that an index of 0.8 or higher indicates a stable spatial pattern. The S Index software also identifies areas in which the change in crime is statistically significant. The process involves taking two point datasets of crime, aggregating each to geographic units (such as jurisdictional areas or grid cells), and comparing the differences between the two spatial distributions.

The following example shows results of an analysis using the S Index for street robberies in six cities in Brazil. Each city has a population of at least 500,000. The S Index values were calculated by comparing robberies in 2015 to robberies in 2016. Robberies increased significantly in each city over this two-year period, from an increase of 25 percent in Rio de Janeiro to an increase of 82 percent in São Gonçalo (Chainey and Monteiro 2019). The application of the S Index would help indicate whether the spatial distribution of robberies remained stable over this two-year period, even though the number of robberies significantly increased. To identify the areas in which the change in crime was significant, small geographic cells of 150 × 150 m were used, with counts of robberies for each grid cell derived from the robbery point data. Following Andresen and Malleson's method (2011), 200 runs of the S Index were performed for each city to determine those grid cells that experienced significant increases in crime (to p <= .01).

Table 4.5. Robbery levels and S Index values for six cities in Brazil

City	Robberies in 2016	Change 2015–16	S Index
Rio de Janeiro	55,149	+25%	0.80
Duque de Caxias	7,938	+63%	0.86
São Gonçalo	12,357	+82%	0.82
Nova Iguaçu	8,310	+69%	0.90
Niteroi	4,629	+37%	0.84
Belford Roxo	2,681	+72%	0.82

Table 4.5 shows that the S Index values for each city were greater than or equal to 0.8. These results suggest the spatial patterns of street robberies in 2016 were like the patterns of robberies observed in 2015, even though the number of robberies increased significantly. Figure 4.7 shows an example for the city of Belford Roxo, in which the changes in the number of robberies were most significant. In total, of the 3,876 150 × 150 m grid cells for the city of Belford Roxo, 202 cells experienced significant increases in crime (p <= .01). As the point pattern of street robberies was stable between the two periods (on the basis of the S Index value of 0.82), the results can be used to deduce that areas in which crime previously was highly concentrated is where crime continues to be highly concentrated, but now with a greater number of crimes.

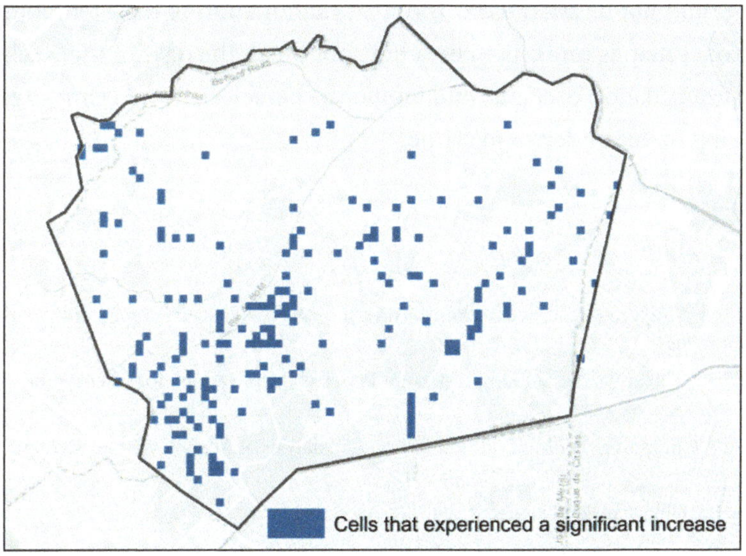

Figure 4.7. Areas that experienced a statistically significant increase in street robberies between 2015 and 2016 in the city of Belford Roxo, Brazil.

Chapter summary

Crime has an inherent temporal component as well as an inherent spatial component. The analyses of temporal patterns of crime are important for understanding crime. Data clocks provide a useful way for examining the temporal profile of crime, by the day of the week and the time of day. These data clocks are particularly useful when applied to specific geographic areas, as they help identify the differences in temporal patterns of crime between areas and can help us consider why these patterns exist in each area. In this manner, data clocks provide a simple but useful means of examining temporal and spatial patterns of crime together. The use of aoristic analysis for data clocks addresses those circumstances in which crime records contain a date and time range.

Analyzing the spatial stability of crime can help us answer the question about whether the places where we have observed crime previously are likely to be suitable for resource targeting because they are unlikely to significantly change soon. The temporal stability index is a useful technique for determining whether the aggregate number of crimes for an area has been consistent over time, or whether the crime in the area relates to only one or a few retrospective periods. If the results from a TSI analysis reveal that the levels of crime in an area, such as in a crime hot spot, have not changed over time, the hot spot then experiences persistence problems with crime. The S Index is a stability measure that can be applied regardless of whether changes in crime have been experienced. The S Index measures whether the overall spatial distribution of crime is similar for two time periods. The S Index, though, also can be used to identify those areas in which the change in crime has been statistically significant.

Spatial patterns of crime should not be analyzed without an examination of their temporal features. For instance, a hot spot on a map is unlikely to be a hot spot on all the days of the week and during all the times of day. An examination of spatial and temporal characteristics of crime together provides a vital means for helping better understand crime.

References

Andresen, M. A. 2009. "An Area-Based Nonparametric Spatial Point Pattern Test: The Test, Its Applications, and the Future." *Methodological Innovations* 9:1–11.

Andresen, M. A., and M. Malleson. 2011. "Testing the Stability of Crime Patterns: Implications for Theory and Policy." *Journal of Research in Crime and Delinquency* 48:58–82.

Ashby, M. P. J., and K. J. Bowers. 2013. "A Comparison of Methods for Temporal Analysis of Aoristic Crime." *Crime Science* 2:1.

Blau, P. 1977. *Inequality and Heterogeneity*. New York, NT: Free Press.

Brantingham, P. J., and Brantingham, P. L. 1981. *Environmental Criminology*. London: Sage.

Chainey, S. P., and J. H. Ratcliffe. 2005. *GIS and Crime Mapping*. Wiley: London.

Chainey, S. P., and J. Monteiro. 2019. "The Dispersion of Crime Concentration during a Period of Crime Increase." *Security Journal* 32 (3): 324–41.

Gibbs, J. P., and W. T. Martin. 1962. "Urbanization, Technology, and the Division of Labor." *American Sociological Review* 27:667–77.

Groff, E. R., D. Weisburd, and N. Morris. 2008. "Where the Action Is at Places: Examining Spatio-Temporal Patterns of Juvenile Crime at Places Using Trajectory Analysis and GIS." In *Putting Crime in Its Place: Units of Analysis in Spatial Crime Research*, edited by D. Weisburd, W. Bernasco, and G. Bruinsma. New York, NY: Springer.

Haberman, C. P., and J. H. Ratcliffe. 2012. "The Predictive Policing Challenges of Near Repeat Armed Street Robberies." *Policing: A Journal of Policy and Practice* 6 (2): 151–66.

Johnson, S. D., and K. J. Bowers. 2004. "The Stability of Space-Time Clusters of Burglary." *British Journal of Criminology* 44 (1): 55–65.

Johnson, S. D., S. P. Lab, and K. J. Bowers. 2008. "Stable and Fluid Hotspots of Crime: Differentiation and Identification." *Built Environment* 34 (1): 32–45.

Johnson, S. D., L. Summers, , and K. Pease. 2009. "Offender as Forager? A Direct Test of the Boost Account of Victimization." *Journal of Quantitative Criminology* 25:181–200.

Ratcliffe, J. H. 2002. "Aoristic Signatures and the Temporal Analysis of High-Volume Crime Patterns." *Journal of Quantitative Criminology* 18:23–43.

Ratcliffe, J. H. 2004. "The Hotspot Matrix: A Framework for the Spatio-Temporal Targeting of Crime Reduction." *Police Practice and Research: An International Journal* 5 (1): 5–23.

Shaw, C. R., and H. D. McKay. 1942. *Juvenile Delinquency and Urban Areas*. Chicago: Chicago University Press.

Weisburd, D., S. Bushway, C. Lum, and S. Yang. 2004. "Trajectories of Crime at Places: A Longitudinal Study of Street Segments in the City of Seattle." *Criminology* 42:283–22.

CHAPTER 5

Risky facilities, and repeat and near repeat victimization

Objectives

Hot spot analysis helps identify where crime spatially concentrates, but often crime can be even more acute in its spatial distribution. In this chapter, we examine three types of acute spatial (and spatiotemporal) patterns of crime: the uneven distribution of crime across *facilities* (often referred to as the *analysis of risky facilities*), repeat victimization, and near repeat victimization. The analysis of crime across facilities (that is, locations that provide a service, such as bars, parks, or supermarkets) involves measuring the distribution of crime across these facilities to identify those that most account for crime. Analysis of repeat victimization can identify the people, or other targets, who experience a significant proportion of all crime. Analysis of near repeat victimization patterns can help identify spates of crime.

In this chapter, we discuss the theory that underpins risky facilities and provide guidance on how to perform the analysis of risky facilities. In modern environmental criminology, the frequently observed patterns of repeat victimization and near repeat victimization have provided an added dimension to the examination of the spatial behavior of offenders and the crime patterns that their offending forms. In this chapter, we discuss the theoretical principles that explain why these patterns occur and how to analyze them. For each of these three patterns, we also explain how police and public safety agencies can use the results from analysis for resource targeting and crime prevention purposes.

Key learning points

- Understand the value in examining the distribution of crime across facilities, and how to perform an analysis of risky facilities. This includes mapping risky facilities, considering how to use denominators to control for differences in facilities, and the practical outcomes of this type of analysis for targeting crime prevention.
- Understand what is repeat and near repeat victimization and the theoretical principles that explain why these patterns form.
- Measure and map repeat victimization and near repeat victimization and determine how these patterns can be used to detect offenders, prevent crime, and assist in crime prediction.

Introduction

In chapter 3, we examined techniques for mapping and analyzing hot spots of crime and discussed the theoretical principles of why crime clusters in these hot spots. The uneven distribution of crime across facilities, the repetitive victimization against people (or other targets), and the spatiotemporal pattern of near repeat victimization provide additional means for examining concentrated patterns of crime. For example, although hot spot analysis may help identify specific areas in which crime concentrates, the reason why these hot spots may be present could be because of the presence of risky facilities in these hot spots. Similarly, a hot spot of crime may be because of a high level of repeat victimization against a person or specific target in the hot spot. Analysis of near repeat victimization may reveal patterns that complement hot spot analysis, identifying areas in which there are short-lived series of crimes, the commission of which may be because of the same returning offender or offenders.

The pattern of risky facilities is well documented, with examples covering many types of crime and many types of facilities. In this chapter, we introduce the concept of risky facilities and how to examine the distribution of crime across facilities. Although the analysis of repeat victimization focuses on people or other targets, we demonstrate in this chapter the value that also comes from examining repeat victimization from a *place* perspective. We begin by discussing the theoretical principles that help explain why repeat victimization occurs, and by doing so illustrate how a consideration of repeat victimization patterns can inform our spatial understanding of crime. An analysis of near repeat victimization can provide a valuable insight into specific spatiotemporal patterns of crime. In this chapter, we discuss the theoretical principles that help explain near repeat patterns of crime and provide details on how this pattern can be analyzed and examined to help improve our spatial and temporal understanding of crime. Throughout the chapter, examples are provided to illustrate how to analyze the patterns of risky facilities, repeat victimization, and near repeat victimization and how to use the results to help target policing and public safety resources.

Risky facilities

The Pareto principle (or the 80/20 law) states that approximately 80 percent of causes come from 20 percent of effects. The principle was originally formed from an observation of landownership, when in 19th-century Italy, Vilfredo Pareto showed that approximately 80 percent of the land in the country was owned by 20 percent of the population. The Pareto principle since has been applied to wealth, economics, sports, health, mathematics, and crime. In many regards, we observe the Pareto principle in how spatial patterns of crime form into hot spots—a large proportion of crime concentrates into a small number of places. But it is also across facilities that we observe this uneven distribution of crime.

Facilities are places that provide a specific public or private function. Examples include bars and nightclubs, shopping malls, specific types of shops (e.g., supermarkets), parking lots for cars, railway stations, recreational parks, schools, and hospitals. Each facility has a location, hence making the study of the uneven distribution of facilities across space a worthwhile topic when seeking to understand the distribution of crime. For crime prevention targeting purposes, the analysis of risky facilities can be extremely useful when we observe that a small proportion of the facilities experience crime, and that most of the facilities experience no crime. That is, rather than thinking that the crime problem must be addressed across all facilities in a city, an analysis of risky facilities can help identify the small number of facilities responsible for most of the crime, which ones to target with crime prevention efforts, and which in turn could result in a significant impact on reducing the crime problem across the city. The observation that a small number of facilities accounts for a large proportion of crime has been shown to be so regular that Wilcox and Eck (2011) describe this pattern as the *iron law of troublesome places*.

Analysis of risky facilities

The analysis of risky facilities involves examining the distribution of crime across a homogeneous set of facilities. For example, rather than examining the distribution of cell phone theft by analyzing all crime records for thefts of cell phones across an entire city, a risky-facilities analysis of cell phone theft would examine the distribution of this crime across a type of facility such as all the cafés in the city or across all bars in the city. Similarly, rather than examining all thefts from cars across a city, a risky-facilities analysis of theft from cars could examine the distribution of this crime across all parking lots in the city. Analysis of this type then would aim to reveal whether the distribution of crime across each type of facility was uneven, and which facilities were most associated with the majority of crime (e.g., the small number of parking lots that accounted for a large proportion of thefts from cars). By doing so, an analysis of crime across facilities then prompts inquiry into why these facilities generate so much crime, such as seeking to determine what it is that these facilities have in common (or not) that makes them vulnerable to crime.

Examples of risky-facilities analysis are numerous. For example, of all the violence and disorder calls for police response across 28 parks in Chula Vista, California, Hilborn (2009) showed that only three parks accounted for 52 percent of all the calls for service. In a study in Merseyside, England, 5 percent of all bus stops experienced nearly 30 percent of vandalism against bus stops (Clark and Eck 2007), and in a study in central London, five bars accounted for 59 percent of all reported incidents of handbag theft across 26 bars that were part of the same commercial chain (Sidebottom and Bowers 2010). For a detailed account of the pattern of risky facilities for many types of crime and disorder, such as robbery, theft, assault, and calls for service, we refer you to the study conducted by Eck, Clark, and Guerette (2007) that examined more than 35 studies of a wide range of facilities.

Analyzing risky facilities

The following steps describe how to perform analysis of the distribution of crime across facilities. The steps are organized into three stages:

- analysis of the distribution of crime across facilities,
- revision of the analysis after considering a denominator, and
- analysis of risk factors associated with the facility.

The steps are illustrated with an example relating to drive-off offenses from gas stations for a study area in England. In England, very few gas stations offer the service of paying for the gas at the pump, and instead require the customer to enter the store of the gas station to pay. One of the main reasons is that most gas stations in England make their money from what they sell in the shop (which can be minisupermarkets) rather than from the gas they sell. Paying at the pump would not entice customers into the shop. Making off without payment for the gas (either intentional or by accident) is a large problem in England. In the study area used for illustration, in a typical year there were more than 1,500 reported incidents of making off without payment for gas. The study area has a population of just over 0.5 million people and has over 300 gas stations. Several major highways cross the study area.

Stage 1: Analysis of the distribution of crime across facilities

Stage 1 involves seven steps.

Step 1: Type of crime and type of facility

The first step involves selecting the type of crime and type of facility for the subject of analysis. In this example, we will examine the crime of making off without payment for gas, and the facility that is the unit of analysis is gas stations.

Step 2: Review of crime data, and data on the facilities

The second step involves reviewing the recorded crime data (or other data about the incidents that are subject to analysis) to determine how the data will allow you to examine the distribution of crime across the facility. For example, the data may include the name of each facility and its address—e.g., BP Station, Acacia Avenue, SW1A 1AA. Each facility must be recorded as unique, with consistency in how the description of the facility is recorded in the data. For example, if two incidents took place at the gas station listed but the address for each incident was entered differently—e.g., BP Station, Acacia Avenue, SW1A 1AA, and British Petroleum Station, Acacia Ave, SW1A 1AA, this inconsistency would cause issues when calculating the number of incidents at each gas station. Therefore, you must check your crime data and perform any necessary data cleaning to ensure the name and address of each facility is entered consistently. An alternative is to use the geographic coordinates of each facility that are recorded in the crime data, as it may solve any inconsistencies in the recording of the name and address of each facility (recorded in the crime data), and could be used to generate a unique identity number (which is useful for a step that then follows). For example, the BP gas station on Acacia Avenue has the x,y coordinates 548623 and 148759. In both cases—BP, Acacia Avenue, SW1A 1AA, and British Petroleum, Acacia Ave, SW1A 1AA—the same coordinates (548623, 148759) have been referenced geographically to the two gas station listings. As a coordinate pair, these coordinates are unique to this gas station. Combining this coordinate pair into a single text string 548623148759 would give it a unique identity number. Repeating the process for all gas stations would result in creating a unique identity number that was repeated only for the same gas stations. This approach assumes the geographic referencing process is suitably precise to ensure that each facility (e.g., gas stations) can be uniquely identified. In the example that will be illustrated, the approach using geographic coordinates was used to identify unique facilities. Then, for simplification purposes, a *unique identity code* from 1 to 357 (relating to the number of gas stations) was assigned to each gas station.

The second part of this step involves reviewing whether a list of all the facilities is available. Although not a showstopper if it is not available, it will help you identify those facilities in which no crime took place.

Step 3: Number of crimes at each facility

The third step involves calculating the number of incidents that took place at each facility. This can be done, for example, by using the Pivot Table function in Microsoft Excel and grouping the unique identity numbers for the facilities under analysis and generating a count of the occurrences of each identity number. For example, if the unique identity number 548623148759, determined from step 2, was listed twice in the crime records data, this gas station would have the number two listed against it. Table 5.1 illustrates an example, listing the identity code of the gas station for the first 10 records in this dataset and the number of making off without payment incidents associated with each gas station. The unique identity numbers calculated from the coordinate pairs are not shown in the table because of the need to sanitize the publication of this precise data. The data showed that 267 gas stations experienced a making off without payment incident. As there were 357 gas stations in the study area, 90 gas stations did not experience any of these incidents. Table 5.1 shows that from a sample of 10 of the gas stations, only five experienced one incident but one experienced 36 incidents.

Table 5.1. Making off without payment incidents associated with gas stations in a sample dataset

Gas station ID code	Number of incidents of making off without payment for gas
187	4
89	1
183	1
52	1
116	36
8	1
257	2
103	1
230	5
107	4

Note: The first 10 records are listed of the 267 gas stations that experienced at least one crime.

Step 4: Ranking the facilities

The fourth step involves ranking the facilities according to the number of incidents, from highest to lowest. This ranking process will generate a table, such as that in table 5.2, listing each facility and the number of incidents that each experienced. Table 5.2 shows that the top-ranked gas station (with an ID code of 223) experienced 236 incidents of making off without payment.

Table 5.2. Gas stations ranked for the number of making off without payment incidents

Rank	Gas station ID code	Number of incidents of making off without payment for gas
1	223	236
2	189	204
3	258	180
4	215	177
5	66	172
6	233	137
7	177	135
8	73	127
9	33	112
10	3	105

Note: Showing the top 10–ranked gas stations of 267 that experienced at least one crime.

Step 5: Figuring the contribution of each facility

The fifth step involves calculating the percentage of incidents that each facility contributes to the total of all incidents, and cumulating the percentages by starting with the *riskiest facility* and adding the percentage contribution of each facility. This tally is illustrated in table 5.3, showing two additional columns of data: proportion of all incidents and cumulative total. This example shows that the top-ranked gas station accounted for 5.2 percent of all making off without payment for gas incidents. The top 10–ranked gas stations accounted for 35.1 percent of all incidents of making off without payment.

Table 5.3. Gas stations ranked for the proportion of making off without payment incidents

Rank	Gas station ID code	Number of incidents	Proportion of all incidents	Cumulative total
1	223	236	5.2%	5.2%
2	189	204	4.5%	9.7%
3	258	180	4.0%	13.7%
4	215	177	3.9%	17.6%
5	66	172	3.8%	21.4%
6	233	137	3.0%	24.5%
7	177	135	3.0%	27.5%
8	73	127	2.8%	30.3%
9	33	112	2.5%	32.8%
10	3	105	2.3%	35.1%

Note: The proportion of all incidents and the cumulative total proportion of all incidents, showing the top 10–ranked gas stations of 267 that experienced at least one crime.

Step 6: Calculating risky facility statistics and visualizing the results on a chart

The cumulate total calculations produced from step 5 can be used to generate useful statistics that illustrate the uneven distribution of incidents across all facilities. For example, using the example of incidents of making off without payment for gas:

- Ten gas stations, representing only 2.8 percent of all gas stations in the study area, accounted for 35.1 percent of all incidents of making off without payment for gas.
- Some 51 percent of all making off without payment incidents were associated with only 18 (5 percent) of the 357 gas stations in the study area.
- A total of 227 gas stations experienced no more than three incidents of making off without payment. Of these 227, 90 gas stations experienced no making off without payment incidents.

The distribution of crime across the facilities also can be usefully shown as a bar chart. Figure 5.1 shows all 357 gas stations and the number of driving off without payment incidents for each, from the riskiest to the least risky, from left to right.

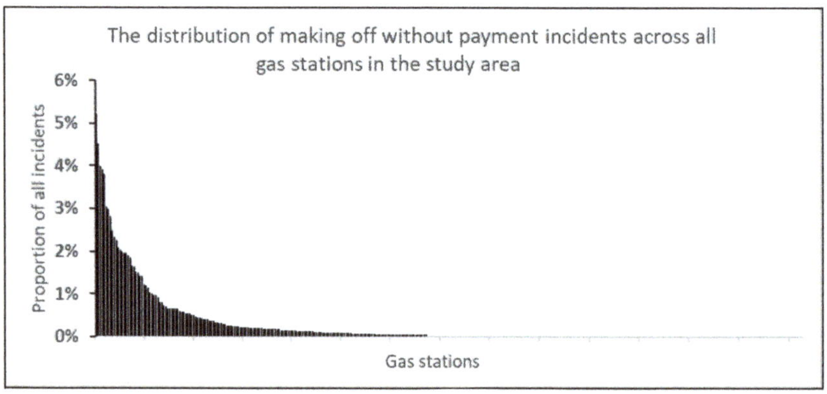

Figure 5.1. Risky facilities chart for the distribution of making off without payment for gas across 357 gas stations.

Step 7: Mapping the results

The final step involves mapping the results to visualize the distribution of risky facilities. Often, a graduated symbol approach is the best method for visualizing the results. Figure 5.2 shows the gas stations identified as the riskiest facilities, each experiencing more than 100 incidents. The map also shows all other gas stations. The gas stations that experienced the most incidents were in the south of the study area. A cluster of gas stations was in a town in the middle of the study area, but none of these were determined to be risky facilities.

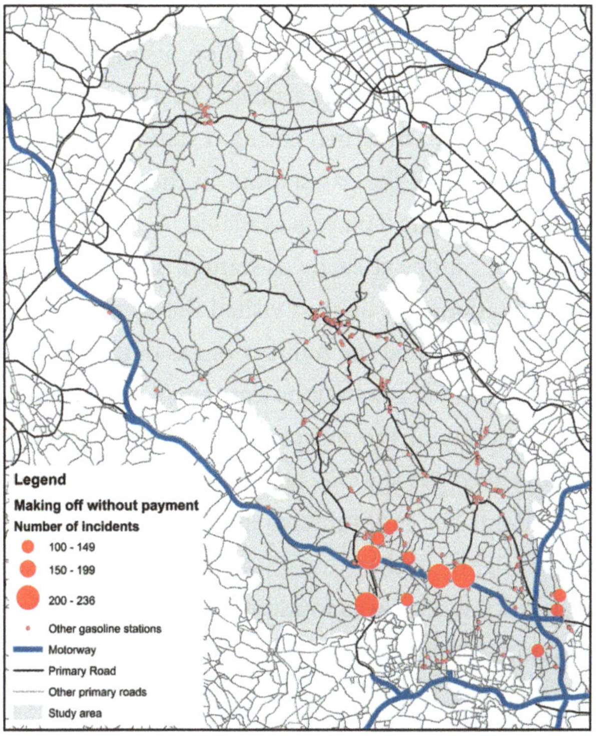

Figure 5.2. A graduated symbol map showing the location of the riskiest facilities—the gas stations that experienced the largest number of making off without payment offenses.

Stage 2: Revision of the analysis after considering the *population at risk (the denominator)*

The results from step 1 are useful in identifying the facilities that experience a large proportion of crime. However, these facilities may be just the large or most frequented facilities and may not necessarily be a true representation of risk. To correct for this requires us to consider a denominator, or population at risk measure, to determine a rate of crime. For example, in the gas station example, the financial transactions of each gas station would provide a good indication of how busy each one was. From this data, theft per $1,000 in transactions could be a meaningful rate to calculate. If this was possible, steps 4 to 7 in stage 1 could be repeated to produce a list of the gas stations with the highest theft per $1,000 in transactions, from which further analysis could be performed. Perhaps, better still, if the value of each making off without payment incident was known, stage 1 could be repeated to identify the gas stations that experienced the highest financial losses from these incidents (rather than just the number of incidents), and from that, the financial loss for every $1,000 in transactions could be calculated as a rate, and an additional measure of risk.

Financial transactions from each gas station, though, would likely be difficult to obtain because this information typically is treated as commercially sensitive by gas stations and petroleum companies. Therefore, generating a rate that relates to transactions for the purchase of gas may not be possible. Similarly, the number of cars that visit each gas station each day would be an alternative, ideal denominator, but it is unlikely to be recorded at all gas stations. In many situations, when analyzing crime data, the ideal data we need, frustratingly, may not be available. An alternative denominator to use in the current example could be the square footage of the gas station, using this measure as a proxy for the size of the gas station. If this data was not easily available from a public records system, land-use data in vector and polygon GIS format, in which each gas station is represented as a single polygon, could be used to calculate the size of the gas station. In the study of making off without payment for gas from gas stations, none of this data was available. A compromise that the analyst came up with was to use the number of pumps at each gas station. This data, however, was not recorded. So, rather than visiting all gas stations, the analyst visited the 20 top-ranked gas stations and the 20 bottom-ranked gas stations and recorded the number of gas pumps at each gas station. The analyst found that many of the bottom-ranked gas stations were small gas stations, with no more than four gas pumps, often in villages. Many of the top-ranked stations were large gas stations with many gas pumps. One of the top-ranked gas stations had 20 gas pumps. Calculating the number of making off without payment incidents per gas pump for the top-ranked gas stations helped revise the analysis by changing the order slightly of these top-ranked facilities. For example, gas station 177, previously ranked the seventh riskiest, now was found to be the riskiest when the number of gas pumps was taken into consideration (table 5.4).

Table 5.4. Gas stations ranked by rate of incidents

Rank	Gas station ID code	Number of incidents	Number of gas pumps	Rate: Incidents per 10 gas pumps	Rank based on rate
1	223	236	12	19.7	3
2	189	204	20	10.2	10
3	258	180	10	18.0	4
4	215	177	10	17.7	5
5	66	172	12	14.3	7
6	233	137	8	17.1	6
7	177	135	6	22.5	1
8	73	127	6	21.2	2
9	33	112	10	11.2	9
10	3	105	8	13.1	8

Note: Stations, listed in rank order, for the number of making off without payment incidents, the number of gas pumps at each station, the number of incidents per 10 gas pumps, and the rank based on rate, showing the top 10–ranked gas stations of 267 that experienced at least one crime.

In practice, examining risky facilities usually involves a combination of the number of incidents and a rate. In the making off without payment of gas example, the analysis initially identified those gas stations responsible for a large proportion of incidents. With this knowledge, efforts to reduce these incidents in the top-ranked facilities could have a sizable impact in reducing the number of incidents overall. Calculating a rate for the top 10–ranked gas stations (as shown in table 5.4) changed their order slightly, identifying which ones to prioritize for any targeted intervention. For instance, the top-ranked gas station, based on the number of incidents, still was ranked in the top three gas stations when the number of pumps was taken into consideration. The calculation of the rate, alongside the calculation of the number of incidents, helped identify the five gas stations to target initially—these five gas stations accounted for one in five of making off without payment incidents across all 357 gas stations.

In many situations, selecting a meaningful rate can be difficult. For example, in the theft of handbags from bars example mentioned earlier, what would be the best denominator to use? It would be the number of handbags in each bar, but many analysts often don't want to spend their Friday and Saturday nights with a clipboard, visiting the town's bars to count the number of handbags in each establishment, and watch how this number changes throughout the evening. In the handbag theft example, what the researchers ended up doing was visiting the bars and counting the number of chairs in each bar and used this as a denominator measure (Sidebottom and Bowers 2010). Figure 5.3 shows an example of the number of cell phone thefts and a rate measure for cell phone thefts in bars in the city of Nottingham, England. The initial risky facilities analysis identified that 10 bars accounted for 40 percent of all cell phone thefts. Bar D accounted for over 300 of these thefts, but as figure 5.3 shows, it is a large venue. In this example, the person capacity of each bar was used as the denominator to calculate a meaningful rate—the number of thefts per 1,000 capacity. This resulted in slightly changing the order of the riskiest facilities—Bar D now was ranked fourth. Bar A initially was ranked sixth considering only the number of incidents, but when the capacity of the venue was considered, it now ranked first. Calculation of the rate in this example was particularly useful for two reasons. First, it identified the bars in which you were at more risk of having your cell phone stolen. If only the number of incidents was considered and used for prioritization, the riskiest bar would have been lower on the priority list for attendance. Second, when the managers of bars A to E were invited to a meeting by the police to discuss how cell phone theft could be reduced on their premises, the manager of Bar D initially reacted by saying, "Well, of course it does not surprise me that my place has the highest number of cell phone thefts. We are the biggest and best place for people to go out to on a Friday or Saturday night, and 10 times the size of the other premises—so, of course, we are likely to have more phone thefts." The police response was, "Well, sir, we have considered the size of your premises in our analysis, and even when taking that into consideration, you are still ranked as the fourth-worst bar in the city for cell phone theft. So, we'd appreciate it if you'd work with us to address the phone theft issues in the city that many of your customers experience."

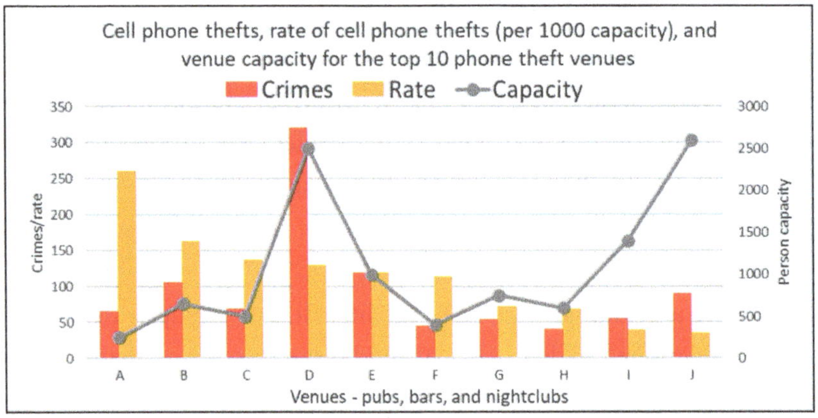

Figure 5.3. Risky bars and nightclubs for cell phone theft in Nottingham, England.

Stage 3: Analysis of risk factors associated with the facility

The final step of a risky facilities analysis involves attempting to determine what it is about the riskiest facilities that make them so vulnerable to crime. Often, this can involve analysis of data that relates to these facilities. For example, a risky facilities analysis of parking lots and parking garages may involve examining risk in relation to the type of parking facility (e.g., open-air lot or multilevel parking garage), the entry system (e.g., entry and exit barriers, or no barriers), and the payment system (e.g., pay and display ticket or pay on exit). Research on risky facilities also has shown that it is often the poor management of facilities that is a key contributing factor to why they are risky facilities. Often, this data for examining risk factors is not recorded, so this investigation may require visiting the risky facilities, conducting a visual audit or survey, and from there determining the factors that make these facilities risky to crime. The POP Center guide on risky facilities listed in the "Further Resources" section at the end of this chapter provides useful additional information about analyzing the factors that may help explain why certain facilities are particularly risky (Clarke and Eck 2007).

Mapping the results of a risky facilities analysis also can help identify whether their location may explain their high level of risk. Have another look at figure 5.2. The riskiest facilities are all in proximity to the major highway that cuts across the south of the study area. The cluster of gas stations in the middle of the study area that are not the riskiest facilities are located far from the major highway. In this example, the results indicate that where facilities are located, such as gas stations having proximity to a major highway, is likely to explain why they are risky facilities for making off without payment for gas incidents.

Repeat victimization

Repeat victimization is when the same type of crime is experienced by the same victim or target. For example, the same person is assaulted once, and then is assaulted again at a later point, or a house is burgled and then burgled again after the first offense. When the focus of examining repeat victimization is on targets, the owner of the target could be a different person (e.g., an apartment is burgled, the resident of the apartment moves out of the apartment, a new resident moves in, and the apartment is burgled again). Typically, analysis of repeat victimization considers incidents for a specific period, such as one year. Repeat victimization refers to the total number of offenses experienced by the victim or target, including the first offense and all subsequent offenses within the period of analysis. The number of *repeats* refers only to the subsequent offenses that took place after the initial offense.

The analysis of repeat victimization is useful for several reasons:

- It helps identify whether certain people or targets experience a large number of offenses against them, and then prompts additional analysis to help understand why.
- It helps determine the extent to which repeat offenses contribute to all crime, and from there directs interventions for crime reduction toward reducing the repeat victimization.
- Additionally, analysis of repeat victimization can inform crime prediction. That is, if the same victim or target has been victimized repeatedly, there is a high risk of further victimization if something is not done to address what makes the victim or target susceptible to victimization. As Skogan (1996) stated a few years ago: "That crime is concentrated on the same people and places and that this has potential for crime control has been described as the most important criminological insight of the decade."

Since the first studies into repeat victimization in the 1990s, there has continued to be a growth in strong empirical evidence for supporting it. Although most studies of repeat victimization have focused on burglary, repeat victimization patterns also have been seen in many other types of crime—from bike theft to domestic violence. Table 5.5 lists examples of crime types and the extent of repeat victimization measured from several studies.

Table 5.5. Repeat victimization for different crime types

Crime type	Repeat victimization (as a percentage of all offenses)	Location
House burglary	32%	Beenleigh, Australia (Townsley, Homel, and Chaseling 2000)
	13%	Merseyside, England (Johnson, Bowers, and Hirschfield 1997)
	9%	Auckland, New Zealand (Chainey et al. 2018)
Domestic violence	42%	West Yorkshire, England (Hanmer, Griffiths, and Jerwood 1999)
Commercial robbery	65%	Indianapolis, Indiana (Weisel 2005)
Gas station robbery	62%	Australia (Taylor 2004)
Bank robbery	58%	England (Matthews et al. 2001)
Sexual offenses	43%	International (ICVS 2000)

For ease of explaining the theoretical principles relating to repeat victimization and how to measure it, we focus on house burglary, but then discuss these principles and the analysis of repeat victimization for other types of crime.

The main findings from the research into repeat victimization show the following:

- If you are burgled, you are at a heightened risk of being burgled again. This is not to say you will be burgled again, but your risk (following the initial incident) of further victimization, in general, has doubled.
- This heightened risk rapidly declines with time, dropping to the same level of risk as other properties, typically within one to two months. The highest period of risk is within a few days of the initial incident.

From these findings, an analysis of repeat victimization can help determine who or what (if the target is property) is at a heightened risk of being a victim of crime again, with this repeat most likely to be soon after the first incident. In this manner, understanding patterns of repeat victimization can help inform the design of crime prevention initiatives, crime risk (which we discuss in chapter 9), and the prediction of crime (repeat victimization is a variable used in many crime prediction software tools). The Kirkholt Burglary Prevention Project (Forrester, Chatterton, and Pease 1988) in Rochdale, England, was one of the first crime prevention initiatives that used repeat victimization patterns to prevent crime. The analysis of crime patterns in Kirkholt identified high levels of repeat victimization, with these results then used to target a crime prevention program to those properties that previously had experienced burglary repeat victimization. The result of the program was an overall reduction in burglary of 53 percent across the Kirkholt estate, and an 80 percent reduction in burglary repeat victimization.

Why does repeat victimization occur?

The reasons why repeat victimization occurs can be explained principally by the boost account and flag account theories.

The boost account

The boost account refers to an offender deciding to victimize the same person again or return to the same target, boosted by the success of previous crime commission (Pease 1998). For example, the offender is boosted to return because of familiarity with the victim and the area in which the victim (or other type of target) is located (following the initial offense), experience from the successful modus operandi from the previous offense (such as the means of breaking into a property), and the means of evading capture (informed by the previous victimization against the victim or target). The boost account, therefore, refers to the same offender returning to commit subsequent offenses, rather than each repeat offense being committed by different offenders. The exception is when the offender passes on his or her experience of crime commission obtained from the first offense to other people with whom he or she may then co-offend, or these other people commit the repeat offenses having been informed by the offender of their successful crime commission from the first offense.

The flag account

The flag account suggests there is some enduring quality about the victim or target that draws an offender's attention to the target's higher level of vulnerability (Pease 1998). This enduring quality, for example, could include features about a target that make it more vulnerable to crime, such as a house being located at the end of a row of houses, with easy access to the rear because of an alley or public path behind the property. Different to the boost account, the flag account suggests a repeat offense just as likely would be committed by a different offender responding to the same cues of high vulnerability associated with a particular target that drew the attentions of the first offender in the commission of a previous offense against that target. The time between offenses against the target may be longer than those influenced by the boost account.

In an analysis of repeat victimization, if the time between incidents is relatively short (i.e., within two weeks), it is more likely that the repeat offense is because of boost account principles, and hence a returning offender. If there is a long time between a first and second offense, it is more likely the second offense is because of flag account principles, with the second offense committed by a different offender than who committed the first offense.

Measuring repeat victimization

Repeat victimization can be measured in two main ways. Like the analysis of risky facilities, one method involves identifying victims or the target by name, and the other involves using geographic coordinates. For both, the data used must be as complete as possible for an accurate analysis of repeat victimization. For example, if the analysis is based on the name of the victim—e.g., Spencer Chainey—the name of the person would need to be entered in the same manner to effectively identify whether the person is recorded multiple times in the crime recording data. The same applies if the address is used for identifying repeat victimization—e.g., 5 Bond Street, London SW1A 1AA—or if geographic coordinates are used to identify repeat locations—e.g., 548623, 148759. This factor, therefore, requires you to check your data and perform any necessary data cleaning to ensure the names, addresses, or coordinates are entered consistently. In the examples of the analysis of repeat victimization that follow, geographic coordinates were used.

The next stage is to generate a count of the number of offenses committed against the same person or target. This can be done using SQL routines in your GIS software or using the Pivot Table function in Microsoft Excel. Like the analysis of risky facilities, the results generated in table 5.6 were calculated from a text string for the pair of geographic coordinates (i.e., concatenating the geographic coordinates to generate an identity number for each record), and then grouping by these identity numbers to generate a count of the number of occurrences of each record's identity number. For those victims or targets that have experienced repeat victimization, it is also useful to generate a table that lists the date of the first offense and the dates of subsequent offenses so that an analysis of the days between offenses also can be performed.

Table 5.6 shows the results of a repeat victimization analysis of residential burglary for a city over a one-year period. The table shows the measures that are useful to calculate when analyzing repeat victimization: the total number of offenses, the number of victims or targets that have experienced more than one offense, the total number of offenses experienced by the victims or targets that experienced more than one offense (represented also as a percentage of all offenses), and the number and proportion of offenses that were repeats (not including the first offense against the victim or target).

Table 5.6. Repeat victimization results of residential burglary in a city for a one-year period

Number of burglaries	Locations experiencing repeat victimization of burglary	Number of burglaries at repeat locations (and % of all offenses)	Number of burglaries that were repeats (and % of all offenses)
1,304	87	200 (15.3%)	113 (8.7%)

Mapping repeat victimization

Measuring repeat victimization is useful for determining the extent of this pattern. Mapping the locations of repeat victimization then shows the locations in which this pattern is most prevalent. Figure 5.4 shows an example for the district of Newcastle (England), using graduated symbol sizes to represent the number of repeat victimization offenses at each location in which repeats were experienced. The figure also shows hot spots of crime, using the procedure for creating KDE hot spot maps, as described in chapter 3. Mapping crime hot spots and the locations of repeat victimization helps show the extent to which the patterns are colocated. In this example, 11.5 percent of all repeat victimization offenses were in the main hot spot (defined by the *high crime density* thematic classification), with these repeat offenses accounting for 25 percent of all the burglaries in this hot spot. This finding suggests that repeat victimization contributes to why an area is a hot spot, but that there were many other offenses (75 percent, in this example) that were single offenses that were also an important factor in the existence of this hot spot. Some 40 percent of burglary repeat victimization offenses were in the two highest thematic classes, meaning that most repeat victimization offenses were not in the areas of highest crime density. Analysis such as this helps indicate how a combination of attention toward all crimes in hot spots and the nature of the spatial distribution of repeat victimization can assist in determining where to target interventions. From the analysis, 36 percent of all the repeats took place within two weeks of the previous offense at the same property. This result suggests that about a third of all repeats likely could be explained by the boost account, and about two-thirds of repeats likely were a result of flag account theoretical principles.

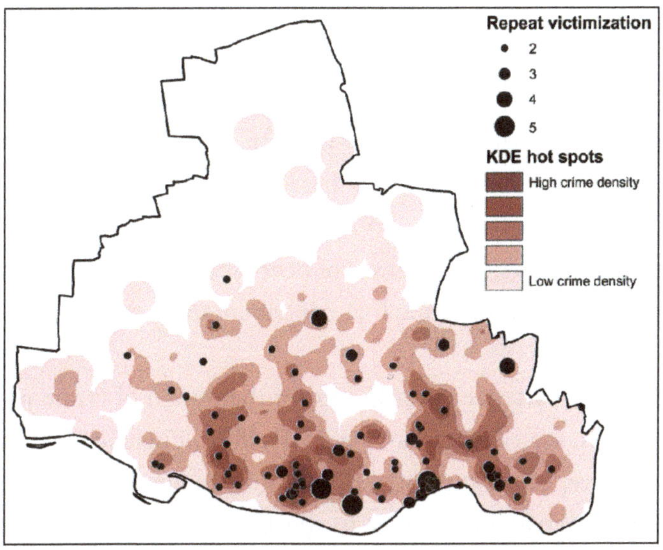

Figure 5.4. Residential burglary repeat victimization, mapped against the KDE hot spots of all burglaries (using a bandwidth of 450 meters and natural breaks [Jenks] classification with five thematic classes) for Newcastle, England.

A similar analysis that compared the location of burglary repeat victimization and the location of burglary hot spots was performed in Wellington, New Zealand, showing that only a third of the repeat incidents took place in hot spots, with the remainder taking place in other locations (Chainey et al. 2018). Performing a combination of hot spot analysis and repeat victimization analysis can help identify the extent to which the two patterns are colocated, and in turn help inform the design of crime prevention initiatives. In situations in which many of the repeats take place inside crime hot spots, a focused intervention to the targets that are repeatedly victimized would help reduce crime overall in the hot spot. In contrast, when many of the repeat offenses take place outside hot spots, interventions that reduce repeat victimization separate to targeted interventions in crime hot spots would be beneficial for crime reduction purposes.

Near repeat victimization

Near repeat victimization is when the same type of crime is experienced by a victim or target nearby soon after the initial offense. For example, the house next door to a recently burgled property is also then burgled. Put another way, if a house is burgled, the neighboring houses now are considered to be at a heightened risk of being burgled. Research on near repeat victimization suggests the risk of victimization to the neighboring targets is lower than the risk of crime to the target already victimized (i.e., the chances of repeat victimization are greater than near repeat victimization), and that this risk declines with distance from the location of the target already victimized. Like repeat victimization, the heightened risk for near repeat victimization declines over time (Johnson and Bowers 2004; Townsley, Homel, and Chaseling 2003). Most research to date on near repeat victimization has focused on burglary, but patterns of near repeats also have been observed for street robbery (Haberman and Ratcliffe 2012), homicide (Chainey and Muggah 2020), and maritime piracy (Marchione and Johnson 2013).

Like the analysis of repeat victimization, analysis of near repeats considers incidents for a specific period, such as one year. Near repeat victimization refers to the total number of offenses experienced by other targets within a set of parameters of distance from and time after an *originator* offense (e.g., within 200 meters and within three days). The originator offense is the crime record identified as the first offense in a near repeat series. The analysis of near repeat victimization is useful, as it can help identify spatiotemporal patterns of crime that contribute to or are distinct from hot spots of crime.

An analysis of near repeat victimization patterns can help inform the design of crime prevention initiatives and where these should be targeted. For example, in Trafford, England, an analysis of residential burglary identified that near repeats contributed significantly to burglary in the town. In response, Greater Manchester Police designed an initiative to specifically counter near repeats. The initiative in Trafford involved neighborhood police officers conducting door-to-door visits to neighboring houses on the day after an initial burglary to help prevent near repeats. The visits involved informing residents about the recent burglary against their neighbor, asking residents to contact the police if they saw someone behaving suspiciously, and advising residents to take practical measures that could reduce their risk of burglary, such as keeping lights on when it was dark and the house was unoccupied. Additionally, a crime prevention officer would visit the burgled property within 24 hours of the incident to identify ways to reduce a repeat victimization (Chainey 2012). The impact in Trafford was a burglary reduction of 42 percent in the areas that were targeted (Fielding and Jones 2012). Similar burglary prevention initiatives to those introduced in Trafford have been implemented by other police agencies in other parts of the world, including Operation Optimal in Leeds, England, that led to a 48 percent reduction in burglary (Professional Security

2012). The pattern of near repeats also often is used for crime prediction purposes (Chainey 2012; Johnson et al. 2009). We discuss its use for mapping and analyzing crime risk in chapter 9.

Why do near repeats occur?

The boost account

Like repeat victimization, the boost account also helps explain why near repeat victimization occurs—the offender is boosted to return to try their luck against nearby targets because of their familiarity with the area (following the initial offense), familiarity with the targets, and the previous success in the modus operandi they used. For example, in the case of burglary, the layout of the neighboring properties is likely to be similar to the property they recently burgled, the means of breaking in is likely to be similar, and the neighbors are likely to have possessions worth stealing, similar to those stolen in the initial burglary (Chainey et al. 2018). Because of these boost principles, near repeats are likely to be committed by the offender who committed the first offense.

Optimal foraging theory

Optimal foraging theory provides a further means for explaining why near repeats occur. This approach likens offenders to foraging animals. As a forager, an offender makes a trade-off between the rewards of crime commission that are most obvious and immediately available (i.e., returning to commit a repeat) and the effort and risks that will be expended in seeking better opportunities. Once an area is exhausted of the best opportunities for crime commission, the forager moves on (i.e., seeking to commit a crime elsewhere). The offender is also conscious of the risks of being caught if the police begin to see a spike of incidents in an area, so he or she decides to move on to commit crime elsewhere and avoid the risk of capture. Near repeats help explain why short spates of several crimes may occur in an area.

Measuring near repeat victimization

Identifying and measuring near repeat victimization is technically a challenge, but thankfully a free tool called the Near Repeat Calculator (Ratcliffe 2009) has been developed to help identify near repeats. The Near Repeat Calculator identifies whether there is a statistically significant near repeat pattern. It also identifies whether a *repeat victimization* pattern is statistically significant. In this section, we describe how to define practical spatial and temporal parameters for near repeat analysis and illustrate the analysis of near repeats with several examples.

Defining spatial and temporal bandwidths for near repeat analysis

The analysis of near repeats requires you to specify the spatial and temporal bandwidths for what you are defining as a near repeat—i.e., how near in distance and near in time to a previous offense. You also must determine the number of bands to apply as multiples of the spatial and temporal bandwidths. Little guidance exists on how to define the spatial and temporal bandwidths, but from experience of performing many tests of near repeats, I suggest the following:

- First, consider the spatial distribution of the data you are analyzing and the spatial distribution of the targets of the crime. For example, if you are analyzing residential burglaries for a city, the spatial bandwidth to choose should consider the distance between the residential properties. Let's say that there is 20 meters between each property. A spatial bandwidth of 100 meters from each property then would include multiple neighboring properties within a bandwidth of this size (figure 5.5). The analysis of near repeat patterns involves examining the distribution of crime across two-dimensional space, so some consideration would need to be given in the interpretation of the results if your study area contained many apartment blocks in which the x,y coordinates for the properties in these apartment blocks do not differ.

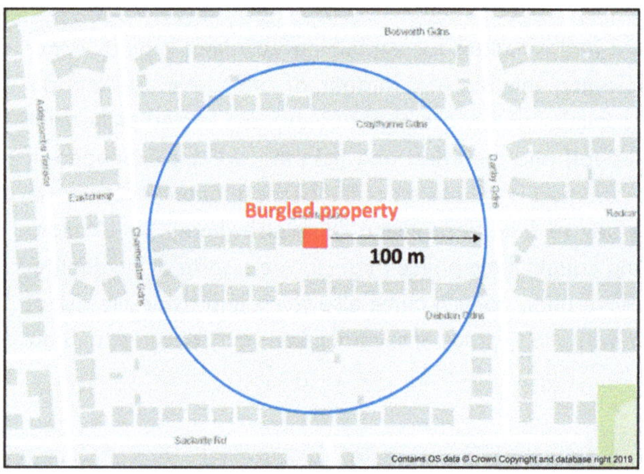

Figure 5.5. A near repeat spatial bandwidth of 100 meters includes several nearby properties to a recently burgled property when these neighboring properties are located reasonably close to each other.

- I find it useful to work with distances that are easy to visualize. A distance of 100 meters is easy to visualize so is often a spatial bandwidth I use in urban areas. For practical purposes, I also find it best to work with no more than five spatial bands. When using a spatial bandwidth of 100 meters, you will analyze whether the presence of near repeats is close to previous offenses (e.g., within 100 meters), and for four additional multiple bands of 100 meters (e.g., 101 to 200 meters, 201 to 300 meters, 301 to 400 meters, and 401 to 500 meters). I have seen some analysts examine near repeats in urban areas for spatial bands of greater than 1,000 meters. Personally, I find this measurement of little practical use. Even if near repeats were found this far away, it is typically too far from an original burglary to instruct a sensible prevention deployment plan.
- I find it useful to experiment with three groupings of temporal bandwidths. First, apply a spatial bandwidth of three days, and apply seven bands. You will examine for the presence

of near repeats within three days, 4 to 6 days, 7 to 9 days, 10 to 12 days, 13 to 15 days, 16 to 18 days, and 19 to 21 days. These calculations will help you initially determine whether near repeats are present in your data. Then, I find it useful to examine for near repeats using one-day bandwidths to determine whether there are any days on which near repeats occur. I find that applying 14 bands is practical, so that you can see if there are any days within two weeks of a previous offense on which near repeats occur. Then, for the third group, analyze for near repeats using seven-day bandwidths and four bands. This will help you determine whether the presence of near repeats is, overall, most significant in the first week following an offense.

Collectively, these three groupings of temporal bandwidths, with the 100-meter spatial bands, will provide you with a thorough assessment of the presence of near repeats and help you effectively communicate the results of your near repeat analysis. An example of near repeat analysis using these spatial and temporal bands is provided below. If you're analyzing crime patterns in less urban or rural areas, the spatial bandwidths likely would need to be larger to reflect the spatial distribution of targets. A useful way to determine the initial size of a spatial bandwidth is to apply the nearest neighbor index (NNI) measure to your data, as described in chapter 2. In calculating the NNI, the average nearest neighbor distance also is calculated and can be used to provide an indication of the distribution of targets that have been victimized previously. Also, for crimes that are rare, temporal bands also may need to be larger, but not too large so that the results are of no practical value.

Table 5.7 shows results of a near repeat analysis of residential burglaries for the city of Wellington, New Zealand. The analysis used five bands of 100-meter spatial bandwidths and the three categories of temporal bandwidths referred to above: (a) three-day bandwidths and seven bands, (b) one-day bandwidths and 14 bands, and (c) seven-day bandwidths and four bands. The results were generated using the Near Repeat Calculator and show those cells in which the presence of near repeats was found to be statistically significant (to $p <= .05$). A result of 4.8 shown in the first column (table 5.7a) suggests that the presence of near repeats is 380 percent greater than chance. The results from Wellington show that, overall, the greatest presence of near repeats (indicated by the higher numbers in each cell) was closest (in space and time) to originator offenses. Personally, when communicating the results of the Near Repeat Calculator, I tend not to refer in detail to the values in each cell but instead use them simply to state the statistical presence of near repeats (when the results are observed to be statistically significant), with these near repeats occurring closer (in space and time) to originator offenses. The results for Wellington show exactly this, with the higher values for the space-time parameters that are closer to originator offenses. The results also show the significant presence of a repeat victimization pattern (indicated by the "Same location" row), with these results showing, and in line with the theory, that offenses at the same location, and soon after a previous offense, are more likely than offenses nearby (indicated by the higher values, in general, for the "same location" in the first temporal column in two of the three tables).

Table 5.7a. The presence of burglary near repeats in Wellington, New Zealand, for three-day bandwidths and seven bands

	0 to 3 days	4 to 6 days	7 to 9 days	10 to 12 days	13 to 15 days	16 to 18 days	19 to 21 days
Same location	4.3		3.6				
1 to 100 m	4.8				1.99		
101 to 200 m	2.2		1.35				1.53
201 to 300 m					1.53	1.43	1.75
301 to 400 m	1.66	1.4	1.4				
401 to 500 m		1.2					1.32

Note: Blank cells represent nonsignificant results.
p <= .05

Table 5.7b. The presence of burglary near repeats in Wellington, New Zealand, for one-day bandwidths and 14 bands

	Within 1 day	Day 2	Day 3	Day 4	Day 5	Day 6	Day 7	Day 8	Day 9	Day 10	Day 11	Day 12	Day 13	Day 14
Same location	11.4					5.8	14.3							
1 to 100 m	7.8	2.7									3.6			
101 to 200 m	4.4	2.0						1.9						
201 to 300 m						1.7								
301 to 400 m	2.2			2.0				2.0			1.7			
401 to 500 m														

Note: Blank cells represent nonsignificant results.
p <= .05

Mapping near repeats

ArcGIS has several functions for near repeat analysis. They include a classification tool to determine the number of crime records that are near repeats (for sets of user-specified spatial and temporal bandwidths) and the ability to map near repeats and their originator offenses. The Near Repeat Calculator also can be used to produce an output file that lists the crime records that were near repeats and the crime records that were *originator* offenses. This output file from the Near Repeat Calculator can be used as an additional data source, imported into a GIS, and mapped to identify where these incidents took place.

A near repeat series can involve more than just a single originator and a single near repeat offense. For example, an offense takes place (the originator), followed by two other offenses two days later and within 100 meters of the originator offense, creating a series of three offenses (figure 5.6a).

Near repeats also can be an originator for a near repeat offense. For example, a near repeat occurs a day after an originator offense, and then on the next day, another offense takes place within 100 meters of the near repeat but farther away than the spatial bandwidth from the originator offense (figure 5.6b). Technically speaking, a near repeat also may have two originators when each originator took place on the same day before the near repeat. Figure 5.6c illustrates this scenario, showing that the near repeat was a subsequent offense to two originator offenses (on the basis of defining a near repeat as being within 100 meters and three days of a previous offense). In reality, one of the originator offenses may have been more of a *boost* to the near repeat, but based purely on spatial and temporal characteristics of the offenses, it is not possible to distinguish between which offense was the more likely originator. Closer examination of the modus operandi of the three offenses, the types of houses targeted, and what was stolen may indicate which originator offense was more likely to be associated with the near repeat. This closer analysis also may lead to identifying that one of the originator offenses was most likely a near repeat of the other originator (e.g., it occurred two hours after the first of the three offenses), and that all three burglaries are part of a near repeat chain that is likely to have been committed by the same offender.

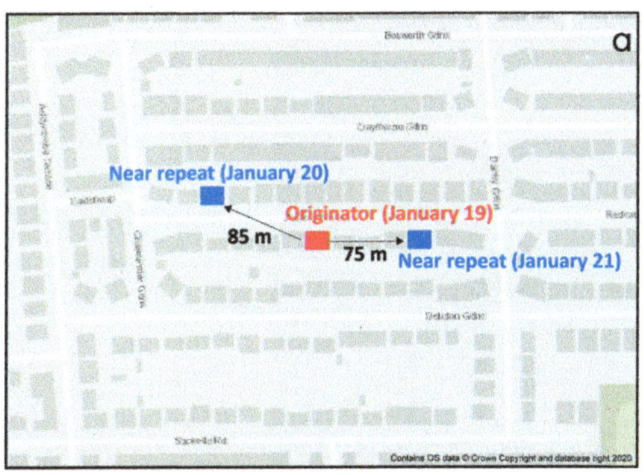

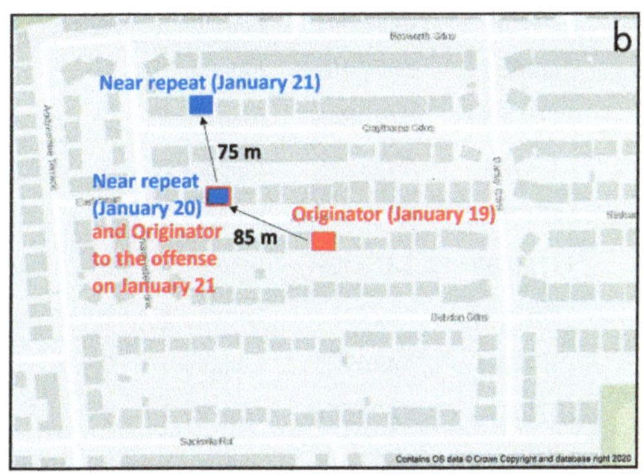

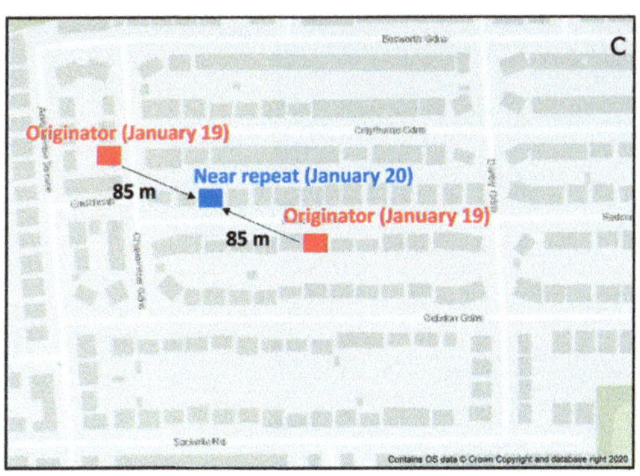

Figure 5.6. (a) Two near repeats (defined as being within 100 meters and three days of a previous offense) associated with a single originator offense, (b) two near repeats, in which one is associated with the originator and the other is a near repeat of the second offense in the near repeat chain, and (c) a near repeat with two originators.

Using the Near Repeat Calculator to produce an output file of near repeats and originator offenses

Table 5.8 shows an extract from the Near Repeat Calculator output file, listing the coordinates of a sample of 20 offenses and columns that indicate whether the offense was an originator for a near repeat or was a near repeat. If the value in the originator column is greater than or equal to one, the offense was an originator. If the value in the near repeat column is greater than or equal to one, the offense was a near repeat. If the value in the originator column is greater than or equal to two, the offense was an originator for two (or more) near repeats (as illustrated in figure 5.6a). If the value in the near repeat column is greater than or equal to two, the offense was a near repeat for two (or more) originator offenses (as illustrated in figure 5.6c). It does not mean it is two near repeats. If a near repeat is also an originator to a subsequent offense (a near repeat), it is indicated by a value of one (or more) in the originator column and a value of one (or more) in the near repeat column (as illustrated in figure 5.6b).

Table 5.8. Near Repeat Calculator output file for originator and near repeat offenses

X	Y	Date	Originator	Near Repeat
1749741	5427131	January 1	0	0
1748553	5429016	January 2	1	0
1750856	5423848	January 2	0	1
1749751	5423576	January 2	0	0
1748830	5424805	January 2	0	0
1750371	5423309	January 2	0	2
1747597	5427036	January 3	0	0
1747815	5426975	January 4	0	0
1752190	5435556	January 5	0	0
1747924	5426744	January 5	1	0
1750799	5434390	January 6	0	0
1748742	5425155	January 6	0	0
1749047	5424648	January 6	0	0
1748415	5426278	January 6	0	0
1749277	5430645	January 6	0	0
1748917	5424797	January 7	0	0
1750868	5423906	January 7	0	0
1747471	5421438	January 7	0	0
1748638	5425105	January 8	1	0
1748767	5427006	January 9	0	0

Analysis of repeats and near repeats: An example from Wellington, New Zealand

Figure 5.7a shows the distribution of residential burglary near repeats in Wellington, defined as being within 200 meters and seven days of a previous offense. During the one-year period of study, of 984 burglaries, 147 were near repeats, accounting for 15 percent of all burglaries. The figure also shows the distribution of burglary repeat victimization, with 137 offenses (14 percent of all burglaries) during the same period. Repeats alone (not including the first offense in each case) accounted for 79 offenses (8 percent of all burglaries). Eighteen of the repeats took place within two weeks of an offense at the same property, and 27 took place within four weeks of a previous offense, whereas the rest (52) took place more than a month after a previous offense at the same property. These results indicate that the majority (two in three repeats) were more likely to be explained by flag account principles, with a third likely to be explained by boost account principles and associated with a returning offender. The distribution of repeats and near repeats occurred in the main urban area of Wellington. Closer inspection of the distribution of the burglary hot spots (see figure 5.7b showing the southern area of Wellington) identified that 42 percent of all near repeats and 32 percent of repeats took place in burglary hot spots. In total, 60 percent of repeats and near repeats did not take place in hot spots. These results, from this example from Wellington, help demonstrate the value in conducting both hot spot analysis, and repeat and near repeat victimization analysis for examining the spatial and spatiotemporal distribution of crime, and where to target crime prevention interventions. Analysis of repeat and near repeat patterns can indicate the extent of these types of offenses, and where they are most likely to take place. However, many additional crimes did not follow repeat and near repeat patterns but instead clustered to form hot spots of crime. An analysis of repeats and near repeats and an analysis of crime concentration in combination therefore can be useful for considering where to target interventions and how to design these interventions. For instance, in areas in which many repeats and near repeats occur, visits by the police offering crime prevention advice to the owners of burgled property and properties nearby soon after the initial burglary may be the most effective way to help reduce further burglaries. In hot spots in which most burglaries are single events, spread over time but that spatially cluster, the conditions that make the area a hot spot would need to be identified, and then a crime prevention plan could be put in place to address these underlying conditions.

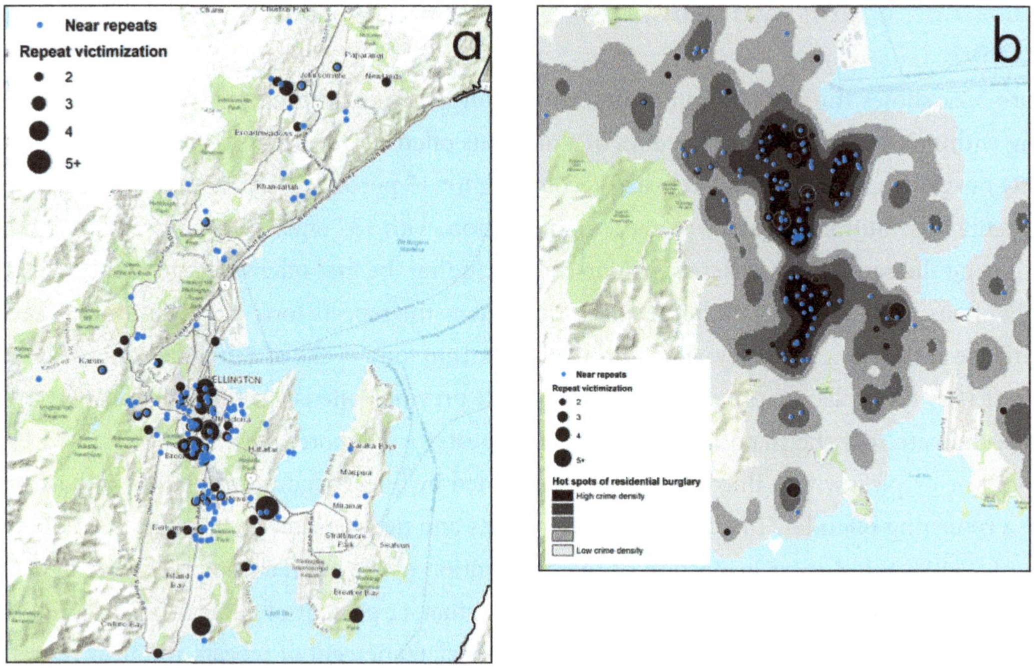

Figure 5.7. (*a*) The distribution of repeats and near repeats in Wellington, shown with (*b*) hot spots of residential burglary for southern Wellington.

Repeat and near repeat victimization analysis applied to other types of crime

Most types of crime are suitable for repeat and near repeat analysis, especially crimes that involve the theft of property and most forms of violent crime. Certain types of crime, however, are not suitable or not relevant for the analysis of repeat victimization and near repeats. For example, many studies have shown the high extent of repeat victimization for domestic violence and have drawn from the boost account to explain the high levels of repeat offending for this type of crime (Hanmer, Griffiths, and Jerwood 1999). Near repeat victimization of domestic violence, however, is unlikely—e.g., a man who violently abuses his wife is unlikely to be *boosted* to go next door and physically abuse the wife of his male neighbor. So, when considering the analysis of near repeat patterns, it is useful first to consider whether the boost account and optimal foraging theoretical principles are likely to apply to the crime type you want to study, as it will help determine whether an analysis of near repeats is suitable. Similarly, with repeat victimization, although homicide repeat victimization cannot occur, research has shown the presence of homicide near repeats. These near repeats of homicide often are associated with retaliation or revenge incidents (Chainey and Muggah 2020).

Repeat and near repeat incidents of street robbery also may occur, although the repeat is not necessarily against the same target but is at the same location to a previous offense (e.g., a street junction). Figure 5.8 illustrates an example of near repeats of street robbery for an area in a city in South America, showing the location of near repeats and their originator offenses, and lines connecting the two. Mapping and examining these *near repeat chains* can help identify whether locations are susceptible to these spates of offending. For example, if near repeat chains are observed to be frequent in an area, when an incident occurs, it can prompt a targeted response in the area to interrupt subsequent offenses from occurring.

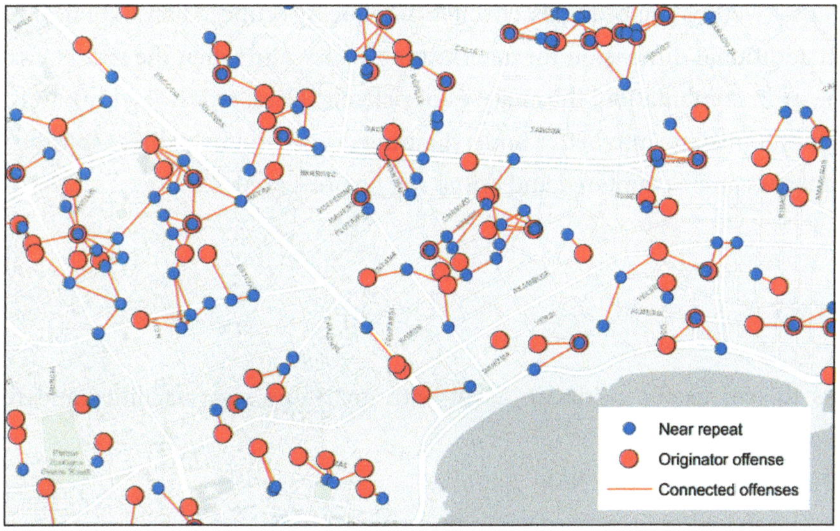

Figure 5.8. Street robbery near repeat chains (for offenses within 300 meters and three days of a previous offense).

Chapter summary

The analysis of risky facilities and repeat victimization provides a means of examining acute concentrations of crime—crimes that take place at the same location, against the same facility, same person, or same target. An examination of the spatial distribution of these offenses then can help identify whether these facilities or repeat victims cluster at certain locations, and the extent to which they contribute to the generation of hot spots. Further analysis prompted from identifying patterns of risky facilities and repeat victimization then may help reveal why these locations and targets experience such acute levels of crime, and inform the design of crime prevention interventions more specifically.

The analysis of near repeats offers additional spatial and temporal insights into the patterns of crime. The locations in which near repeats take place are where there are spates of crime, often committed by the same returning offender. The locations in which near repeats take place may not necessarily be in crime hot spots, and hence can provide another means for determining how to prevent offenses.

Places in which crime previously concentrated are often where crime is likely to concentrate in the future, hence why the analysis of crime hot spots is useful. Places that experience acute levels of crime concentration, such as risky facilities, and targets that are repeatedly victimized are also where crime is likely to take place in the future. Hence, targeting crime prevention efforts to the riskiest facilities and repeat victims can be highly effective in reducing crime. Analysis of near repeat victimization offers an additional dimension for determining where and when the risk of crime is greatest. On their own, or in combination, the analyses of risky facilities, repeat victimization, and near repeats can offer useful insights into better understanding crime, and help direct specific ways in how these crime patterns can be countered and crime can be prevented.

Further resources

Listed below are links to two useful guides to assist with analyzing risky facilities and repeat victimization.

POP Tool Guide No. 6: *Understanding Risky Facilities*, https://popcenter.asu.edu/content/understanding-risky-facilities.

POP Tool Guide No. 4: *Analyzing Repeat Victimization*, https://popcenter.asu.edu/content/analyzing-repeat-victimization.

References

Chainey, S. P. 2012. Predictive Mapping (Predictive Policing). JDI Briefs Series. London, England: University College London.

Chainey, S. P., S. J. Curtis-Ham, R. M. Evans, and G. J. Burns. 2018. "Examining the Extent to Which Repeat and Near Repeat Patterns Can Prevent Crime." *Policing: An International Journal* 41 (5): 608–22.

Chainey, S. P., and R. Muggah. 2020. "Homicide Concentration and Homicide Near Repeats in a Latin American Urban Context." *The Police Journal.*

Clarke, R. V., and J. E. Eck. 2007. *Understanding Risky Facilities.* POP Center Tool Guide No. 6. Arizona State University Center for Problem-Oriented Policing. https://popcenter.asu.edu/content/understanding-risky-facilities.

Eck, J. E., R. V. Clarke, and R. T. Guerette. 2007. "Risky Facilities: Crime Concentration in Homogeneous Sets of Establishments and Facilities." In *Imagination in Crime Prevention*, Crime Prevention Studies *21*, edited by G. Farrell, K. Bowers, S. D. Johnson, and M. Townsley. Cullompton, England: Willan.

Fielding, M., and V. Jones. 2012. "Disrupting the Optimal Forager: Predictive Risk Mapping and Domestic Burglary Reduction in Trafford, Greater Manchester." *International Journal of Police Science and Management* 14 (1): 30–41.

Forrester, D., M. Chatterton, and K. Pease. 1988. *The Kirkholt Burglary Prevention Project, Rochdale.* Crime Prevention Unit Paper 13. London: Home Office.

Haberman, C. P., and J. H. Ratcliffe. 2012. "The Predictive Policing Challenges of Near Repeat Armed Street Robberies." *Policing: A Journal of Policy and Practice* 6 (2): 151–66.

Hanmer J., S. Griffiths, and D. Jerwood. 1999. *Arresting Evidence: Domestic Violence and Repeat Victimization*. Police Research Series Paper 104. London: Home Office.

Hilborn, J. 2009. *Dealing with Crime and Disorder in Urban Parks*. Problem-Oriented Guides for Police, Response Guides Series No. 9. Washington, DC: US Department of Justice.

ICVS 2000. *International Crime Victims Survey*. European Union. http://www.unicri.it/services/library_documentation/publications/icvs/.

Johnson, S. D., and K. J. Bowers. 2004. "The Burglary as Clue to the Future: The Beginnings of Prospective Hotspotting." *European Journal of Criminology* 1 (2): 237–55.

Johnson, S., K. J. Bowers., D. Birks, and K. Pease. 2009. "Predictive Mapping: Accuracy, Units of Analysis, and the Environmental Backcloth." In *Putting Crime in Its Place: Units of Analysis in Geographical Criminology*, edited by D. Weisburd, W. Bernasco, and G. Bruinsma. New York, NY: Springer.

Johnson S. D., K. Bowers, and A. Hirschfield. 1997. "New Insights into the Spatial and Temporal Distribution of Repeat Victimization." *The British Journal of Criminology* 37 (2): 224–41.

Marchione, E., and S. D. Johnson. 2013. "Spatial, Temporal, and Spatio-Temporal Patterns of Maritime Piracy." *Journal of Research on Crime and Delinquency* 50 (4): 504–24.

Matthews R., C. Pease, and K. Pease. 2001. "Repeated Bank Robbery: Themes and Variations." In *Repeat Victimization*, edited by G. Farrell and K. Pease. Crime Prevention Studies 12. Monsey, NY: Criminal Justice Press.

Pease, K. 1998. *Repeat Victimization: Taking Stock*. Home Office Police Research Group, Crime Detection and Prevention Series, paper 90. London: Home Office.

Professional Security. 2012. *Tackling Burglary*. Accessed May 20, 2020. http://www.professionalsecurity.co.uk/news/case-studies/tackling-burglary/.

Ratcliffe, J. H. 2009. *Near Repeat Calculator*. Accessed November 24, 2019.Available at https://www.cla.temple.edu/center-for-security-and-crime-science/projects/#near-repeat-calculator.

Sidebottom, A., and K. Bowers. 2010. "Bag Theft in Bars: An Analysis of Relative Risk, Perceived Risk, and Modus Operandi." *Security Journal* 23:206–24.

Townsley, M., R. Homel, and J. Chaseling. 2000. "Repeat Burglary Victimization: Spatial and Temporal Patterns." *Australian and New Zealand Journal of Criminology* 33 (1): 37–63.

Townsley, M., R. Homel, and J. Chaseling. 2003. "Infectious Burglaries. A Test of the Near Repeat Hypothesis." *British Journal of Criminology* 43 (3): 615–33.

Skogan, W. G. 1996. *The Decade's Most Important Criminological Insight*. National Institute of Justice: Research in Action. Washington: Department of Justice.

Taylor, N. 2004. "Petrol Service Stations as Victims of Crime: Their Risks and Vulnerabilities." *Crime Prevention and Community Safety: An International Journal* 6 (1): 31–41.

Weisel, D. L. 2005. *Analyzing Repeat Victimization*. Problem-Oriented Guides for Police, Problem-Solving Tools Series No. 4. Washington, DC: US Department of Justice.

Wilcox, P., and J. E. Eck. 2011. "Criminology of the Unpopular: Implications for Policy Aimed at Payday Lending Facilities." *Criminology and Public Policy* 10 (2): 473–82.

CHAPTER 6

Persistent, emerging, and dispersed spatial patterns of crime

Objectives

In chapter 4, we discussed techniques for examining the spatial stability of crime—the temporal stability index and the S Index. In this chapter, we build on these spatial techniques for analyzing crime patterns over time by examining the spatial persistence of crime, identifying emerging problem areas, and analyzing the dispersion of crime concentration during a period of crime increase.

When we examine crime in an area, such as observing a crime hot spot, we often want to know if the crime levels in this area are persistent, or if crime is increasing. We also may want to know if other areas that have yet to get our attention are *emerging hot spots*, and distinguish from these areas the places in which crime is declining. In this chapter, we describe three techniques that, collectively, help us answer these questions. Each complements the other, which we will see in the examples that illustrate their use. These techniques are particularly useful when an area has experienced an increase in crime, as the results we generate using these techniques can help us be even more specific in any resource targeting. These techniques can help us identify those areas that are contributing the most to a crime problem.

We illustrate the use of these techniques using two examples—one of an area in which crime has declined and another of an area in which crime has increased—to show the value in using each technique.

Key learning points

- Understand how to examine the spatial persistence of crime, and in doing so also identify areas in which crime levels are intensifying or diminishing. This will involve use of the *Space Time Cube* and the *Emerging Hot Spot Analysis* tools in ArcGIS.
- Identify areas that have contributed most to a recent increase in crime. This process involves generating global statistical index measures—the Offense Dispersion Index (ODI) and the Non-Contributory Dispersion Index (NCDI)—that help identify if only a small number of areas are responsible for the crime increase across a wider study area.
- Understand how to determine whether, during a period of crime increase, it is the areas in which crime has previously concentrated that are responsible for the increase, or if the increase is associated with new emerging hot spots of crime. This includes describing the use of the Crime Concentration Dispersion Index (CCDI).

Introduction

When we identify a hot spot or we are focusing our analytical attention on a specific area, we often want to know if the spatial pattern of crime we are observing is persistent, historical, diminishing, or increasing. Typically, when we analyze crime data, we use a snapshot of data for a specific period (e.g., one month or one year). A map showing the spatial patterns of this snapshot of crime can help us identify high-crime areas and low-crime areas but hide how crime has changed over the period the data is used. For example, figure 6.1 shows hot spots of robberies and thefts against pedestrians for an area in London. It is clear that several hot spots exist, but are these hot spots persistent areas of high crime, is one hot spot in particular an emerging area of high crime that must be prioritized, and is one hot spot an area in which (even though high levels of crime remain) the numbers of offenses have been declining?

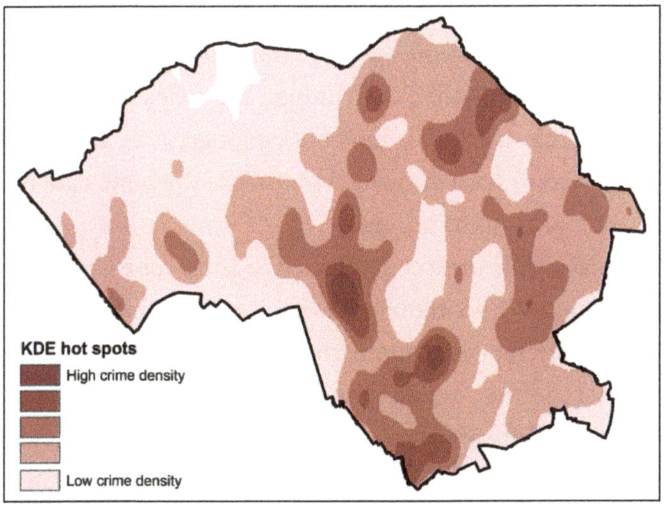

Figure 6.1. Hot spots of robberies and thefts against pedestrians for an area in London.

In this chapter, we describe three analysis techniques that help answer these questions. We choose these techniques because each offers something that complements the other. This analysis includes global statistics that provide simple yet instructive information that relates to the direction of the change in crime and how it relates to areas within our study area, as well as mapping output that helps us visualize these changes over time.

So far, virtually all the content of this book has been about discussing and illustrating examples of useful spatial (and temporal) analysis techniques that are applicable to all researchers and analysts of crime, regardless of the technology they use. However, in this chapter, we begin with two analysis tools that are specific to ArcGIS, the creation of space-time cubes and the use of the Emerging Hot Spot Analysis function, because they are extremely useful for identifying persistent problem areas, alongside several other categories for describing how crime has changed over time in a study area. We describe how the calculations generated from the Space Time Cube and Emerging Hot Spot Analysis tools are performed in ArcGIS in generic terms so the technique can be replicated anywhere.

The two other techniques we describe make use of free software and can be calculated in Excel (or some other spreadsheet software) for import into a GIS for mapping and further analysis.

Identifying persistent spatial patterns of crime

Areas that experience persistent patterns of crime are areas in which crime levels have remained high (or low) for some time—*time* referring to the period covered by the data that is subject to the analysis. As a rule, it is useful to use data for at least a one-year period to examine the persistence of crime, as it helps overcome any variation in crime related to seasonality. Figure 6.1 was generated using two years of data and shows the areas in which the levels of robberies and thefts against pedestrians in this area of London were highest. In this section, we describe an analysis technique that identifies areas in which high crime levels have persisted. The technique also helps us identify areas in which the high level of crime is historical (i.e., used to be high but no longer is as high), is intensifying, is diminishing, or is persistently low in its level of crime. We will use the same data used to create figure 6.1 to illustrate the value of this analysis technique.

The analysis technique uses two tools available in ArcGIS: the space-time cube Creation function and the Emerging Hot Spot Analysis tool. We do, though, describe each technique in generic terms, so if you do not use ArcGIS, you can attempt to replicate the techniques, or at least parts of the techniques, in your own GIS. In the next section, we describe another measure for identifying emerging problem areas that makes use of free software, produces output that can be imported into any GIS, and produces several useful global statistics about the spatial pattern of a change in crime over time.

Over the two-year period covered by the data in London (used for producing figure 6.1), robberies and thefts against pedestrians dropped from 3,482 in year 1 to 3,029 in year 2, a reduction of 13 percent. Our goal is to continue to achieve these reductions. Therefore, it would be useful to determine where our efforts should now be targeted and where reductions have been most observed. An approach for doing this is illustrated in figure 6.2, where, for each area (represented by a 250 × 250 m grid cell), the differences in crime between year 1 and year 2 are mapped. This shows that several areas (a total number of 128 grid cells, representing 19 percent of all grid cells) have experienced reductions of at least three or more crimes, many areas have experienced no or little change, and some areas have increased in their level of crime (12 percent of all grid cells) over the period when these crimes, overall, have declined.

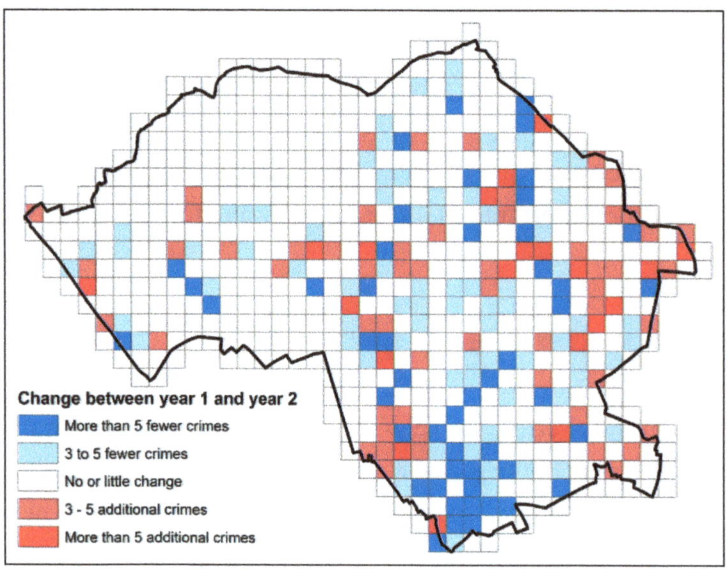

Figure 6.2. Change in robberies and thefts against pedestrians for an area in London over a period in which these crimes decreased by 13 percent.

Although the map in figure 6.2 is useful for identifying where further efforts can be targeted to help reduce robberies and thefts against pedestrians in this area of London, the map indicates only numerical change. Some areas shown in white may be areas in which high levels of crime persist, whereas some areas shown in red may be areas in which crime has increased from a low level (e.g., of only one crime) to what is still a relatively low level (e.g., four crimes—a difference of three additional crimes). A more useful map to produce may be that which shows the areas in which high levels of crime have persisted (i.e., persistent hot spots), alongside areas in which hot spots have intensified, and others in which hot spots have diminished in crime.

The Space Time Cube and Emerging Hot Spot Analysis tools in ArcGIS can generate mapping output that shows areas in which high crime levels persist, together with 16 other categories of change over time. Seventeen categories is a large number to interpret, but from experience, we often observe only half of these categorical interpretations of change over time in our data. An illustration is provided in the example that follows.

The Emerging Hot Spot Analysis tool determines whether hot spots are present by using a space-time implementation of the Gi* statistic (the singular *spatial implementation* of the Gi* statistic is described in chapter 3). The analysis involves a statistical significance test for determining the presence of hot spots, alongside an examination of how these hot spots change over time. The analysis process involves organizing the data into a series of time steps and bins that relate to equal temporal periods and geographic units of analysis across the period of study. A *time step* is the temporal unit for how the crime data is organized (e.g., 12 time steps for a one-year period, each representing one month). *Bins* are represented by geographic units, to which crime for a time step is aggregated (e.g., a grid cell for an area, with a count of crime for a single time step interval). The categories for the output from the Emerging Hot Spot Analysis tool are as follows:

- Persistent hot spot: a location that has been a statistically significant hot spot for 90 percent of the time step intervals, with no discernible trend indicating an increase or decrease in the intensity of clustering over time.
- New hot spot: a statistically significant hot spot for the final time step, and which never before has been a statistically significant hot spot.
- Consecutive hot spot: a location with a single uninterrupted run of statistically significant hot spot bins in the final time step intervals.
- Intensifying hot spot: a location that has been a statistically significant hot spot for 90 percent of the time step intervals, including the final time step. In addition, the intensity of clustering of high counts in each time step is increasing overall, and that increase is statistically significant.
- Diminishing hot spot: a location that has been a statistically significant hot spot for 90 percent of the time step intervals, including the final time step. In addition, the intensity

of clustering in each time step is decreasing overall, and that decrease is statistically significant.

- Sporadic hot spot: a location that interchanges between being a hot spot and not being a hot spot. Less than 90 percent of the time step intervals have been statistically significant hot spots, and none of the time step intervals have been statistically significant cold spots.
- Oscillating hot spot: a statistically significant hot spot for the final time step interval that has a history of also being a statistically significant cold spot during a prior time step. Less than 90 percent of the time step intervals have been statistically significant hot spots.
- Historical hot spot: the most recent time period is not hot, but at least 90 percent of the time step intervals have been statistically significant hot spots.
- Persistent cold spot: a location that has been a statistically significant cold spot for 90 percent of the time step intervals, with no discernible trend indicating an increase or decrease in the intensity of clustering of counts over time.
- New cold spot: a statistically significant cold spot for the final time step and never has been a statistically significant cold spot before.
- Consecutive cold spot: a location with a single uninterrupted run of statistically significant cold spot bins in the final time step intervals.
- Intensifying cold spot: a location that has been a statistically significant cold spot for 90 percent of the time step intervals, including the final time step. In addition, the intensity of clustering of low counts in each time step is increasing overall, and that increase is statistically significant.
- Diminishing cold spot: a location that has been a statistically significant cold spot for 90 percent of the time step intervals, including the final time step. In addition, the intensity of clustering of low counts in each time step is decreasing overall, and that decrease is statistically significant.
- Sporadic cold spot: a location that interchanges between being a cold spot and not being a cold spot. Less than 90 percent of the time step intervals have been statistically significant cold spots, and none of the time step intervals have been statistically significant hot spots.
- Oscillating cold spot: a statistically significant cold spot for the final time step interval that has a history of also being a statistically significant hot spot during a previous time step. Less than 90 percent of the time step intervals have been statistically significant cold spots.
- Historical cold spot: the most recent period is not cold, but at least 90 percent of the time step intervals have been statistically significant cold spots.
- No pattern detected: a location in which the pattern of crime does not fall into any of the hot spot or cold spot patterns defined in this list.

Figure 6.3 shows the output of an analysis using the Space Time Cube and Emerging Hot Spot Analysis tools using the same data from London used for creating the maps in figures 6.1 and 6.2. The analysis involved a two-step process:

- The creation of the space-time cube—this involved selecting the point dataset of robberies and thefts against pedestrians of London (covering a two-year period) and specifying the Time field as the date on which each offense occurred. Optional fields that could be entered into the tool were left blank, except for the *Aggregation Shape Type* being specified as hexagons rather than square grids (in part, to illustrate the visualizations you can achieve when using hexagons rather than square grids), and the *Distance Interval* being set to 250 meters. This resulted in the creation of 12 time step intervals, each two months in length, with data aggregated into geographic units, hexagonal in shape, and with a width of 250 meters[1]. The analysis included in the creation of the space-time cube detected a statistically significant decreasing crime trend.
- The generation of Emerging Hot Spot Analysis output—this involved selecting the space-time cube file that was created in the previous step as the input file, and selecting the *Analysis variable* as being the aggregate count of crime for each bin. The optional entries for *Neighborhood Distance* and *Neighborhood Time Step* were left blank (so that the analysis examined each hexagonal grid cell only on its own, rather than neighboring cells).

From the analysis output shown in figure 6.3, 51 (7 percent) of the 756 grid cells were identified as persistent hot spots (see also table 6.1), 97 cells (13 percent) were identified as persistent cold spots, no grid cells were identified as new or intensifying hot spots, and 13 cells (2 percent) were identified as diminishing hot spots. Most of the diminishing hot spots were in the northeast of the study area. A large area consisting of 14 grid cells toward the south of the study area was identified as a historical hot spot—i.e., in the most recent period, the area was *not* a statistically significant hot spot, when previously it had been. Several grid cells were identified as sporadic hot spots, as a single area in the east of the study area or distributed around persistent hot spots or diminishing hot spots. These findings importantly add to the results generated in figures 6.1. and 6.2, as they show the area in the central part of the map (in figure 6.3) as being, perhaps, the area to prioritize for attention—it is an area in which high levels of crime have persisted over the last two years, even when overall levels of robberies and thefts against pedestrians have significantly decreased by 13 percent in the study area.

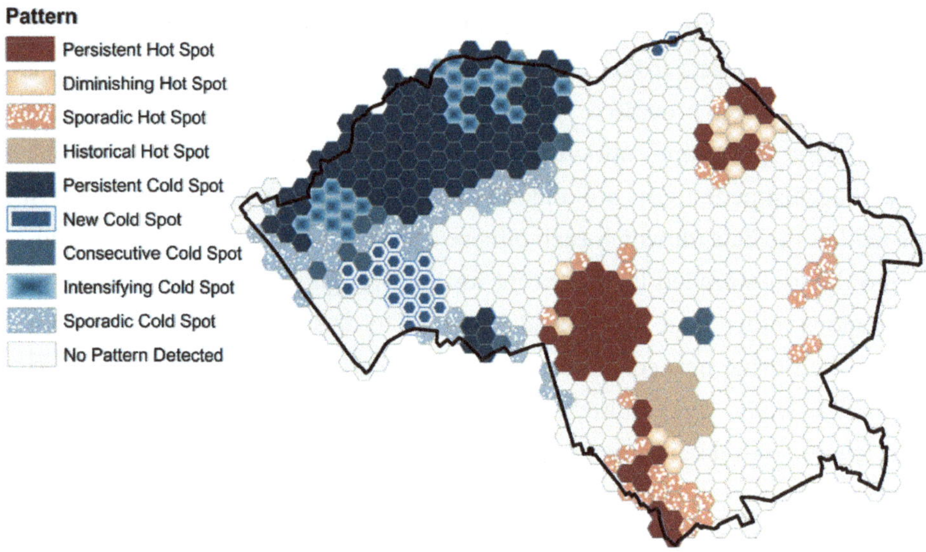

Figure 6.3. Patterns of the change in crime over time for an area in London (showing only those patterns that were present).

Table 6.1. Number and proportion of grid cells (shown in figure 6.3) by hot spot pattern type

Pattern type	Number (and %) of grid cells
Persistent hot spots	51 (6.7%)
New hot spots	0
Intensifying hot spots	0
Diminishing hot spots	13 (1.7%)
Historical hot spots	14 (1.9%)
Sporadic hot spots	31 (4.1%)

Identifying emerging problem areas

In the study area used as the example in figures 6.1, 6.2, and 6.3, robberies and thefts against pedestrians had been declining over the period of interest. What about when crime increases? One option would be to use the technique that produced the map in figure 6.2, showing the areas in which the crime level had been greatest. Although useful, this technique is sensitive to small numeric changes for areas in which the levels of crime are relatively low (e.g., an increase from one crime to four crimes). The techniques used for creating the map in figure 6.3 also would be useful, but perhaps of more use would be a technique that identifies those areas that are most responsible for an overall increase. The technique we describe next produces this result. The technique we describe uses the Dispersion Calculator[1] software, produced by Jerry Ratcliffe (2010).

To illustrate the technique for identifying emerging problem areas, we use data from three cities in Brazil: Rio de Janeiro, Duque de Caxias, and São João de Meriti. São João de Meriti has a population of about 500,000, the population for Duque de Caxias is about 1 million, and for Rio de Janeiro, the population is about 6.5 million. Data on street robberies was used for a two-year period. Over the two-year period, each of the three cities saw significant increases in robberies (table 6.2).

Table 6.2. Change in robbery for three Brazilian cities

City	Population (2016)	Population density (pop per km²)	*n* robbery year 1	*n* robbery year 2	Change year 1 to year 2
Rio de Janeiro	6,498,837	5,306	44,096	55,149	+25%
Duque de Caxias	1,044,058	2,248	4,869	7,938	+63%
São João de Meriti	460,541	13,219	2,996	5,293	+77%

Not all areas in each of these three cities in Brazil are likely to have experienced an increase in robberies to the level of the overall city increase. Some areas likely have contributed more than other areas to this increase. To conduct an analysis to identify those areas that have contributed the most to an increase requires us to aggregate counts of our crime data into small geographic units (using a GIS to do so). These geographic units could be census areas (e.g., census blocks or tracts) or grid cells. For each of the three cities, 150 × 150 m grid cells were created, and a count of robberies in year 1 and in year 2 were calculated for each grid cell. The Dispersion Calculator requires data to be input in comma-separated format, with three fields (columns) of data: an identifier for each geographic unit (e.g., the grid cell identification number), the count of crime for year 1, and the count of crime for year 2. Inclusion of the grid cell identification number also means the results produced by the Dispersion Calculator can be imported into a GIS and joined to the polygons for each grid cell.

The Dispersion Calculator generates two useful global index measures. The Offense Dispersion Index determines the proportion of areas that alone have contributed to a study area–wide equivalent increase in crime. The ODI ranges from zero to one. If the ODI value is close to zero, it indicates that crime has increased in only a few areas. The Non-Contributory Dispersion Index, ranging from zero to one, indicates the proportion of other areas that are a concern. An NCDI close to zero, especially in conjunction with a low ODI, suggests the increase in crime has occurred in only a few areas, whereas a value close to one indicates the crime increase has spread to other areas (Ratcliffe 2010). Table 6.3 lists the ODI and NCDI values that were calculated for each city. For Rio de Janeiro and Duque de Caxias, the ODI values were extremely low (.034 and .048, respectively), suggesting

1 Dispersion Calculator software: https://www.jratcliffe.net/post/calculating-the-dispersion-when-crime-increases

that the robbery increase in these two cities was experienced in only a small number of areas. In Duque de Caxias, the NCDI was also extremely low (.082), suggesting the crime increase was contained within only a few areas and that very few additional areas had experienced an increase. In Rio de Janeiro, the NCDI was slightly higher (.134) than for Duque de Caxias but still relatively low, suggesting that relatively few other areas experienced any increase, but the number of areas that experienced an increase was greater than in Duque de Caxias. In São João de Meriti, the ODI and NCDI values were still relatively low (.233 and .206), but the results suggested the crime increase was more dispersed in this city than in the other two cities.

Table 6.3. Dispersion Calculator values

City	ODI	NCDI	*n* of grid cells that contributed to the equivalent citywide increase (and percentage of all grid cells)
Rio de Janeiro	.034	.134	1,427 (3.4%)
Duque de Caxias	.048	.082	577 (4.8%)
São João de Meriti	.233	.206	387 (23.3%)

The Dispersion Calculator also lists the contribution that each area (in our example, grid cells) has made to the overall increase but does so by ranking in order those that contributed the most to the increase. From this list, we can identify the areas that, alone, have contributed to an equivalent increase in crime experienced across the entire study region. For example, if the study area increase was 25 percent, the process identifies the areas that, alone, have contributed to a 25 percent increase in crime across the entire study area. We can refer to them as our emerging problem areas, or EPAs. Table 6.3 also lists some of the other results from the Dispersion Calculator for each of the three cities. The table shows, for example, that in Rio de Janeiro, 1,427 grid cells (out of 41,943) contributed to the equivalent increase in robberies of 25 percent experienced across the entire study region. That is, only 3.4 percent of all grid cells were responsible for a citywide increase in robbery of 25 percent. Other cells experienced a marginal increase, but they were offset by cells that experienced a decrease in crime. The 1,427 grid cells identified from this analysis were the main contributors to the increase—in this case, the increase in these cells being equivalent to a citywide increase of 25 percent. Thinking of this result in practice, if robberies could be reduced in the 1,427 cells (3.4 percent of the area) that alone contributed to a citywide increase of 25 percent, there is potential that robberies could be reduced significantly across the whole city (note, this would require reducing robberies to previous levels in these grid cells, rather than eliminating robberies completely). Similar, albeit more significant, results were generated for Duque de Caxias (only 4.8 percent of the area was responsible for the citywide increase of 63 percent), but in São João de Meriti, almost a quarter of all grid cells contributed to the equivalent citywide increase of 77 percent. This is a large area in

São João de Meriti, relative to the results generated for the two other cities. We take this up further in the next section when we describe our third technique.

The areas that contributed to the equivalent citywide increase can be mapped. Mapping requires us to export the Dispersion Calculator results from MS Excel (or another spreadsheet software application in which the Dispersion Calculator results were reviewed) and join these results for each grid cell to the polygons for each cell in our GIS. Figure 6.4 shows examples for each of the three cities, alongside KDE hot spot maps for comparison. In each city, hot spots of robberies can be observed, and although in each case the EPAs are observed in hot spots, they also are observed in other areas. In Rio de Janeiro and Duque de Caxias, the EPAs represented less than 5 percent of the whole city area. In São João de Meriti, the ODI was much higher, as illustrated in figure 6.4f, with almost a quarter of the cells in the city identified as EPAs.

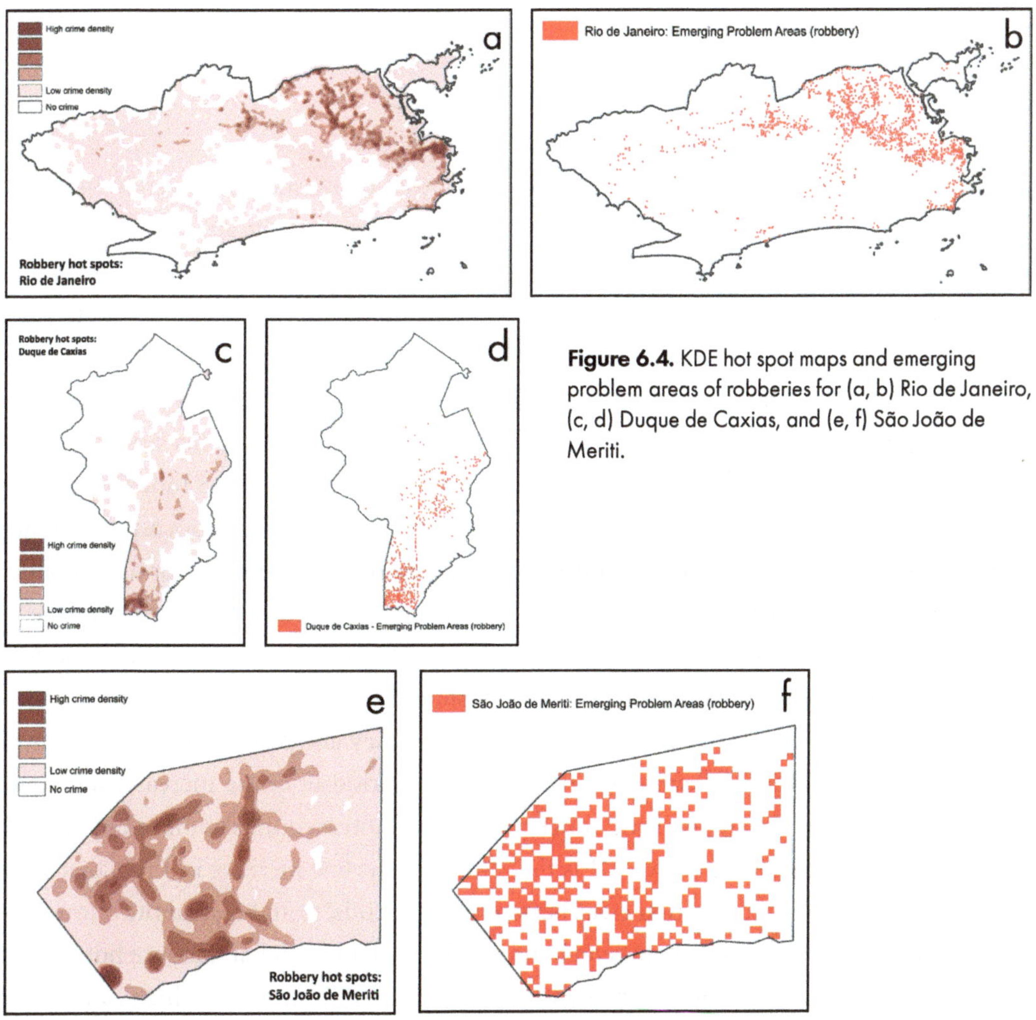

Figure 6.4. KDE hot spot maps and emerging problem areas of robberies for (a, b) Rio de Janeiro, (c, d) Duque de Caxias, and (e, f) São João de Meriti.

The space-time cubes and emerging hot spot analysis technique described in this chapter can be used regardless of whether there is an increase or decrease in crime. The Dispersion Calculator is most suitable when there is an increase and builds on the use of space-time cubes and emerging hot spot analysis by identifying the areas that are most responsible for the increase, and whether this increase is contained or dispersed. The Dispersion Calculator, though, still can result in identifying many areas, when often a key purpose of crime analysis is to focus attention and determine the areas to prioritize. In the example from Rio de Janeiro, the EPAs represented only 3.4 percent of the area, but as can be seen, the EPAs were still expansive. One approach for prioritizing areas for resource targeting would be to focus on the EPAs that contributed most to the increase (i.e., the highest-ranked EPAs). In the next section, we introduce an alternative technique for determining which areas to prioritize, by considering together patterns of crime concentration and crime dispersion during a period of crime increase.

The Crime Concentration Dispersion Index

In chapter 3, we illustrated that crime concentration is a well-established observation in the spatial analysis of crime. At the micro-place level, crime concentrates at significant levels, relative to the distribution of crime across micro-place units. But what about when crime increases? Is it areas in which crime previously concentrated that we see the biggest increases (i.e., hot spots have just become *hotter*), or is the increase associated with new hot spots emerging? For resource targeting purposes, knowing whether the areas in which crime has increased are the areas in which crime has previously concentrated or whether the crime increase is associated with other areas is important for informing prioritization strategies. In chapter 4, we described Andresen's S Index, a spatial point pattern test for measuring the degree to which one pattern of crime differs from another. Ratcliffe's Dispersion Calculator can help identify those areas that are most responsible for an overall crime increase. However, neither technique provides a direct means of determining whether the crime increase is more associated with hot spots becoming hotter, or if the increase in crime is more associated with increases in other areas (i.e., new hot spots are responsible for the crime increase).

In this section, we describe how to use the CCDI (Chainey and Monteiro 2019) and illustrate how this measure can help determine whether, during a period of crime increase, interventions should be targeted at areas of high-crime concentration or if interventions additionally should consider other emerging hot spots. The CCDI is the ratio of the crime increase in EPAs that were *not* identified as high-crime concentration areas between two periods, t1 and t2, and the crime increase in high-crime concentration areas between t1 and t2. The CCDI is a global measure but helps determine whether the focus should be toward areas of high-crime concentration, which can be mapped following the instructions provided in chapter 3, or if the focus should be toward EPAs.

To calculate the CCDI first requires an analysis that uses the Dispersion Calculator to identify those areas that were EPAs. The EPAs are then compared with the high-crime concentration areas (i.e., areas that account for 25 percent of crime, the analysis for which is described in chapter 3) for period t2. Areas that are identified as EPAs and those that are identified as crime concentration areas are then compared. The following steps explain this process:

- Step 1 involves determining the high-crime concentration areas that were responsible for 25 percent of all crime in the study area. Next, EPAs are identified using the Dispersion Calculator.
- In step 2, the results from step 1 are compared by determining which EPAs were not identified as high-crime concentration areas in t2. From this point, we will refer to these areas as *non-CC EPAs*. The number of non-CC EPAs is noted. For the non-CC EPAs, the total increase in crime experienced between t1 and t2 is calculated. An average increase in crime per non-CC EPA is then calculated (i.e., by dividing the crime increase across all non-CC EPAs by the number of non-CC EPAs).
- Step 3 involves calculating the total increase in crime experienced between t1 and t2 for all high-crime concentration areas (areas accounting for 25 percent of crime). We will refer to these areas as *crime concentration areas* (CCAs). The number of CCAs is noted, and the average increase per CCA is calculated (i.e., by dividing the crime increase across all CCAs by the number of CCAs).
- For step 4, the CCDI is the ratio between the increase in crime per non-CC EPA and the increase in crime per CCA.

$$\text{CCDI} = \frac{\text{Crime increase between t1 and t2 per non-CC EPA}}{\text{Crime increase between t1 and t2 per CCA}}$$

An example of the calculation of the CCDI follows:

- Crime increase in whole study area (between t1 and t2) = 1,000
- *n* of geographic units in study area = 40,000
- *n* of geographic units that are CCAs (representing 25 percent of all crime) = 400
- *n* of geographic units that are EPAs = 1,550
- *n* of EPAs that are CCAs = 350
- *n* of non-CC EPAs = 1,200
- Crime increase in CCAs (between t1 and t2) = 400
- Crime increase per CCA = 1.0
- Crime increase in non-CC EPAs (between t1 and t2) = 400
- Crime increase per non-CC EPA = 0.33
- CCDI = 0.33 (i.e., 0.33/1.0)

A CCDI of less than one indicates that high-crime concentration areas contributed more to the increase than other emerging problem areas (i.e., the non-CC EPAs) (figure 6.5). A CCDI of one indicates that high-crime concentration areas and other EPAs equally contributed to the increase. In practice, the closer the CCDI is to zero, the less the need for targeting resources to areas *other* than high-crime concentration areas when a crime increase is experienced. That is, the increase is because the hot spots have become hotter. CCDI values close to one indicate that other areas are also responsible for the crime increase, suggesting that these new emerging areas also require attention if crime is to be reduced. A CCDI of greater than one indicates that other EPAs contributed more to the increase than high-crime concentration areas. We now illustrate an example using data on robberies for the three cities in Brazil used in previous examples.

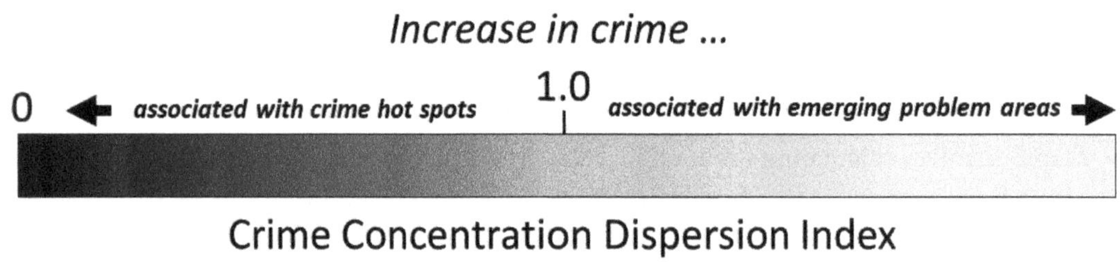

Figure 6.5. The Crime Concentration Dispersion Index.

Recall that each of the cities of Rio de Janeiro, Duque de Caxias, and São João de Meriti experienced increases in crime over a two-year period. Identifying EPAs is useful, but so, too, is the analysis of crime concentration at micro-places. So, which should be used to best assist resource targeting? Is it that areas in which crime previously concentrated have just become *hotter* over the period of crime increase, or have other hot spots emerged? In following the steps described above, the CCDI for each city was calculated (table 6.4). Table 6.4 also shows S Index, ODI, and NCDI values for each city. The S Index values indicate that the spatial pattern of robberies between year 1 and year 2 was similar in both Rio de Janeiro and Duque de Caxias (indicated by an S Index value of at least 0.8), but dissimilar in São João de Meriti. The ODI and NCDI values for Rio de Janeiro and Duque de Caxias suggest the crime increase was contained, and associated with only a small number of areas, but in São João de Meriti, the crime increase was more (spatially) disperse. The CCDI results provide another valuable perspective. In São João de Meriti and Duque de Caxias, the CCDI values were 0.382 and 0.281, respectively, suggesting the increase was associated mainly with hot spots (the areas of highest crime concentration) getting hotter, whereas in Rio de Janeiro (with a CCDI of 0.702), hot spots became hotter, but other new hot spots (new areas of high crime concentration) also have emerged, contributing to the increase. In practice, in the cities of São João de Meriti and Duque de Caxias, a strategy for reducing robberies that targets hot spots is likely to be most effective, whereas in Rio de Janeiro, efforts to reduce robbery will be more challenging—it is not only hot spots that must be targeted, but new areas also have emerged as hot spots and will require attention.

Table 6.4. Crime Concentration Dispersion Index values

City	CCDI	S Index	ODI	NCDI
Rio de Janeiro	0.702	.80	.034	.134
Duque de Caxias	0.281	.86	.048	.082
São João de Meriti	0.382	.54	.233	.206

For each of these three cities in Brazil, additional analysis was performed to compare differences in the number of cells of interest, which were identified using the S Index, Dispersion Calculator, and CCDI approaches, and the increase in crime in these cells. The S Index and Dispersion Calculator techniques identified approximately 5 percent of grid cells in Rio de Janeiro and Duque de Caxias as experiencing either a significant increase in robberies ($p <= .01$) or as emerging problem areas (table 6.5). In São João de Meriti, more than 15 percent of the area was identified as being of interest when using these two techniques. Consider that 15 percent, and even 5 percent, of any city is still a large area, and typically too large for any targeted resource deployment. The areas of high-crime concentration for each city were much smaller in coverage—high-crime concentration areas covered less than 1 percent of the city area in Rio de Janeiro and Duque de Caxias, and 3 percent of the city area in São João de Meriti. Identifying high-crime concentration areas may offer a better initial means of selecting areas to prioritize, and in the cases in which the CCDI was less than one (as in the case of all three cities), targeting resources to areas of high-crime concentration likely would be of most benefit. Comparison of average change in crime for the cells selected using the three techniques also showed that the largest increases were in those cells identified as high-crime concentration areas. For example, robberies increased on average by 10 offenses in high-crime concentration areas compared with increases of five offenses on average in areas identified using the S Index and Dispersion Calculator approaches (see table 6.5).

Table 6.5. Average increases in robbery per grid cell for S Index, Dispersion Calculator, and crime concentration measures

City	S Index—cells identified with statistically significant increase		Dispersion Calculator—cells identified as EPAs		Crime concentration—cells identified as CCAs	
	n (proportion of all cells)	Average increase in robberies per cell	*n* (proportion of all cells)	Average increase in robberies per cell	*n* (proportion of all cells)	Average increase in robberies per cell
Rio de Janeiro	2,425 (5.8%)	5	1,427 (3.4%)	7	350 (0.8%)	8
Duque de Caxias	442 (3.7%)	4	577 (4.8%)	4	60 (0.5%)	11
São João de Meriti	258 (15.6%)	5	387 (23.3%)	5	51 (3.1%)	10

Additionally, in the calculation of the CCDI, each grid cell is ranked using the Dispersion Calculator results, and in turn can further assist the prioritization of areas to target. Table 6.6 provides an illustration of this additional prioritization, using the example of Duque de Caxias. Sixty cells were identified in Duque de Caxias as high-crime concentration areas—these cells covered 0.5 percent of the city area and accounted for 31 percent of all robberies in year 2. The top 10 high-crime concentration areas that were ranked by the Dispersion Calculator as experiencing the highest increases in robbery covered an area of 0.08 percent of the city and accounted for 6.2 percent of all robberies. These 10 highest-ranked high-crime concentration areas experienced a 142 percent increase in robbery compared with the 116 percent increase across all high-crime concentration areas in the city, with these highest-ranked cells each accounting for an average of 40 robberies in year 2. Furthermore, the robbery increase in just these 10 cells was equivalent to a citywide 7.6 percent robbery increase. Targeting high-crime concentration cells by the order of their Dispersion Calculator ranking can provide an effective additional means of prioritization for areas in which the CCDI value is less than one, and particularly when it's close to zero.

Table 6.6. Robbery levels in Duque de Caxias for the top 10 CC EPAs, all CCAs, and all other areas

City: Duque de Caxias	n of cells (%)	n robbery year 2 (%)	Average robberies per cell (Year 2)	Change in robberies for year 1 – year 2 (%)	Contribution to citywide increase of 63%
Top 10 CC EPAs	10 (0.08%)	400 (6.2%)	40	235 (+142%)	7.6%
All CCAs	60 (0.5%)	1,985 (31%)	33	1,067 (+116%)	30%
All other areas	11,917 (99.5%)	4,421 (69%)	0.4	1,400 (+46%)	25.6%

Chapter summary

The analysis of the spatial distribution of crime often involves taking a snapshot of a period (e.g., one year) and analyzing the overall patterns. Although useful, this analysis can hide variation over the period of study. In this chapter, we have illustrated how to use techniques that identify areas in which high (and low) crime levels persist, intensify, and diminish; identify the specific areas that are most responsible for an overall crime increase; and explain how these observations relate to crime concentration. If we decide to plan a new crime reduction intervention that targets hot spots because we have observed an increase in crime across our jurisdiction, we want to be able to determine the extent to which the hot spots are responsible for the crime increase, or if new areas of crime concentration appear to be emerging. Each technique we have described in this chapter complements the other rather than being the optimal technique to use in these circumstances. Collectively, they can help us better determine where to prioritize resource targeting. If an area persists with high levels of crime, our next line of inquiry is to help understand why.

To understand crime requires an examination of data beyond the simple production of a hot spot map. Hot spot maps provide a useful initial point of inquiry, from which we then should examine patterns within them in more detail. In previous chapters, we have considered how an analysis of spatial stability, repeat victimization, and near repeat victimization contributes to helping us understand spatial patterns of crime. All these techniques illustrate the inherent temporal qualities, alongside the inherent spatial qualities, observed in crime data. In this chapter, highlighting these temporal qualities further has been a key objective, so that our analysis of crime helps determine the extent to which our spatial patterns persist, are emerging, or are dispersing over time.

Further resources

Space Time Cube: including details about time steps and bins, https://pro.arcgis.com/en/pro-app/tool-reference/space-time-pattern-mining/create-space-time-cube.htm.

Jerry Ratcliffe's Dispersion Calculator: https://www.jratcliffe.net/post/calculating-the-dispersion-when-crime-increases.

References

Chainey, S. P., and J. Monteiro. 2019. "The Dispersion of Crime Concentration during a Period of Crime Increase." *Security Journal* 32 (3): 324–41.

Ratcliffe, J. H. 2010. "The Spatial Dependency of Crime Increase Dispersion." *Security Journal* 23:18–36.

CHAPTER 7

Geographic profiling analysis: Supporting criminal investigation

Objectives

Geographic profiling is an investigative method that uses the locations associated with a series of crimes to determine the most probable area of offender residence, or some other location from which the offender anchors their activities. Geographic profiling can be applied to all types of serial crime, and to certain types of crime that involve only a single incident.

In this chapter, we introduce geographic profiling. Geographic profiling involves much more than entering crime locations into a spatial analysis software tool and generating results. To do geographic profiling effectively to support criminal investigation requires specialist knowledge and skills that cannot be covered in a single chapter. However, geographic profiling is an important method to be familiar with to better understand crime. In this chapter, we describe the key principles underpinning geographic profiling, discuss spatial analysis techniques that support geographic profiling, and illustrate its application with several examples. The aim is to provide sufficient information that will help you incorporate principles of geographic profiling into your analysis and research and allow you to consider ways you can further develop professional skills in geographic profiling.

Key learning points

- Improve awareness of geographic profiling to understand how it can support criminal investigation.
- Understand the principles underpinning geographic profiling, including theoretical perspectives from environmental criminology, offender spatial mobility, and offender search

and attack methods. In addition, recognize the importance of good linkage analysis as an essential input to geographic profiling analysis and consider the types of investigative strategies that geographic profiling can support.

- Become familiar with a range of spatial analysis techniques for supporting geographic profiling and understand the differences between them.

Introduction

I could write a whole book to explain how to use geographic profiling, which is not necessary because it has been done by Kim Rossmo in his definitive book on the subject (2000). Geographic profiling is a proven, invaluable tool for supporting criminal investigation. Geographic profiling involves analyzing the geography of a linked series of crimes and using the analysis results to help inform a criminal investigation. It can help identify and lead to the arrest of the offender but can also support other geographically targeted investigative strategies, such as intelligence gathering and surveillance.

In this chapter, we illustrate how you can use certain spatial analysis techniques that help you apply geographic profiling and give examples of its use. We come short of leaving you thinking, by the end of the chapter, that you are now an expert on the subject, and effectively can apply geographic profiling to support the investigation of a murder or rape series. Geographic profiling requires specialist skills to apply it professionally, but it would be remiss not to refer to this method in a book on analyzing the geography of crime. We cover the topic in enough detail in this chapter to help you better apply geographic profiling principles in your work.

Geographic profiling is more than just inputting data into a computer and using the results for criminal investigation. Geographic profiling involves looking for patterns that can be extracted from an analysis of a crime series and requires careful consideration of the information used and diligent and thoughtful interpretation of the analysis results. To do it effectively requires experience from a range of investigative scenarios and cases to determine how the spatial anchors identified relate to the offender's behaviors.

Geographic profiling does not solve crime. There are only three ways a crime can be solved, in that an offender is charged for an offense: through a witness statement, from a confession by the offender, or from physical evidence. Geographic profiling assists the investigative process that can result in a crime being solved. Geographic profiling also is not an "x marks the spot" routine. Instead, it is better to think of geographic profiling as identifying what is *most likely*.

Geographic profiling uses the locations of a connected group or series of crimes to determine the most probable area of offender residence, or some other location from which the offender anchors their activities (Rossmo 2000). These anchors could include a place of work, or the location from which the offender conducts a *hunt* for their victims. Geographic profiling initially was used to investigate stranger violent crime (such as murder and rape) but is now applied to the full range of crime types considered to have serial qualities. Geographic profiling also can be applied to single offenses, in which additional geographic information exists beyond where the offense took place. This last point recognizes that geographic profiling involves using information not only on the location of offenses but other geographic information related to the offense, such as the last location where the victim was seen, and where items of the victim's property were found or used (e.g., where a stolen credit card or cell phone was used or found). Geographic profiling seeks to use the geographic traces left by the offender to assist the police in their investigation.

In the next section of this chapter, we briefly describe the theoretical principles underpinning geographic profiling, explain the importance of good linkage analysis as an input to geographic profiling, and describe the types of investigative strategies that geographic profiling can support. We then describe several spatial analysis techniques that support geographic profiling, including the use of centrographic statistics and the criminal geographic targeting (CGT) equation. We use examples throughout to help illustrate these principles, the use of spatial analysis techniques, and the application of investigative strategies. I am limited with the examples I can show in this book because all of my examples are based on real cases. Behind each point that is mapped is a person—a person who has been a victim of crime, and sometimes a serious crime. The examples used in this chapter are from real cases and often are detailed, but for some cases, details have been removed to avoid revealing the identities of the victims who have suffered these crimes.

Professional geographic profilers and geographic profiling analysts

There are two levels of certification in geographic profiling: professional geographic profilers and geographic profiling analysts.

Professional geographic profilers receive training in criminal behavioral psychology, mathematics and statistics, and environmental criminology and undergo extensive practical training, involving visiting many crime sites and practicing on past cases. Training takes a minimum of one year. Professional geographic profilers typically are sworn officers who work for large regional or national police agencies and are assigned to complete geographic profiles full time, or at least most of the time. Many professional geographic profilers also have a serious crime investigation background that provides them with the additional qualification to work effectively on murder and rape serial cases. In 2020, there were fewer than a dozen professional geographic profilers in the world.

Geographic profiling analysts receive training in environmental criminology, offender spatial behavior, crime linkage, investigative strategies, and spatial analysis techniques for geographic profiling, especially the use of the CGT equation and analysis. Training involves a two-week course and includes practice and additional skills development from working through several real cases. The training is most suited to police crime and intelligence analysts, technically competent investigative officers, and others working in applications in which geographic profiling can be applied (e.g., military analysts). Geographic profiling analysts complete geographic profiles on a part-time basis, applying geographic profiling when necessary in their analysis and investigative tasks.

Geographic profiling analysis training courses are provided by trained experts from the UK, USA, and Canada. See https://www.ucl.ac.uk/short-courses/search-courses/geographic-profiling-analysis-crime-and-security.

Key principles and inputs to geographic profiling

Theoretical principles underpinning geographic profiling

Geographic profiling draws from four key theoretical perspectives: crime pattern theory, the least-effort principle, the routine-activity perspective, and rational choice. We describe these at length in chapter 1 (with references) so here we provide only a quick recap but framed in relation to geographic profiling.

- The *routine-activity perspective* states that for a crime to occur, three components are necessary: a likely offender, a suitable target, and the absence of a capable guardian. These three components must meet in time and space to provide the necessary chemistry for crime. This meeting in time and space is not random but dictated by the natural rhythm of daily life, with people going about their routine activities. The combination of these three components dictates that the risk of crime changes over time with the movement of people throughout the daily routine activities of their lives. That is, as these meetings in time and space are not random, the crime patterns we can observe relate to people's routine activities.
- The *rational-choice perspective* provides a framework to consider offender decision-making when a crime opportunity is presented. It also can be used to consider likely strategies that will influence the decision-making of the offender. Most offenders are known to make some sort of decision to commit a crime by weighing the risks and effort involved against the rewards they may gain. This decision-making suggests committing a crime is a (fairly) rational decision. On its own, the theoretical principles of rational choice do not necessarily determine the spatial patterning of an offender's behavior, but when considered alongside the influence of routine activities, they can help inform a model for the interaction between offenders and victims—crime pattern theory.
- *Crime pattern theory* can help explain where and when crime is likely to occur. Crime pattern theory brings together the two areas of offender spatial distribution and offense

spatial distribution by examining the relationship of the offense to the offender's use of space. Crime pattern theory suggests that offenders are influenced by the daily activities and routines of their lives, so that when they search for a criminal opportunity, they will tend to steer toward areas that are known or most familiar to them. These areas make up the offender's awareness space. In an offender's day-to-day activities, they will be watching for targets that have no guardians (i.e., the areas of opportunity), in areas in which they feel comfortable and are familiar.

- The *least-effort principle* dictates that there are spatial limits to the geographic coverage of an offender's awareness space, and it influences why an offender chooses a location to commit crime. As increased distance to commit crime increases the effort, it also likely increases the risks associated with their involvement in crime commission and increases the possibility the offender will stray into an unknown area. This theoretical concept of minimizing the distance traveled to commit crime is observed in studies on the journey to crime, which indicate that most offender journeys to crime tend to be short (as described in chapter 2). The least-effort principle, therefore, is a useful mechanism for thinking about the geographic extent of an offender's spatial behavior and the crime patterns that may result. Furthermore, an offender's choice in where to commit a crime and the patterns that result additionally can be explained in terms of how opportunities to commit crime concentrate. These opportunities may be influenced by the large numbers of people or other targets present at certain locations, or by places that create specific criminal opportunities, and in doing so attract offenders.

The journey to crime and the search to crime

Results from numerous studies on offenders' journeys to crime show that these journeys tend to be short (see chapter 2 for a longer discussion on the journey to crime). However, it would be too simplistic to think that the offender travels from the location in which they are anchored straight to the location in which the crime takes place. Instead, the offender must conduct some form of search to find their potential victim. This search may take time, requiring the offender to be patient by remaining stationary at a point and waiting, or by traveling around the area in search of potential targets. The distance that an offender travels in their commission of crime, and the time it takes to find suitable targets, are likely to be much longer than indicated from simple journey to crime metrics (figure 7.1).

Figure 7.1. Search to crime distance compared with journey to crime distance.

Knowing that offenders tend *not* to travel long distances to commit crime is useful, as it indicates the offender is likely to be local. Thinking in terms of the *search to crime* is also useful because it suggests, from an investigative point of view, that there is potential for the offender to be identified from their suspicious activity, or from targeted surveillance in the area in which the offender is most likely searching for his (or her) victims.

Crime pattern theory provides a model to help explain why an offender chooses a location to commit a crime. It illustrates that this decision-making is informed by the offender's awareness of the area; the availability of, and their familiarity with, crime opportunities; and their previous experience with crime commission. Crime pattern theory, therefore, provides a theoretical explanation of why the spatial patterns of crime associated with an offender are not random and can be predicted. However, the activities performed in spaces are not static, meaning that different social, economic, and institutional structures that operate in these spaces, and hence the routine activities that operate within them, result in changes over time in the pattern of opportunities to commit crime. In turn, these conditions influence when offenders commit crime. The motivations of the offender also are bounded by a degree of rationality in their decision-making. Offenders prefer to commit offenses in areas in which they are comfortable, in which risks are lower, and which suit their preference to minimize effort. The combination of these theoretical explanations helps us consider space in terms of a criminal geographic landscape, which in turn provides the source for determining characteristics about the offender's spatial behavior that could be used to assist in their apprehension.

From an investigator's point of view, the theoretical principles underpinning geographic profiling can help provide a path *outward* from the crime scene. The investigator can consider why the offender chose this location to commit their crime, and chose this person (or target) to commit their crime against (i.e., was it just a case of the victim being in the wrong place at the wrong time, or was it something about them that made them a suitable target?). The location acts as a clue, because it is likely that the offender's spatial behavior has been influenced by their awareness space, the principle of least effort, and an assessment of the risks that helped them determine where to commit a criminal act.

Offender hunting methods

In geographic profiling, a consideration of the offender's hunting behavior can help in the interpretation of the offender's spatial behavior. Examining the offender's hunting method requires consideration of how the offender searches for a suitable victim or target, as well as the attack method the offender uses.

The search for a suitable victim or target can be grouped into four categories:

- *Hunter*: the offender sets out specifically to search for a victim. The search often begins from the offender's home, and then proceeds through areas in their awareness space that they believe contain suitable targets (figure 7.2a).
- *Commuter* (or *poacher*): the offender sets out specifically to search for a victim, but the search is based from a location other than the offender's residence. That is, it involves the offender traveling to an area from which the search process commences (figure 7.2b).
- *Troller*: the offender, while involved in other noncriminal activities, opportunistically encounters a victim (or target). These other noncriminal activities relate to the offender's normal routine activities at the time of the encounter, such as driving home from work and spotting an opportunity to commit crime (figure 7.2c). Troller crimes are often spontaneous.
- Trapper: the offender has an occupation in which potential victims come to him or her or uses subterfuge to entice victims into the offender's home or some other location for which the offender has some control over (figure 7.2d) (Rossmo 2000).

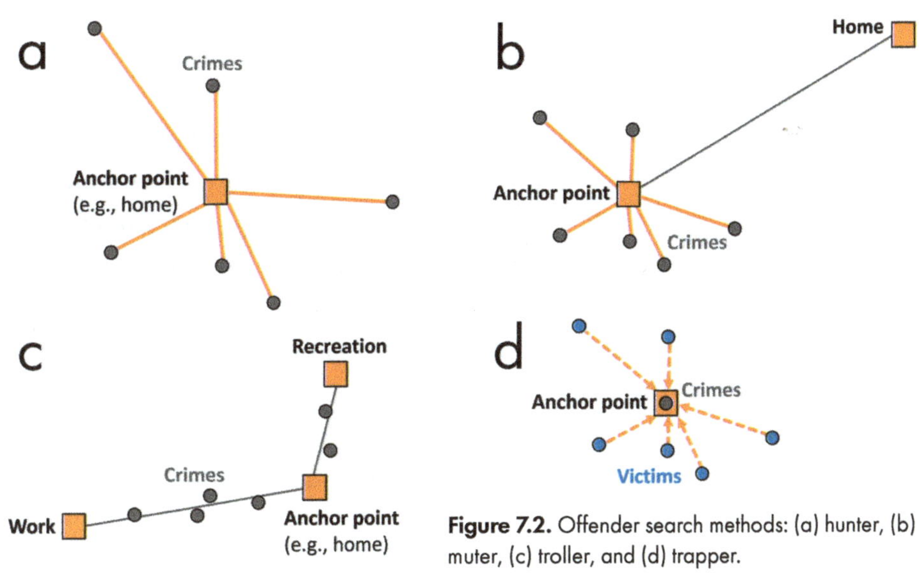

Figure 7.2. Offender search methods: (a) hunter, (b) commuter, (c) troller, and (d) trapper.

For each search behavior, the offender may use a variety of attack methods. The raptor style of attack is when the offender attacks on encounter. The stalker form of attack is when the offender follows a victim and then attacks (Rossmo 2000). This type of attack typically involves the offender moving into the victim's activity space, and possibly away from the offender's own awareness space, and waiting for an opportune moment to strike. The third type is the ambush style of attack (Rossmo 2000), in which the offender attacks a victim that he (or she) has enticed to a location. For each of the search methods, and the different styles of attack that follow, different spatial patterns can result, and different interpretations of the geographic anchor of the offender are required. Determining when the crime pattern relates to a hunter who uses the stalking form of attack or a commuter who follows the same style of attack or a troller who attacks a victim they opportunistically encounter (in the raptor style) takes a careful interpretation of the crime series and experience from working cases. There is no simple statistical test to indicate whether the crime refers to one hunting method over another, and hence is one of the reasons why effective geographic profiling involves much more than entering data into a software tool, producing a map, and using the results to determine where to start knocking on doors to find the offender.

Linkage analysis

A starting point for any geographic profile is to determine the crimes that are linked and form the crime series. Crime linkage analysis (also referred to as *comparative case analysis*) involves comparing features about each crime to determine whether they were committed by the same offender. The process is subject to error, but by following certain analytical steps, errors can be avoided.

The first stage in crime linkage analysis involves assessing whether a crime series is occurring. Crime series often can be identified because they increase the number of crimes that an area experiences. This inquiry initially requires establishing the *baseline*, which relates to the number of crimes typical in this area (for the crime type that is subject to the analysis) for a specific period. For example, if on average, there are two sexual assaults in an area in a typical month, and then four occur within a week of each other, it may indicate the presence of a crime series. Assessment against the baseline can be difficult for certain crimes that are more frequent (e.g., theft from the person), but the fact that there are many of them should prompt you to consider examining the serial features of these crimes as worthwhile. When a crime is occurring relatively frequently, it is likely that many of the offenses are associated with several individuals committing separate series of crimes rather than all crimes being committed by different individuals who never go on to commit further crimes.

Crime linkage involves comparing similarities and differences about the crimes to determine the ones that are related and those that are unrelated. To be successful at crime linkage analysis involves following two simple principles:

- Connected crimes should show more similarities than differences.
- Unconnected crimes should show more differences than similarities.

Three methods can determine whether crimes are linked: use of physical evidence, an assessment of the description of the offender and the description of the victim (or target), and an assessment of crime scene behavior. From these methods, crime linkage variables are established and used to analyze comparisons between the crimes. A crime linkage variable is any factor that possibly links a series, and can include DNA, footprints, and fingerprints; the gender, skin color, hair color, estimated age, height, and build of the offender; the offender's accent (and if they did or did not speak); the gender, skin color, hair color, estimated age, height, and build of the victim (or other descriptive features about the target); the modus operandi of the offender (e.g., if they threatened or used violence against the victim, how they broke into a property, and how they escaped the property); and the proximity in space and time between offenses. With regard to the latter, if a residential burglary took place on a weekday, at about the same time, close to where a burglary recently occurred, the inclusion of location and time variables can help determine whether a crime is linked to others in the series. Variables associated with modus operandi can be one of the best ways to determine crime linkage, and this approach works best when care is taken in separating the features of the modus operandi and treating each as a separate variable (e.g., for a theft that involved violence, variables could include the offender's use of a weapon, how they approached the victim, the type of violence used, and the injuries the victim sustained). The presence of any unique, *signature* characteristics of the crime or crime scene also can be valuable to linkage analysis. The Irvine Chair Burglar in the city of Irvine, California, was so named because he often would use a chair, positioned against the garden wall at the rear of the properties he burgled, to assist in his escape (Rossmo and Velarde 2008).

Common linkage errors

Crime linkage analysis requires a careful, methodic approach, which can go wrong if the principles in crime linkage are not followed. There are three common errors made in crime linkage analysis that can result in unrelated crimes being included in the series and related crimes being omitted:

- The failure to take the baseline into account to identify that a crime series is or is not present.
- Too much emphasis is placed on differences between the crime incidents rather than similarities.
- The failure to recognize that modus operandi is not constant but can change because of the circumstances before, during, and after the crime occurs and the experience the offender develops from committing previous crimes. The examination of modus operandi variables should not be too strict. A useful rule to follow is that the modus operandi is more likely to be similar for crimes that are close in space and time.

Offenders are creatures of habit, so it is useful to focus on the similarities in the factors involved in each criminal act, while also recognizing that the situation in which the offender was operating influences differences that appear in their offending commission. When conducting linkage analysis, avoid using variables about the offender's description that can easily change, such as hair style, facial hair, or clothing, and be conscious that witness memory can be faulty in their description of the offender.

When conducting linkage analysis, it is important to consider how the offender may have adapted their behavior over time, such as evolving their offending by learning from mistakes. Also, it is important to assess the precision of the recorded crime data and the level of detail (or not) entered by the person who completed the crime record. Crime reporting of levels of crimes also bears consideration. Not all crimes are reported, so it is important to consider how the underreporting of crime (i.e., what is not known) influences the results from a geographic profile.

To be professional in geographic profiling requires skills in crime linkage analysis. Even if a series of crimes has been given to a police crime analyst to analyze, it is the responsibility of the analyst to review the linkage to ensure it is correct, and question whether crimes have been omitted. If unrelated crimes are included in a geographic profile, you effectively are introducing the spatial decision-making behavior of a different person into the geographic profile, and hence the results and their interpretation will be inaccurate and misleading.

Questions that an effective geographic profiler aims to answer

To illustrate that geographic profiling involves more than using only a spatial analysis technique to examine crime locations from a crime series, following are some of the questions that an effective geographic profiling analysis attempts to answer:

- Did the routine activities of all (offender, target, guardian) produce this intersection in time and space?
- What do the locations of the crime sites tell us about the choices the offender may have considered?
- What characteristics are present in the crime series that indicate the offender is likely to live locally, or have local knowledge?
- Is the offender choosing specific types of targets?
- Are the temporal features of the offenses based on the target opportunity structure or the offender's own availability?
- Are there multiple locations that relate to the crime event—e.g., an encounter site that is separate from the attack site?
- Are there any other locations associated with the crime event, and how might they assist the geographic profile—e.g., use of a stolen credit card or location where an item of clothing from the victim was found?
- What search and attack method is the offender employing?
- Is the journey to crime likely to start from the offender's home or some other location?
- What is the offender's likely mode of travel, and how is it likely to influence their spatial offending pattern?
- Does the offender have a preference to return to a location multiple times (to commit crime), and how may this help in understanding the offender's spatial decision-making behavior?

- How confident are you that the linkage is accurate and complete—can different scenarios relating to the confidence in linkage be tested to compare spatial analysis results?
- Is there any evidence of pre- or post-offense movement by the offender, and how may it be useful in interpreting the offender's spatial behavior?

Considering and answering these questions is central to effective geographic profiling.

Using the results from a geographic profile to inform investigative strategies

The results of a geographic profile can inform a wide range of investigative strategies. The investigative strategy implemented depends on the interpretation of the results of the geographic profile. The results of geographic profiling are best used in the following investigative strategies:

- Suspect prioritization. This strategy involves narrowing down the list of suspects by identifying the neighborhood in which the offender most probably lives, and then reassessing and reprioritizing suspects on the basis of cross-referencing of the geographic profile with a list of suspects.
- Analysis against known offender registries (e.g., criminal records and parole or prison records). This analysis can help generate additional investigative leads from recording systems based on the results of the geographic profile. It can involve the police searching their own records of offenders to determine whether the geographic features of the series match with those known to the police or help identify other crimes that may be linked to the series. It also can include the search of other systems such as vehicle license databases (for suspect vehicle descriptions).
- Patrol saturation and surveillance. The results of a geographic profile may indicate where to direct saturation patrols and surveillance. This approach is particularly effective if offenses occur during specific times of day, on certain days of the week. This approach also can be effective when it is suspected that the offender is required to spend time searching for targets or waiting for the right circumstances to commit a crime. It gives police the opportunity to observe any people who appear to be acting suspiciously or loitering in the area that the geographic profile has highlighted as an area of interest.
- Targeted police response plans. Many offenders return home after committing a crime. This knowledge can be used to deploy quick response police units to the area of probable offender residence or to focus on the most logical routes from the crime site to where it is suspected the offender resides (or has a key anchor). Roadblocks may be an option along the offender's probable escape route. The strategy also can be useful in identifying locations for possible witnesses and gathering information from them about any activity of interest they observed before or after a crime incident.

- Public awareness campaigns. This strategy can include the mailing of information about the offender to residents and businesses in an area the geographic profile has identified as being of interest so that they can be alert and report any suspicious activity. This approach, however, may alert the offender that the police are attentive to the offender's movements and may result in the offender displacing their activities. This investigative tactic is most viable if the offender description information is adequate and reliable.
- Neighborhood canvassing by the police. This strategy can include door-to-door canvassing, interviews, signs, and targeted media campaigns in the areas identified from a geographic profile.
- DNA screening. This approach involves DNA testing in the areas identified from a geographic profile of targeted groups in a population (e.g., all men between the ages of 18 and 30) or the key suspects in the geographic profile peak area (Rossmo 2000).

In the examples that follow, the applications of several of these investigative strategies are illustrated.

Use of geographic profiling in other applications: From counterterrorism to disease prevention—and revealing the identity of Banksy

Since geographic profiling emerged as a criminal investigative method in the 1990s, it since has been applied to cold cases (Rossmo and Walton 2006); other areas of security (e.g., counterterrorism [Rossmo and Harries 2011] and border security [Rossmo et al. 2008]); and other applications, including ecology (e.g., the predatory attacks of sharks [Martin, Rossmo, and Hammerschlag 2009]) and epidemiology (e.g., infectious sources of disease [Le Comber et al. 2011]). All these applications are based on a central geographic profiling principle that where incidents happen (that, in some way, are connected or linked) provides clues that can assist investigation. Geographic profiling also can support prevention, by preventing the source (e.g., criminals) from creating events (e.g., crimes) by limiting or controlling the source's activities in some way.

Geographic profiling also has been used to reveal the identity of the graffiti artist Banksy. I will not give it away here, but all I will say is that he likely lives in central east London, and has another geographic anchor in Bristol, England. You can find out who Banksy (most likely) is from the full article by Hauge et al. (2016) or the numerous online news stories (and the saga of Banksy's lawyers trying to suppress his identify) when the original research article was published.

Spatial analysis techniques for geographic profiling

The application of geographic profiling is associated with numerous spatial analysis techniques. Although specialist software for geographic profiling is what tends to be used by those certified to conduct geographic profiling, other techniques and tools are available. In this section, we discuss the main techniques and tools for examining the spatial distribution of offenses committed as part of a series. We begin by considering the use of centrographic statistics (building on the description of these techniques from chapter 2) and discuss the application of circle theory, set theory, and journey to crime circles. We then describe two specialist software applications for geographic profiling—Dragnet and Rigel—and then refer to packages available in R for supporting geographic profiling.

Centrographic techniques

As described in chapter 2, the spatial mean is a measure of central tendency in the spatial distribution of data and was used in some of the earliest attempts at examining the spatial behavior of offenders who had committed a series of crimes, such as in the case of a serial rapist in San Diego, California, in the 1980s (LeBeau 1987). Figure 7.3a shows a more recent example of the spatial mean technique applied to a series of eight sexual assaults in Bath, England (also previously illustrated in chapter 2). This example from Bath will be used in this chapter to help illustrate the application of several spatial analysis techniques for supporting geographic profiling. Figure 7.3a also shows the location of the weighted spatial mean for the eight offenses. Both the spatial mean and weighted spatial mean draw attention to the city center of Bath, with each point positioned on either side of George Street. Geographic profiling is not an "x marks the spot" routine, although the result from these two analysis techniques may give that impression.

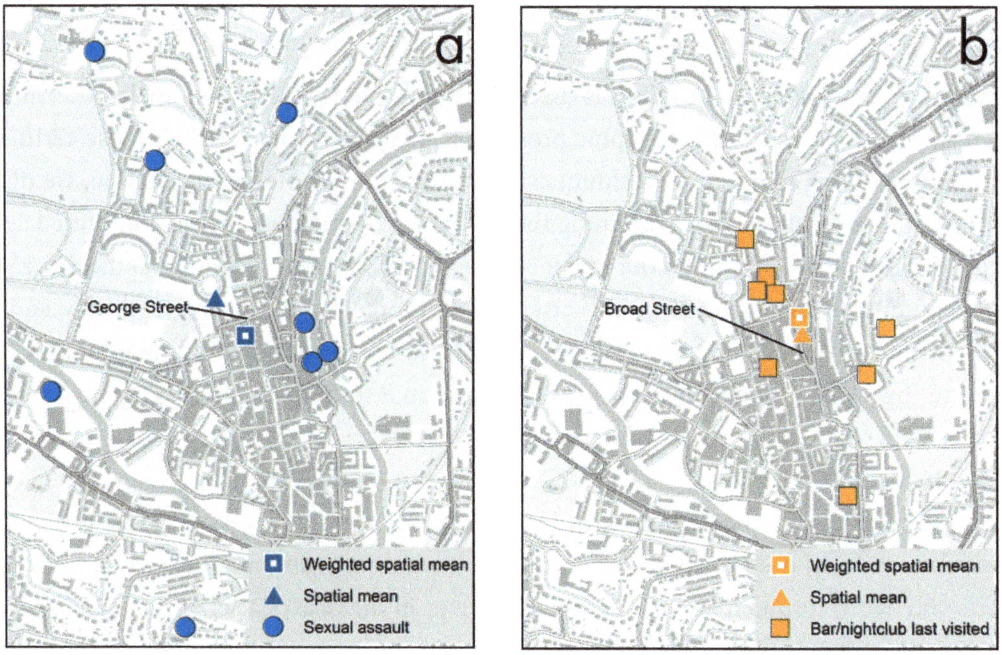

Figure 7.3. The spatial mean and weighted spatial mean for a series of sexual assaults in Bath, England: (a) based on the location of the assaults, and (b) based on the last bar or nightclub the victim visited on the night of their assault.

"Geographic profiling simply involves inputting data into a spatial analysis tool"

By this stage, and assuming you have read the previous sections of this chapter, I hope you will appreciate that geographic profiling involves more than just the analysis of serial crime points on a map. It requires an understanding of the theory that underpins the method, a careful consideration of the data used as inputs to the geographic profile, knowledge drawn from cases that have applied geographic profiling, and experience in criminal investigation. Mapping the crime points from a crime series in a GIS, and applying a spatial analysis technique to these points, helps in the examination and interpretation of the series but on its own does not make a geographic profile.

The spatial mean and weighted spatial mean results shown in figure 7.3a were based on the locations in which the assaults took place. Yet another set of locations were recorded that related to these offenses—the locations of the bar or nightclub in which each woman who was assaulted last visited with friends on the night of their assault. Figure 7.3b shows the spatial mean and weighted spatial mean for the bars and nightclubs in which each victim last visited. The locations of the spatial mean and weighted spatial mean are different, this time centered on Broad Street—albeit still identifying a similar area in the city center to the spatial mean and weighted spatial mean for the assault locations.

Which locations then should we use to inform the geographic profile? The locations of the assaults, the locations of the bars and nightclubs in which each person visited, or all these locations? The answer is all of them, for the purposes of an accurate assessment and interpretation of the spatial behavior of the offender, the opportunity space, the likely hunting method, and the target backdrop (i.e., the temporal and geographic distribution of suitable crime targets) for determining an appropriate investigative strategy. However, for the purposes of using a spatial analysis technique to help determine the likely location of the offender's anchor point, the answer is, perhaps, not all of them. In this example from Bath, only the assault locations should be used because these locations were *chosen* by the offender, meaning that the location of each assault is more likely to exhibit something useful about the offender's spatial decision-making behavior, rather than the location associated with the choice of the bar or nightclub that each woman decided to go to with her friends that night. If all the women, or several of them, had visited the same bar or nightclub on the night they were assaulted, that would be different because this common location could suggest other aspects of the offender's spatial decision-making behavior. For example, the offender visited this venue purposely to seek potential victims.

This example illustrates an important component of geographic profiling—consideration of the spatial data to include in the geographic profile, and whether to include or exclude this data in a spatial analysis technique that assists the geographic profile. Geographic profiling involves more than just producing a map based on the location of crime sites from a crime series. The spatial analysis of location data associated with a crime series can provide an important contribution to a geographic profile, but care must be taken in the selection of location data to include, and in what the results mean. The results from the spatial mean and weighted spatial mean examples could be useful, but it is important to recognize that both these measures are only synthetic points. As Jim LeBeau (1987) stated, the spatial mean (and weighted spatial mean) represent "the average location of a phenomenon, and not the average of the characteristics of the phenomenon at that location."

Circle theory

Circle theory, and its development into a technique for examining the spatial behavior of serial offenders, emerged in the 1990s from research conducted by David Canter and colleagues at the University of Liverpool, England (Canter and Larkin 1993). The approach involves drawing a circle and using the two most distant crime events in an offender's crime series as the limits of the circle's diameter. This technique is illustrated in figure 7.4 for the series of sexual assaults in Bath, England. The original research, based on solved rape series, found that 87 percent of the offenders had a residence within the circle (Canter and Larkin 1993). Although interesting, the technique was of limited practical application because of the large sizes of some circles covering geographic areas consisting of whole cities or large areas between several cities. Additionally, the research drew attention only to the offender's residence, when in practical terms, geographic profiling recognizes that the area of interest for targeting an investigative strategy could be many other locations (e.g., the anchor point relating to where an offender bases a search for victims).

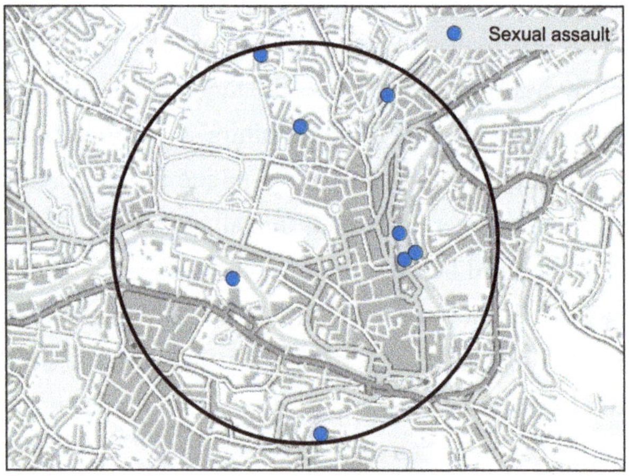

Figure 7.4. An application of circle theory to a series of sexual assaults in Bath, England.

Journey to crime circles

Journey to crime circles are a geographic profiling technique that uses the likely distance an offender travels to commit a crime to draw circles around each offense and examine where these circles overlap to help determine the likely area in which the offender lives (or has some other geographic anchor). For example, figure 7.5 shows three circles drawn around three points. Each point represents a residential burglary. Research findings on journey to crime indicate that, on average, offenders who committed residential burglaries traveled 1.88 miles from their home to their crimes (Wiles and Costello 2000). Circles with a 1.88-mile radius are then drawn around each crime location. The lune between all the circles for each crime considered to make up the crime series is the area with the highest probability of being where the offender lives (or is the location of some other spatial anchor of the offender).

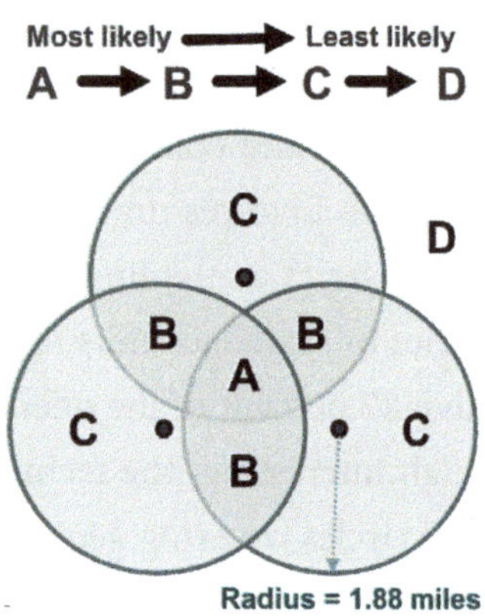

Figure 7.5. Journey to crime circles, with the lune between all three circles indicating the most likely offender anchor point.

Very few examples of the use of journey to crime circles have been published. Figure 7.6 shows the application of this technique to the eight sexual assaults in Bath, using a radius of 0.5 miles for each circle. The radius was determined from previous research on the journey to crime of offenders committing sexual assaults (Rossmo 2000). Figure 7.6a shows circles of 0.5 miles around each assault. Figure 7.6b shows the areas in which seven, six, or five circles overlapped. There was no area in which all eight circles overlapped. Based on the journey to crime circle's principles, the offender most likely lives (or has an anchor) in the area in which most of the circles overlap. The *most likely* area determined using journey to crime circles was slightly farther north than the areas shown by the spatial mean and weighted spatial mean.

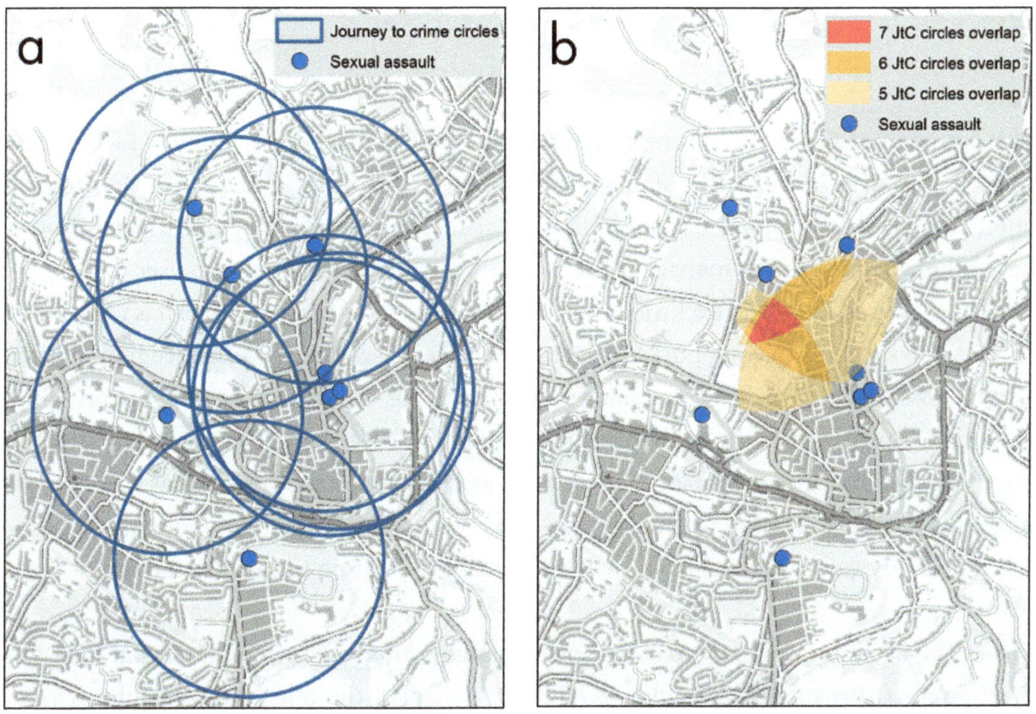

Figure 7.6. (a) Journey to crime circles, and (b) the area in which most circles overlap for a series of sexual assaults in Bath, England.

Dragnet

Dragnet is a spatial analysis technique developed by David Canter and colleagues, following their extensive research on the analysis of serial murderers and serial rapists in the United States (Canter and Larkin 1993; Canter and Gregory 1994; Canter et al. 2000; Canter 2003). The technique in many ways builds on journey to crime circles by applying a distance decay function from the center of each circle (and the location of the crime) to the edge of the circle. The function assigns higher values to the center of the circle (one) and lower values (approaching zero) at the edge of the circle. The circle is based on the likely journey to crime distance that offenders take to commit their crimes. Once multiple circles are drawn, and the values in each circle have been calculated, the value at each location is summed and converted into a probability estimate relating to where the offender lives or has some other anchor. For example, for a location covered by three circles, and where the values from each circle for this specific location are: Circle 1 = 0.80, Circle 2 = 0.15, and Circle 3 = 0.25, the sum at this location would be 1.2. Once values for all locations are calculated, these numbers can be converted to index values, ranging between zero and one in relation to all values, and used as a probability estimate. It was not possible to illustrate the application of Dragnet for the Bath series of sexual assaults. An example that illustrates the application of Dragnet can be found in Canter et al. 2000.

Rigel

Rigel is a professional geographic profiling tool developed by Canadian company ECRI. Rigel supports the inputting and geocoding of data used for a geographic profile (i.e., crime locations, other locations of interest that are associated with each crime, and the locations in which suspects live or have other anchors), helps manage this data into case files, and allows the user to test different scenarios for the inclusion and exclusion of location data in their analysis. Rigel has tools for simple temporal analysis, tools that support linkage analysis, and uses the criminal geographic targeting equation (Rossmo 2000) as its spatial analysis technique for supporting geographic profiling. The CGT equation incorporates the principles of crime pattern theory, hunting methods, and offender journey to crime to calculate probability scores on where the offender lives or has a key geographic anchor. This process works by creating a grid matrix of the study area and populating each grid cell with a probability value (derived from the CGT equation) relating to the distribution of the crime series.

The CGT equation works by making several assumptions: the series contains at least five geographic points (to ensure the results are valid and reliable); the offender is operating as a local hunter; if the series involves more than one offender, the offenders reside together; and the offender's search base has not changed over the duration of time the series has run (e.g., the offender has not moved) (Rossmo 2000). Although there is some flexibility in the consideration of these assumptions, it requires good knowledge of geographic profiling to determine how the results are affected by relaxing some of these assumptions. For example, although there is an assumption that the offender is a local hunter, an offender who travels to an area and commits crime may have an anchor that relates to where they begin their search for crime commission in the place they travel to. The Irvine Chair Burglar provides an example of an offender who commuted to a location, and then operated from an anchor point in that location in his commission of a residential burglary series (reported in Rossmo and Velarde 2008).

A police investigation for a serial sex offender, Operation Lynx, was the first application of geographic profiling in the UK and used one of the early versions of Rigel to assist detectives in the investigation. Between 1982 and 1995, three rapes, one attempted murder, four kidnappings, one serious sexual assault, one indecent assault, and one other assault, all against women, were committed in north England. The crimes were considered linked, even though they were committed over a large geographic area that covered the jurisdiction of three separate UK police forces. The police investigation had collected DNA from one of the incidents and a partial fingerprint. Analysis of the DNA did not result in a person being identified on the UK's national database, and the print fragment was too small to search using the UK's Automated Fingerprint Identification System. In 1995, 13 years after the first offense, the case remained unsolved.

In 1995, an analysis of the offenses suggested that the offender was likely to have a prior offending history of robbery, and fraud, because of the violent way he initially behaved against his victims, and that he often would steal credit and bank cards from his victims, using the cards to extract money from their accounts. The offender, therefore, was likely to have had some previous contact with the police, but on its own, this information was not that useful to the detectives working on the case. A sketch of the offender also had been drawn, based on information provided by one of the victims. Police investigation efforts were focused on suspects in the metropolitan area of the city of Leeds and its surrounding areas because it is where many of the offenses took place. However, it was a large area, in which the population was several million.

To help prioritize the suspect search strategy, the police investigating the case decided to use geographic profiling, and to do so called on assistance from the UK's National Crime Agency and Kim Rossmo, who was a detective at that time in Vancouver, Canada. Prior to 1995, the police in the UK never had used geographic profiling, had begun to consider its use, and decided that Operation Lynx provided an ideal opportunity to evaluate its use for assisting in serious crime investigations in the UK. Figure 7.7 shows the geographic profile map produced from the Rigel software system after an analysis of the crime series. In this example, the geographic spread of where the offenses were committed meant that specific streets could not be identified. Instead, geographic neighborhoods were identified. The dark-orange and red areas in the map indicate where the offender was most likely to reside, or anchor their activities, based on the probability values calculated using the CGT equation.

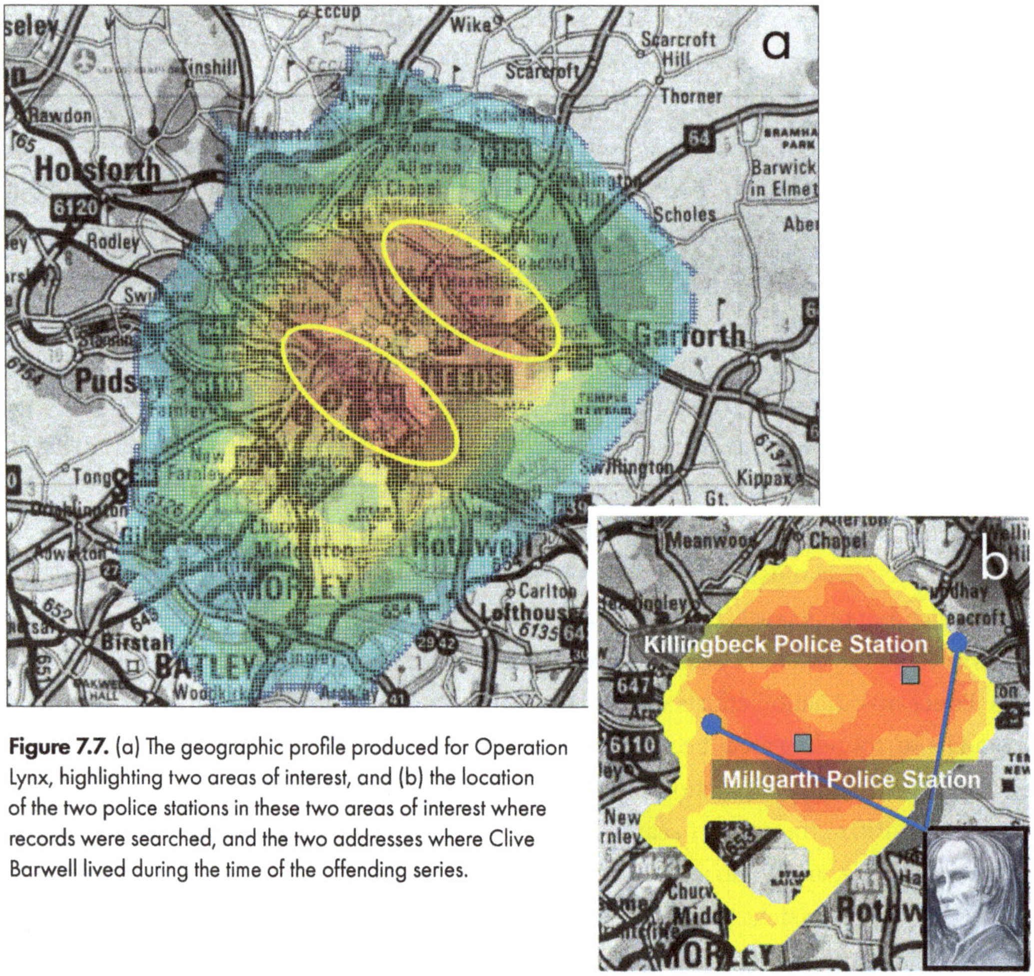

Figure 7.7. (a) The geographic profile produced for Operation Lynx, highlighting two areas of interest, and (b) the location of the two police stations in these two areas of interest where records were searched, and the two addresses where Clive Barwell lived during the time of the offending series.

This example is an illustration that the precision of the geographic profile is closely related to the geographic distribution of the offense sites. If the offense sites are close together, individual streets or a collection of streets may be identified as the area of focus in a geographic profile map, but when the distribution of the offenses covers a much wider area, the geographic profile is more likely to identify neighborhoods, rather than streets, to target. In figure 7.7a, two specific areas of Leeds were highlighted for attention. These *peak profile* areas still covered reasonably large areas in which many people lived. However, based on the hypothesis that the offender probably was known to the police because of previous criminal activity, the investigative strategy the police employed was to focus on examining offender records that were stored in the two police stations located in the peak profile areas—Millgarth and Killingbeck police stations (figure 7.7b). This strategy was on the basis that records for offenses of fraud or robbery committed by the offender and fingerprint records of the offender were likely to be stored in these two police stations, more so than in any other police station in Leeds. The search began, examining records of fingerprints from all robbery and fraud offenders against the single partial fingerprint that had been obtained from one of the crime scenes.

On the second day of the search (recall, the offenses had remained unsolved for 13 years), a match was made between the partial fingerprint and a crime record at Millgarth police station, identifying Clive Barwell as the suspect. Barwell was arrested. His DNA was taken and matched to the DNA of the offender from one of the rape offenses. Barwell was subsequently charged and convicted for the offenses. Barwell had lived at two addresses over the duration of the crime series, with each address being close to the peak profile areas generated from Rigel and shown in figure 7.7b.

This test case in the use of geographic profiling proved to the UK police authorities that it was a method worth investing in, which resulted in the UK National Crime Agency establishing a specialist geographic profiling unit consisting of four professionally trained geographic profilers to work on other cases of serial crime investigation. Since Operation Lynx, geographic profiling has been used in the UK to support investigations associated with murder, other cases of rape, kidnapping, child abduction, sexual assaults, and bank robbery. Expanding the capacity for geographic profiling in the UK (and internationally) is supported by the Geographic Profiling Analysis training program. This training program has resulted in many other applications in the use of geographic profiling in the UK and Europe for residential burglary, burglary of commercial premises, street robbery, criminal damage, arson, metal theft (chiefly copper), fuel theft, vehicle theft, and assaults.

Figure 7.8 shows the results from Rigel for the series of sexual assaults in Bath, England—the results from the CGT equation for the whole study area, and the peak profile area. The peak profile output identified the area of highest probability as being specifically around Milsom Street, George Street, and Broad Street—this area being the likely geographic anchor point for the offender. The interpretation from the times of day and the days of the week of the offenses (most took place on a Friday or Saturday night, during the late evening hours, or the early hours of the next day) and the modus operandi of the offender suggested that the peak profile area most likely identified where the offender was searching for victims rather than this location being the offender's place of residence. It was suspected that the offender was following his victims for some distance before assaulting them: identifying them in the city center, following them as they walked from the city center, and attacking them in quieter areas in which there were no people on the street, and in which there was little formal surveillance (e.g., CCTV cameras). The investigative strategy the police decided to implement involved setting up a decoy operation on each of the next weekend evenings and using a young female police officer (dressed in plain clothes) as the decoy, walking along George Street, Milsom Street, and Broad Street, supported by a team of police colleagues providing surveillance to ensure she would be safe. Her police colleagues also would be on the lookout for anyone behaving suspiciously in the area the CGT spatial analysis results identified as the most likely geographic anchor point of the offender. The sexual assault series had begun several months before, and the police had no leads, so they hoped this operation would assist them in some way.

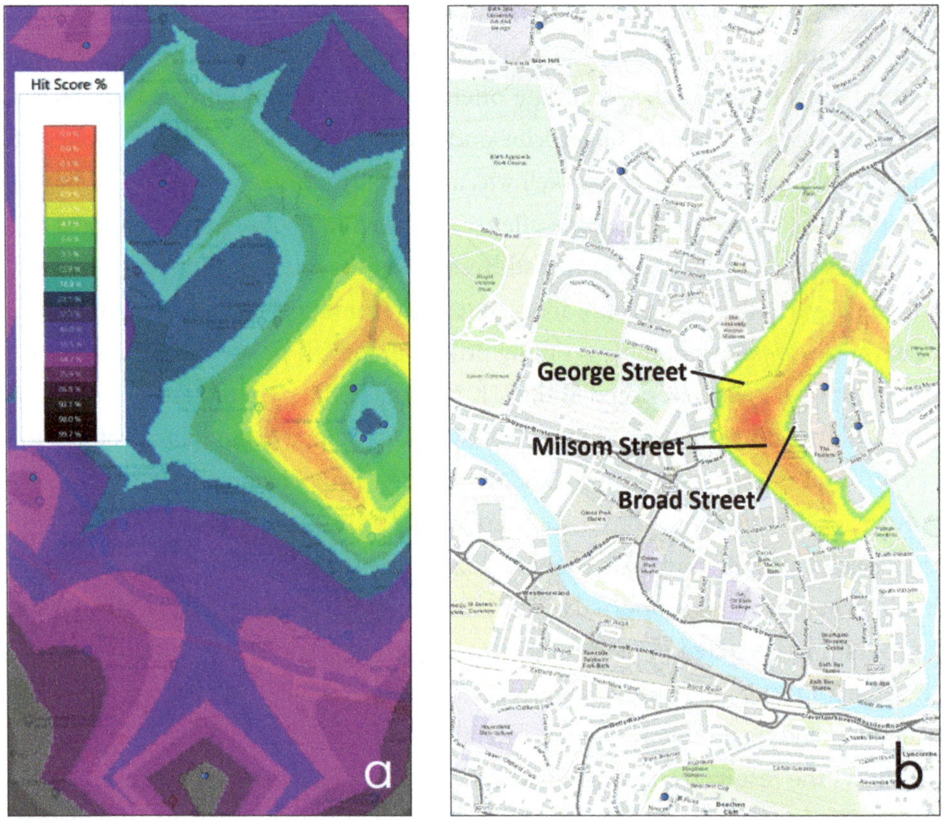

Figure 7.8. Criminal Geographic Targeting equation output from Rigel for a series of sexual assaults in Bath, England: (a) showing the output for the full geographic coverage of the series and (b) the peak profile area.

On the first night of the police operation, at 3 a.m., a man was seen driving slowly around the area. The undercover female police officer was informed and started to walk north from the city center. The man under police surveillance watched her, and then followed her. He watched her enter a house nearby, and then tried to gain entry, unaware this was the police safe house, with several police officers waiting inside. He was challenged by police officers, arrested, and made a full admission to the sexual assaults. He subsequently was sentenced to nine years' imprisonment and is now registered indefinitely on the UK's sex offender register (Daniell 2008).

From a spatial data point of view, the Bath sexual assault series was reasonably straightforward—there were few offenses, all offenses displayed similar features that resulted in a high degree of confidence that they were linked, and the geographic coverage of the series was contained in a relatively small area. As particular streets in the city center consistently were identified as being the likely area in which the offender anchored their activities, and this was predominantly a commercial area, it was felt that the area that comprised George Street, Milsom Street, and Broad Street was the offender's hunting ground—the location in which he most likely would search for suitable targets.

It was likely that on the evenings he was motivated to commit an assault, he would have identified a target and followed them, but the woman then went into a bar, got into a taxi, or was protected by the presence of other people on the street at the moment when the offender was contemplating an attack. It can take an offender time to search for and find a suitable target, which the police investigators in this case realized, took advantage of, and then caught him in the process.

Geographic profiling applied to co-offending and criminal groups

Geographic profiling spatial analysis techniques assume the crimes are committed by one person, or if committed by a group, the group lives together. Many crimes involve more than one offender (e.g., co-offending in a robbery), and it's unlikely that all the co-offenders live together. This situation imposes a restriction on using geographic profiling spatial analysis techniques on a lot of crime that is committed because the results it may generate when more than one offender is involved are likely to show a meaningless anchor point resulting from a number of people's spatial decision-making behavior. For example, this profiling would include a mixture of individuals' awareness spaces and the group's collective assessment of risk and effort rather than those of a single individual.

Geographic profiling *principles*, however, can be applied to groups, but to do so requires care in extracting the activities of individuals from these groups. For example, a group involved in committing a bank robbery is likely to involve each offender playing a different role (e.g., the lead offender who plans the robbery, the driver, and supporting roles played by accomplices). Once the activities of individuals in a group are extracted, a spatial analysis of each person, using geographic profiling principles, may help identify interesting spatial anchors. Research on the application of geographic profiling applied to groups (including organized-crime groups) is ongoing but shows promise, and I anticipate will be published soon after the publication of this book.

More than just the spatial center of the crime series

The spatial analysis techniques applied in this chapter to the Bath sexual assaults series all indicated an anchor point on or close to George Street. Most spatial analysis techniques used for geographic profiling tend to identify the areas that represent the *center of gravity* of the crime series. It is important to recognize that simply mapping the locations associated with a crime series and determining some central area as the area of investigative focus is insufficient for geographic profiling and can produce misleading results. In turn, this can have serious consequences on an investigation, and on people's lives (leading to further victimization and targeting the wrong suspects). Geographic profiling involves significantly more than the use of spatial analysis techniques for identifying possible areas in which police investigative efforts should be targeted.

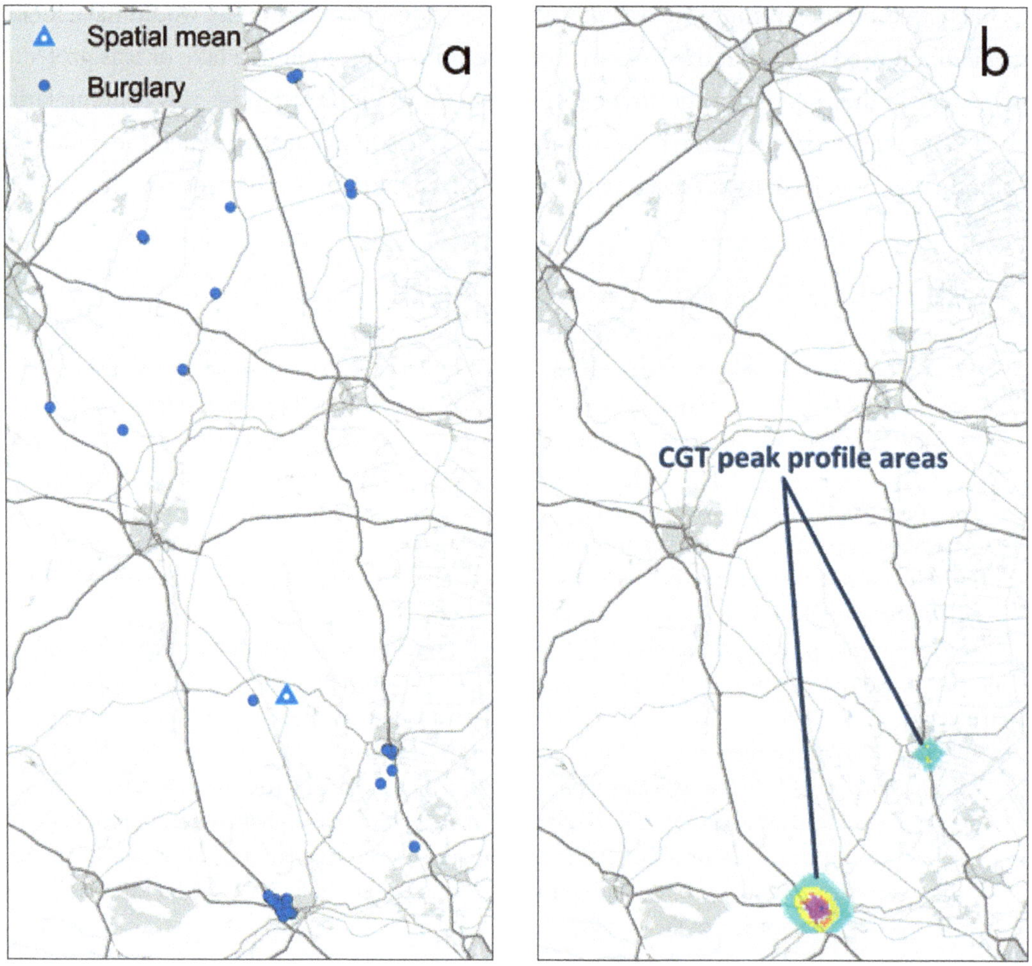

Figure 7.9. (a) The spatial mean (and crime locations), and (b) the CGT equation peak profile areas for a series of burglaries.

Figure 7.9 shows an example of a residential burglary series, linked as being committed by the same offender. Detailed street mapping data and labels for the cities are not included because of the sensitivity of the crime data. The series involved 46 offenses, committed over a two-year period, covering an area of 65 kilometers from north to south. Figure 7.9a shows the distribution of the series and the spatial mean. Figure 7.9b shows the CGT peak profile output (based on all 46 crime locations), highlighting two areas different from the spatial mean—a small peak profile area in the southeast of the study area and a larger peak profile area in the south. Both areas are located over two small towns. In this example, use of the spatial mean would lead to a misleading result. The spatial mean highlights a field, nine kilometers from the nearest settlement. The only building nearby is a farmhouse that has been burgled by the offender. So, is the CGT result correct? Not necessarily.

When conducting a spatial analysis of a crime series for geographic profiling, it is essential to consider certain aspects of the series that determine whether a crime location should be included in the spatial analysis. Thinking that the application of a spatial analysis technique to all the crime locations in a series is the only thing required can have potentially damaging effects on the effectiveness of the criminal investigation.

Excluding and including locations in the spatial analysis of a crime series

Often, the spatial analysis of the crime locations in a geographic profile requires certain crime locations to be excluded. Excluded offenses for a spatial analysis of the series would remove the following:

- Repeat locations and other crimes in proximity to a previous crime in the series. We already know the offender has been to this location, so we already recognize it as a location that features in their awareness space. An analogy I like to give to analysts and investigators I train in geographic profiling is to think of this information as though the offender has placed a flag in the ground of where he has been (i.e., where he has committed crime). By doing so, he has signaled to us, "I know this place." Having two flags in the ground at the same location tells us nothing more about the places he knows. In technical terms, repeat offenses are not independent of previous offenses, and this is a consideration that results in their exclusion when data is used for the spatial analysis of an offender's spatial decision-making behavior.
- A crime that took place soon after a previous failed attempt to commit a crime. This also often is excluded from the spatial analysis of an offender's spatial decision-making behavior. Again, it is because of the second incident not being independent of the first. That is, the second offense may not have happened if the first incident had been successful.
- Spatial outliers, as these offenses differ significantly from the other offenses in the series.
- Opportunistic offenses. In the burglary series illustrated in figure 7.9, it was suspected that some of the burglaries were committed only because the offender came across an opportunity that he decided to take advantage of while going about his other noncriminal activities (i.e., offenses that fit the troller search method). The troller search behavior was unlikely to be the main search method the offender used, so on occasions, these types of offenses are not included as inputs to the spatial analysis technique being used.

Although spatial data on the locations of some crimes often is excluded from the technical spatial analysis of a crime series, it is *not* excluded from the geographic interpretation of the crime series. For example, the fact that an offender returned to a property, or to a property close to where they previously committed a crime, can be a useful indicator of the preferences in their spatial decision-making. Similarly, crimes that appear to be opportunistic or spontaneous, in that the offender came across an opportunity while doing something else, can tell us useful things about their other likely routine activities. For example, the routes they regularly travel for work commuting purposes. In practice, determining locations to include and exclude in the spatial analysis of a crime series requires a careful examination of the locations associated with the crime series, and experimentation with several scenarios that include and exclude data. Using different scenarios helps analysts, researchers, and investigators test hypotheses about the offender's spatial behavior, and is an important component in effective geographic profiling.

A detailed examination of the data on each of the 46 crimes linked in the previous burglary series example resulted in using only 12 offenses in the south area of the series as inputs to the geographic profiling spatial analysis of the series. In addition to the offense locations, the location of where a car was recovered that had been stolen from a burglary victim's house the previous day was added to the spatial analysis. On this occasion, the offender stole a flat-screen TV, an item he had not stolen in any of the other 45 offenses. It was suspected he walked to most of the houses he burgled in the southern area of the crime series. To take the TV, he took the victim's car, using the car keys that were on a small table in the hallway. The next morning, the car was discovered seven blocks from the burgled property. The location of the deposition of stolen property, which is then later recovered, often can be a useful spatial indicator that relates to an offender's spatial decision-making, where an offender lives, or where the offender has another anchor. The inclusion of only 12 burglary locations and the inclusion of the location in which the stolen car was recovered produced the peak profile map shown in figure 7.10, highlighting an area covering only a few streets in the southern part of the study area, rather than a field in the middle of nowhere (figure 7.9a). The result from the CGT spatial analysis then was used in an investigative strategy targeted to this area, which resulted in the arrest and prosecution of an offender who had been prolific in committing burglaries over the previous three years. The offender had a full-time job that required him to travel throughout the local region. He committed most of his burglaries close to home but also would commit others around his work activities.

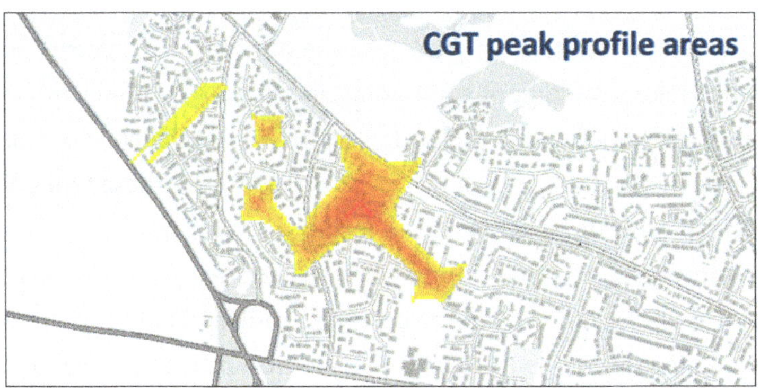

Figure 7.10. CGT peak profile areas of a series of burglaries.

In the Bath sexual assault series example, one valuable piece of information that contributed to the geographic profile was a map of the routes that each victim walked on the night they were assaulted—from the bar or nightclub they last visited to the location of the assault. This information (figure 7.11) showed that five of the eight victims had walked down George Street on their way home from the city center on the night of their assault, adding even more to the likelihood from the technical spatial analysis results that George Street, and its immediate surrounding streets, represented an area of significance to the offender's spatial criminal behavior. In addition, the victims were asked by investigators the direction in which the offender approached them, and the direction he then fled after the assault. On each occasion, the offender approached the victim from behind, and then fled in the direction the victim had walked from the city center. This was a possible indication of the offender following his tracks, retreating to his area of comfort and familiarity with the city center.

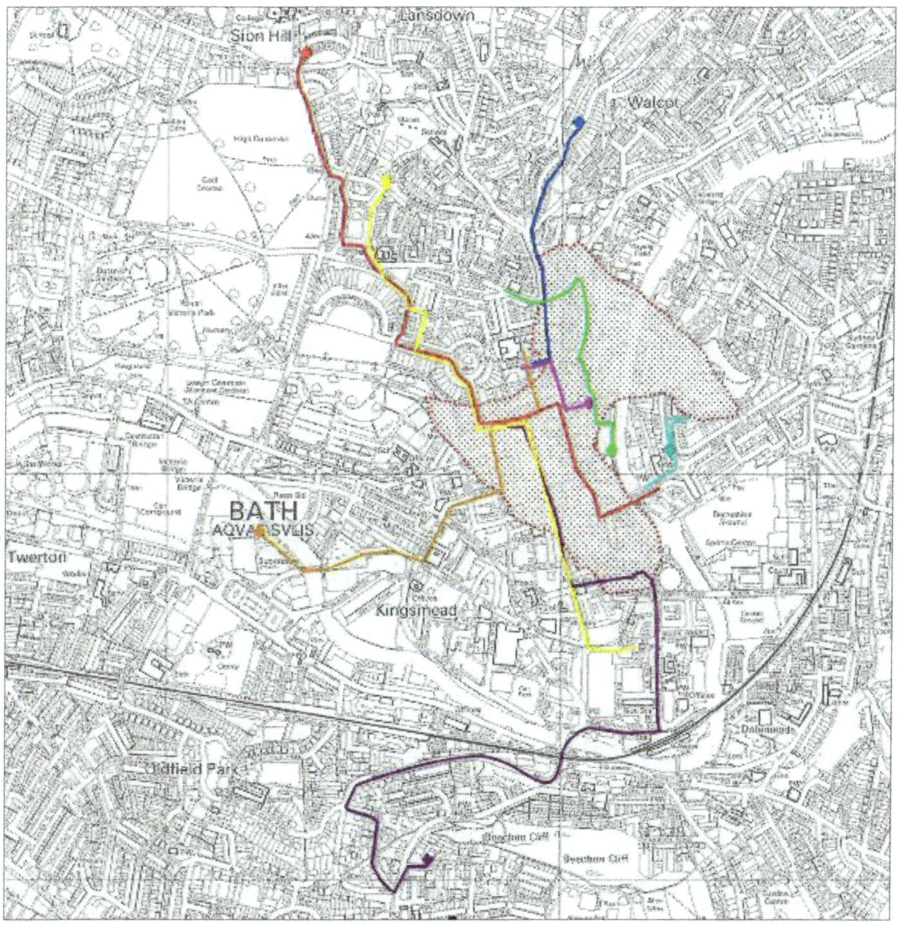

Figure 7.11. The routes that each victim walked on the night they were assaulted—from the bar or nightclub they last visited to the location of the assault. Source: Chainey and Tompson 2008.

Spatial analysis techniques for geographic profiling using R

Several packages are available in R that support spatial analysis techniques used for geographic profiling. One of these is the R package developed by a team of researchers from Queen Mary University, London (QMUL) (https://evolve.sbcs.qmul.ac.uk/lecomber/sample-page/geographic-profiling/geographic-profiling-in-r/). The team at QMUL has used geographic profiling for a range of epidemiology and ecology purposes. The R geographic profiling package produced by the QMUL team uses a Dirichlet process mixture (DPM) model to compute and visualize the source location from a series of connected input locations. If applied to a crime series, the source locations would be the locations of suspects, and the input locations would be the crime locations of the series. The DPM model is more suited for identifying the source when more than one source location exists. In an epidemiological sense, this relates to situations in which more than one source location of a disease exists. In a criminological sense, it relates to situations in which a suspect may have more than one location from which they commit their crimes (e.g., from home and from work or school). For an example of the use of the DPM model in epidemiology, see Verity et al. 2014. Another R package is geoprofiling_w_map.R, available on GitHub (https://gist.github.com/colemanja91/4184683), but this package does not come with any documentation support, and no peer-reviewed examples of its use have been published.

Those unfamiliar with R (and even some of those familiar with R) will find the R packages (Rgeoprofile 1.2; geoprofiling_w_map.R) a little cumbersome. I have tested the Rgeoprofile package on several occasions. It involves working through the multistep process the QMUL team describes in its user guide and requires the use of supporting R packages. It also requires diligence in working out the errors in the syntax for running functions and figuring out how the parameters in functions influence the results for the data being analyzed. The time it takes to work through the series of steps is short (i.e., a couple of minutes, including processing time, on a 10-point dataset) once you have figured out how to use it. Unless you are familiar with R, I would suggest you opt for other, easier-to-apply spatial analysis techniques for geographic profiling.

Chapter summary

Geographic profiling is a proven technique for supporting criminal investigation. Geographic profiling involves more than the use of a spatial analysis technique for examining the patterns of crimes connected as part of a series. Geographic profiling requires a careful examination of the data, care in the choice of locations to include in the spatial analysis, and care in the analytical interpretation of the crime series. Several spatial analysis techniques can inform a geographic profile, but again, care is needed in how they are used, to ensure the results avoid generating misleading interpretations. The objective of this chapter is not to leave you thinking you now have the expertise to do geographic profiling to support a serious crime investigation, but instead aims to provide several insights that will help you apply geographic profiling principles in your analytical work. A key aim is to demonstrate that to do geographic profiling effectively involves more than just inputting crime locations into some software and generating results that will lead you directly to the offender. An appreciation of what effective geographic profiling involves also should help you consider ways to improve your skills in this useful investigative method.

Further resources

Wired Magazine article: "How Geographic Profiling Helps Find Serial Criminals," https://www.wired.co.uk/article/mapping-murder.

Police Foundation article: "The Use of Geographic Profiling in Crime Analysis," https://crimemapping.info/article/use-geographic-profiling-crime-analysis/.

Open-access article: "Recent Developments in Geographic Profiling" (Rossmo 2012), https://academic.oup.com/policing/article/6/2/144/1521464.

Geographic Profiling Analysis training, https://www.ucl.ac.uk/short-courses/search-courses/geographic-profiling-analysis-crime-and-security.

References

Canter, D. V. 2003. "Offender Profiling and Investigative Psychology." *Journal of Investigative Psychology and Offender Profiling*, 1 (1): 1-15.

Canter, D. V., and A. Gregory. 1994. "Identifying the Residential Location of Rapists." *Journal of the Forensic Science Society* 34:169–75.

Canter, D. V., and P. Larkin. 1993. "The Environmental Range of Serial Rapists." *Journal of Environmental Psychology* 13:63–69.

Canter, D. V., T. Coffey, M. Huntley, and C. Missen. 2000. "Predicting Serial Killers' Home Base Using a Decision Support System." *Journal of Quantitative Criminology* 16:457–78.

Chainey, S. P., and L. Tompson, eds. 2008. *Crime Mapping Case Studies: Practice and Research*. Chichester: Wiley.

Daniell, C. 2008. "Geographic Profiling in an Operational Setting: The Challenges and Practical Considerations, with Reference to a Series of Sexual Assaults in Bath, England." In *Crime Mapping Case Studies: Practice and Research*, edited by S. P. Chainey and L. Tompson. Chichester: Wiley.

Hauge, M. V., M. D. Stevenson, D. K. Rossmo, and S. C. Le Comber. 2016. "Tagging Banksy: Using Geographic Profiling to Investigate a Modern Art Mystery." *Journal of Spatial Science* 61 (1): 185–90.

LeBeau, J. L. 1987. "The Methods and Measures of Centrography and the Spatial Dynamics of Rape." *Journal of Quantitative Criminology* 3:125–41.

Le Comber, S. C., D. K. Rossmo, A. N. Hassan, D. O. Fuller, and J. C. Beier. 2011. "Geographic Profiling as a Novel Spatial Tool for Targeting Infectious Disease Control." *International Journal of Health Geographics* 10:35–42.

Martin, R. A., D. K. Rossmo, and N. Hammerschlag. 2009. "Hunting Patterns and Geographic Profiling of White Shark Predation." *Journal of Zoology* 279:111–18.

Rossmo, D. K. 2000. *Geographic Profiling*. Boca Raton, FL: CRC Press.

Rossmo, D. K. 2012. "Recent Developments in Geographic Profiling." *Policing: A Journal of Policy and Practice* 6 (2): 144–50.

Rossmo, D. K., and K. D. Harries. 2011. "The Geospatial Structure of Terrorist Cells." *Justice Quarterly* 28:221–48.

Rossmo, D. K., and L. Velarde. 2008. "Geographic Profiling Analysis: Principles, Methods, and Applications." In *Crime Mapping Case Studies: Practice and Research*, edited by S. P. Chainey and L. Tompson. Chichester: Wiley.

Rossmo, D. K., and R. Walton. 2006. *Geographic Profiling in Cold Case Investigations, Cold Case Homicides: Practical Investigative Techniques*. Boca Raton, FL: CRC Press 537-560.

Rossmo, D. K., Q. C. Thurman, J. D. Jamieson, and K. Egan. 2008. "Geographic Patterns and Profiling of Illegal Crossings of the Southern U.S. Border." *Security Journal* 21:29–57.

Verity, R., M. D. Stevenson, D. K. Rossmo, R. A. Nichols, and S. C. Le Comber. 2014. "Spatial Targeting of Infectious Disease Control: Identifying Multiple, Unknown Sources." *Methods in Ecology and Evolution* 5:647–55.

Wiles, P., and A. Costello. 2000. *The "Road to Nowhere": The Evidence for Traveling Criminals*. London: Home Office.

CHAPTER 8

Applying spatial regression analysis to crime data

Objectives

Regression analysis involves testing the relationship between a subject of interest (i.e., crime) and the factors that cause it. Rather than thinking of a regression analysis technique as being a single process, generating a single result, regression analysis involves a set of statistical processes from which the results are used collectively to interpret the relationships being tested. Regression analysis, particularly of spatial data, can be daunting, and many texts on the subject can be difficult to make sense of for practitioners and those not blessed with advanced mathematical skills. Spatial regression is useful but requires thought and diligence. There are numerous spatial regression techniques, many of which are available in GIS and other spatial data analysis software, but often it is not clear which one to use, how to use it, and what all the results mean.

In this chapter, we introduce you to how to use spatial regression analysis for understanding crime. For those familiar with regression analysis, we hope the material we provide on spatial analysis techniques encourages greater use of these techniques to understand crime. Spatial regression analysis uses many technical terms and statistical tests. In this chapter, we attempt to demystify spatial regression and the terms and tests used, by putting them in plain English. The use of spatial regression techniques, things to look out for, and how to interpret the results are illustrated with several examples. We also describe how the results generated from spatial regression analysis can help inform the design of programs to reduce crime and improve public safety, how spatial regression helps us consider *what-if* scenarios, and how it also can be useful for assisting evaluation.

Although correlation does not equal causation, in this chapter we show how a careful recursive specification of a spatial regression model and the results it generates can provide unique insights that help us understand patterns of crime that we cannot extract from other spatial analysis techniques. In this chapter, we focus on three spatial regression techniques that are of most practical use for crime analysis. We begin by describing the use of ordinary least squares regression, before focusing on spatial lag regression, spatial error regression, and geographically weighted regression.

Key learning points

- Understand the merits of ordinary least squares (OLS) regression for informing spatial regression analysis but also recognize the limitations of OLS and that it comes with several assumptions that are usually violated when examining spatial relationships.
- Become familiar with the many statistical terms and statistical tests used in spatial regression analysis and understand what these mean and what you need to look out for. This includes understanding terms such as "spatial dependence" and "heteroscedasticity" and statistical tests that indicate whether the regression model performs well or has problems, such as statistical bias and unreliability.
- Understand the differences between spatial lag and spatial error models, when to use one over the other, and what they can help reveal.
- Understand how to use geographically weighted regression (GWR) and generate GWR mapping output that can help you better understand spatial patterns of crime and how the results can inform the design and evaluation of crime-reduction programs.
- Recognize that spatial regression analysis involves a recursive process of model specification, which requires care and diligence.

Introduction

So far, most of this book has been about identifying geographic and temporal patterns of crime and examining characteristics about these patterns that can help us understand crime. Although all these techniques are useful, in a statistical analytical sense, they do not necessarily help explain the external factors that may influence why these spatial patterns exist. An important part of crime analysis is to be diagnostic and attempt to reveal the underlying causes of crime. Understanding these causes requires knowledge of appropriate theory and an iterative interchange between analysis and theory to help make judgments on why the crime problem exists. We can do this partly by making inferences from the patterns we observe in crime data. For example, suppose robberies cluster near a school, on school days, very soon after the end of the school day, with most victims being schoolchildren—the inferences we can make from these patterns is that the problem likely is related to schoolchildren being highly vulnerable to these robberies and that offenders are exploiting the opportunities presented when and where favorable conditions to commit robbery exist.

Additionally, we can use regression analysis to make inferences about the crime patterns we observe and provide insights we cannot extract solely from the examination of crime data.

Regression analysis involves testing the relationship between one phenomenon (e.g., a crime problem) and factors that are considered to cause the phenomenon. The result of the regression analysis determines which factors can be used to explain the phenomenon and determines the strength in the relationship. The results also can identify whether other factors not included in the model should be identified and examined for their causal influence. Regression analysis models seek to examine the relationship between *variables*—a *dependent variable* and *explanatory variables*—to determine the presence, direction, and strength of any relationships.

The most basic regression model is ordinary least squares. OLS assumes that the relationships between a dependent variable and one or more explanatory variables are consistent geographically. That is, the reasons that explain the presence of a phenomenon in one location are the same as those in another location. For example, if two locations in a study area experience high levels of burglary, an OLS regression would assume that the reasons for these burglary hot spots are the same, influenced equally by the factors causing burglary. However, this similarity is unlikely to be the case. When using spatial data in OLS regression analysis, other issues associated with spatial autocorrelation (i.e., nearby observations being like each other) can generate misleading results. These issues are discussed further in the next section.

Analysis techniques for spatial regression examine the spatial relationship between variables. That is, spatial regression analysis is the process of identifying variables that explain the spatial variation in the phenomenon under study. Spatial regression techniques also seek to account for the spatial autocorrelation between dependent and explanatory variables to ensure the results are reliable. Outputs generated from a spatial regression analysis can tell us something about the variables and their *spatial effects* and produce results that can be mapped. These include the *parameter estimates* that describe the relationship between explanatory variables and the dependent variable for each observation on the map, the *statistical significance* of these parameter estimates, and the level of *error* (often referred to as *error terms*) or unexplained component of the relationship between the explanatory variable and the dependent variable in the model.

In this chapter, we already have introduced several terms, such as spatial effects, parameter estimates, the dependent variable, and error terms. More terms are to come. Often, people can be put off from doing spatial regression analysis because they are not clear about what these terms mean. At the end of this chapter, a reference box lists and describes all these terms in plain English and describes statistical tests that are important in spatial regression analysis. Any word, term, or statistical test that is written in *italics* the first time it appears in the chapter is included in this list. I purposely have included this list to help you become more confident in doing spatial regression analysis because, to date, it is underused in crime analysis.

Types of regression methods

In most cases, linear regression is the type of regression method used for crime data. The reason is because crime data, and the explanatory variables we want to examine in relation to crime, exists on a continuous scale. The data can be counts of events (e.g., the number of crimes, the number of people of a particular age, the number of liquor stores) or some other continuous value (e.g., a rate of crime, the density of crime, the percentage of the population that is unemployed). Poisson regression is often applied when using count data. We discuss Poisson regression later in this chapter to detail the complications in applying it to crime data. Other types of regression that are often used include the following:

- Logistic regression (also referred to as "binary regression"): when the dependent variable is binary (i.e., it has two possible outcomes, such as the presence of crime: yes or no). The explanatory variables can be continuous or binary.
- Ordinal logistic regression: when the dependent variable's values are represented across an ordinal scale of more than two categories (e.g., 1 = never been victimized, 2 = has been victimized fewer than 3 times, 3 = has been victimized more than 3 times).
- Multinomial logistic regression: when the dependent variable relates to categories, of which there are more than one (e.g., 1 = robbery, 2 = theft, 3 = assault, 4 = harassment).

Several approaches can be used in regression analysis for capturing the *spatial effects* that may be present in an examination of spatial data. They include incorporating spatial effects through an error term in the regression analysis (spatial error models), or the spatial effects are incorporated by including a spatially lagged variable as an additional predictor, alongside other explanatory variables (spatial lag models). Other spatial regression approaches that have been developed and tested to address issues in the use of OLS regression on spatial data include the spatial Durbin model, which incorporates both spatial lag and spatial error effects (Anselin 1980; Durbin 1960); the generalized linear model (Littell et al. 1996; Schabenberger and Pierce 2002) and its nonparametric extension, the generalized additive model (Guisan, Edwards, and Hastie 2002; Hastie and Tibshirani 1990); artificial neural networks (Fischer 1997; Shin and Goel 2000); and Bayesian spatial regression (LeSage 1997; Wheeler and Calder 2007). These types of models have been applied to a wide range of spatial applications, including house and land prices (Can 1990; Dubin 1992; Dubin 1998; Gao, Asami, and Chung 2006), forestry (Austin and Meyers 1996; Foody 2003; Haiganoush, Rappaport, and Wood 1997; Wang, Ni, and Tenhunen 2005; Zhang et al. 2004; Zhang and Shi 2004; Zhang, Ma, and Guo 2009), child poverty (Voss et al. 2006), epidemiology (Mohebbi et al. 2011), demography (Loftin and Ward 1983), and crime (Cahill and Mulligan 2007; Wheeler and Waller 2009; Waller et al. 2007). In this chapter, we focus on describing the more commonly used spatial error and spatial lag models, also because these two techniques provide a useful foundation for examining the use of other spatial regression techniques.

Geographically weighted regression is an alternative approach to addressing the limitations of OLS regression applied to spatial data. GWR is a local version of spatial regression that involves measuring the variation in relationships between variables by performing localized regression equations over the entire study area. The final section of this chapter describes and illustrates the application of GWR using two examples.

A starting point for explaining spatial regression is to review the use of OLS regression as a means for exploring possible causal factors. OLS is also a vital ingredient in the spatial regression analysis process, because typically, spatial regression begins with an OLS regression analysis to assist in the specification of the explanatory variables to include in a spatial regression analysis and to determine whether any spatial effects require further examination.

Examining spatially varying relationships: Hypothesis-driven approach and the exploratory approach

There are two approaches for examining spatially varying relationships—a hypothesis-driven approach and an exploratory approach.

The hypothesis-driven approach requires you to consider, on the basis of sound theoretical grounds, the reason why a phenomenon may be present. For example, violent-crime hot spots may cluster within a city center and around entertainment malls, with analysis showing that these problems tend to primarily exist on Friday and Saturday evenings. From these results, it would be plausible to suggest that the reason these violent-crime hot spots are present is because these are the areas in which bars, pubs, and nightclubs are located—the logic being that the high concentration of violent crime in these hot spots is related to the location of premises where people gather to consume alcohol. It also could be plausible to suggest that other violent-crime hot spots observed close to schools and transportation hubs on school days are related to altercations between students from rival schools, with these altercations occurring where children from the schools come together at the main transportation hubs on their way home from school. These two plausible hypotheses are based on reasoned logical and theoretical grounds. Explanatory variables that capture these factors then can be included in the regression model to explore, statistically, whether they are related to crime, and whether these relationships vary geographically. The hypothesis-driven approach, therefore, involves determining a plausible explanation at the outset, determining the data that can be used to test the hypothesis, applying a regression model, and then using the results to identify whether there is evidence that supports this hypothesized reasoning.

The exploratory approach involves choosing a range of variables that are then tested in the regression model, and where correlations are found, attempting to interpret the results on the basis of plausible logic or some theoretical basis—i.e., fitting a theory to the results to explain the relationship between the dependent and explanatory variables. The exploratory approach can be used when you are not initially confident on the theoretical grounds to determine plausible hypotheses, or you have limited access to data (e.g., data may not exist that is perfectly associated with a hypothesis. So instead, you rely on a proxy measure for exploring whether a relationship exists).

Ordinary least squares

OLS regression is a widely used technique for identifying relationships between a *dependent* variable (i.e., the phenomenon that is under study, such as crime) and one or more *explanatory* variables (often referred to as the "independent variables"). An OLS regression produces a global model, explaining the relationships (and their significance) between the dependent and explanatory variables. Figure 8.1 illustrates an example, where y is the dependent variable and x is an explanatory variable. The figure shows four points plotted on the basis of their x,y values. A straight line is calculated from the location of these points on the graph that minimizes the *variance of the error term* about this fitted line. The value at which x is zero and the starting position of the line on the y-axis is called the "*intercept.*" A value relating to the relationship between x and y is known as the *regression coefficient*. The regression coefficient represents the rate of change of the dependent variable (y) as a function of change in the explanatory variable (x). In other words, it is the slope of the regression line.

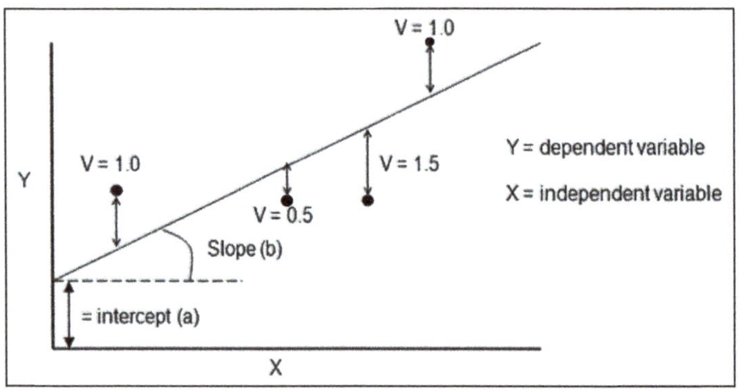

Figure 8.1. Characteristics of the OLS linear regression line.

A useful place to start with OLS is to produce a scatter plot between the phenomenon being examined and the variable being used to explain its presence. To illustrate the use of regression analysis techniques, we begin by using data that examines variables between countries. As we progress through this chapter, we focus mainly on using data for smaller geographic areas (such as census block areas). In the first example, homicide is the dependent variable, and unemployment and income inequality are two explanatory variables. Homicide rates (per 100,000 population) were obtained from the World Bank Open Data database for a sample of 38 countries for the year 2012. Unemployment data for 2012 was obtained from the International Labor Organization Department of Statistics' ILOSTAT database for the 38 countries. Data on income inequality for 2012 was obtained from the United Nations database. Definitions for each of the data used can be reviewed on the respective websites in which this data was obtained—web links to these websites are provided in the "Further Resources" section at the end of the chapter.

Figure 8.2 shows scatter plots for (a) homicides and unemployment and (b) homicides and income inequality. Software such as Excel can be used to create these scatter plots and, with the use of the Excel Analysis Toolpak, can be used for OLS regression analysis (ArcGIS and other spatial analysis software also can be used for OLS regression and is described in later sections). The figures are fitted with a goodness-of-fit line, derived from the OLS equation for these variables. Table 8.1 lists the coefficients, *standard error* values, *t-statistics*, and *p-values* for each relationship (remember, a description of *statistical measures, tests, and terms* is provided at the end of this chapter). The table includes the R^2 *statistic*. The results from an OLS regression analysis of homicides and unemployment suggest there was no statistically significant relationship between these variables (the p-value was greater than .05), and the strength of the relationship was weak (R^2 = .019). That is, there is no correlation between homicide and unemployment based on the data sample used. The results from the OLS regression analysis of homicides and income inequality suggest there was a statistically significant positive relationship between these variables ($t = 3.35$; $p < .01$). The R^2 of .238 indicates that about a quarter of the observed variation in homicides is explained by income inequality (i.e., in percentage terms, the R^2 of .2381 is equivalent to 24 percent). The coefficient value of 1.11 for income inequality suggests that for every one-unit increase in income inequality, homicide rates increase by 1.11 per 100,000 population (i.e., homicides per 100,000 population is the unit by which homicide rates were measured).

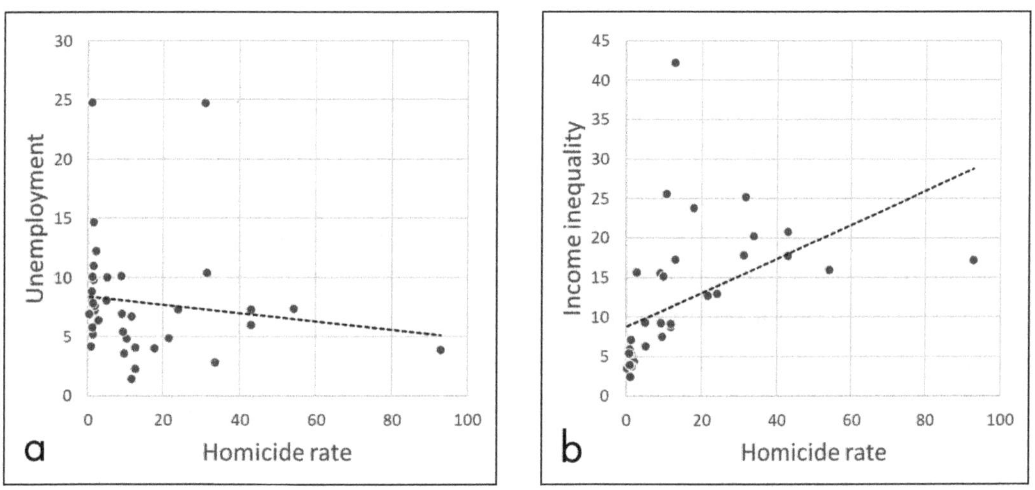

Figure 8.2. Scatter plots with fitted trend lines for (a) homicide and unemployment and (b) homicide and income inequality for a sample of 38 countries.

Table 8.1. OLS regression analysis results for homicide and unemployment, and homicides and income inequality for a sample of 38 countries

	Coefficient	Standard error	t-statistic	p-value	R^2
Unemployment	−0.537	0.647	−0.830	.412	.019
Income inequality	1.106	0.330	3.354	.002	.238

Although we should consider several other things to ensure our OLS results are reliable and applicable to all types of regression, we will consider these tests as we progress through this chapter. An assumption we make when performing an OLS regression is that the relationships we have identified are equal across space. For example, the OLS results for homicide and income inequality indicate there is a significant positive relationship between these variables, and the coefficient for this relationship is 1.11. Our general understanding of spatial patterns of crime, however, would suspect that this relationship is unlikely to be exact for all locations. In some countries, the coefficient may be greater or less than 1.11, and indeed in some cases, we may find the odd location in which the relationship is negative (i.e., opposite to what was found from the global OLS regression analysis). That is, although the overall relationship between homicides and income inequality is positive, in a small number of countries, the relationship could be negative—as income inequality increases, homicides decrease. The relationship between dependent and explanatory variables is likely to vary spatially. This variance, therefore, suggests that any analysis that aims to examine relationships between spatial data variables should be sensitive to exploring whether relationships between variables vary spatially.

Additionally, OLS regression assumes there to be zero or negligible errors in the explanatory variable, and that the *residuals* are independent of one another. When dealing with spatial data, the assumption that residuals are independent of each other is often violated because of spatial autocorrelations in data; that is, nearby observations are like each other. This can lead to a biased estimation of the standard errors of the model's variables, and consequently result in misleading significance results (Anselin and Griffith 1988; Fox, Ades, and Bi 2001). All this means is that when performing regression analysis on spatial data, we must consider giving this data special treatment. As I often have said to my students, spatial data is special, and much more interesting to work with than the nonspatial stuff.

Examining spatial relationships

In the next sections, we examine the use of three spatial regression techniques. Although many others exist, these three are the most used in practice in crime analysis, and also often outperform other regression techniques for their accuracy (Nakaya et al. 2005; Wang, Ni, and Tenhunen 2005; Zhang, Gove, and Heath 2005; Zhang, Ma, and Guo 2009): spatial lag models, spatial error models, and geographically weighted regression. They also provide an excellent foundation for you to consider the application of other spatial regression techniques. As we progress through the chapter, the examples we illustrate also help determine which technique to use when and why.

If we first assume that *spatial dependence* is likely to be generating the processes we observe (e.g., the presence of crime in one location is influenced by the presence of crime in neighboring locations), and that we can attempt to identify the reasons why the presence of events in one area are a function of the presence of events in neighboring areas, we then must consider the type of spatial dependence model that best introduces these neighboring effects. If we think that, in spatial regression terms, what may happen in one location depends on the values observed at neighboring locations, we should opt for applying the spatial lag model rather than the spatial error model. If we approach our spatial regression task with uncertainty on why spatial dependence is present, a spatial error modeling approach can indicate the extent to which we have failed to include other variables in our model, which if not addressed would affect the inference we make from the model. A spatial error model provides an indication that the relationship between the dependent and explanatory variables is influenced by unmeasured variables (Messner and Anselin 2004). We also will describe a statistical process that helps determine whether a spatial lag regression or a spatial error regression approach is the way to proceed. GWR is an alternative spatial regression approach to these two other models; it involves measuring the variation in relationships between variables by performing localized regression equations over the entire study area. We also describe when we have a good candidate for following a GWR approach.

Where to start?

So far in this book, we have described and discussed a wide range of techniques for analyzing the geography of crime. The application of spatial regression does not start with a technique for spatial regression, but with an examination of the data (the dependent and explanatory variables) to be analyzed. This examination helps develop familiarity with the data and its patterns, and effectively provides several of the preliminary diagnostic checks for a spatial regression analysis to progress. These checks include testing for clustering, analysis of spatial autocorrelation, and visualizing on a map the distribution of crime (i.e., the application of techniques already described in this book) and the other data to be included as explanatory variables. Scatter plots, as already illustrated, provide a useful technique for exploring relationships between crime and other spatial data and can help identify any outliers that may need consideration.

Before progressing with the next section, we assume you already have performed some analysis of your crime data, have knowledge of your crime data and its patterns, and have some idea of why you think the patterns may vary spatially. Although an exploratory approach provides an alternative to a hypothesis-driven approach to spatial regression analysis, you still must commence the exploratory approach with some idea of the data you want to include in your analysis, rather than loading every dataset you can get your hands on. With this prior knowledge, you then can progress more confidently in examining spatial relationships between your crime data and other variables.

Spatial lag regression

Definition and use

The spatial lag model adds to the standard OLS regression model equation by considering how the dependent variable for each observation is influenced by the dependent variable in neighboring areas. The spatial lag model implies that the presence of crime, for example (as our dependent variable), in one area is influenced by the presence of crime in neighboring areas (i.e., there is a lag effect from these neighboring areas).

The output from a spatial lag model consists of the coefficient values describing the relationship between the dependent variable and each explanatory variable included in the model (similar to OLS) as well as a coefficient value for the influence that the dependent variable in neighboring areas has on the dependent variable for the area that considers these other areas as its neighbors. By doing so, the spatial lag technique helps us determine how the effect of neighboring values of crime influence crime in each area and identifies other factors that help explain the distribution of crime (our other explanatory variables) but corrects the coefficient values for these other explanatory variables once the spatial lag effect of the dependent variable (i.e., crime) is considered. It is important to recognize that the coefficients of the explanatory variables from a spatial lag model refer to the relationship with the dependent variable across the whole study area rather than different coefficients being produced for each observation (i.e., for each geographic unit for which we have data). Parameter estimates are produced from the spatial lag model for each observation, which can be mapped to show how these values vary across space. Coefficient values for each observation are an output that the GWR technique generates, the third technique we examine in this chapter.

Spatial lag regression analysis is available in a wide range of software applications. The spatial lag regression example included in this section was performed using the free software GeoDa (Anselin 2005). GeoDa accepts files in shapefile format, so it is easy to load data into GeoDa for analysis, complete the regression analysis steps, and if required, import the results and output files from the regression analysis into ArcGIS or other GIS software. Other software that can be used for performing spatial lag regression analysis includes R.

Defining the neighborhood using a spatial weights matrix

A key step in any spatial regression analysis involves defining what constitutes a neighborhood—i.e., defining which neighboring areas to include in the analysis. This involves creating a weights matrix that defines this spatial neighborhood. A weights neighborhood matrix can be defined in terms of distance or contiguity.

Defining a weights matrix using distance involves specifying the neighborhood in distance terms, from an observation to another observation. For example, if the distance was defined as 2,000 meters, all observations within 2,000 meters would be included in the analysis of each observation.

Spatial contiguity is typically defined in either queen or rook terms. These terms refer to how these pieces are moved around a chessboard. Rook's case contiguity will include only neighbors that are forward, back, and on each side of the geographic unit of interest (figure 8.3). The queen on a chessboard can move in any direction, so all geographic units that share a boundary with the geographic unit of interest are included in the neighborhood defined when a queen's case contiguity weights matrix is used. Typically, for almost all scenarios when examining spatial patterns of crime, we would use the queen's case contiguity to define the weights matrix, but experimentation with other weights matrixes created under different definitions is useful to further examine relationships in data.

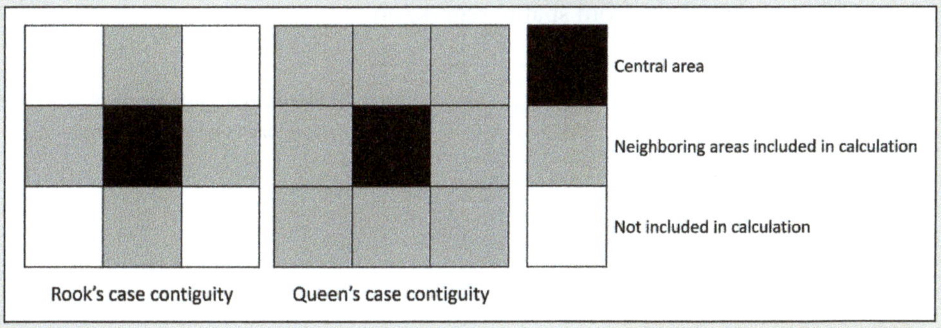

Figure 8.3. Rook's case and queen's case contiguity.

Data and regression diagnostics

Data for the district of Newcastle, England, is used in the examples that follow. When performing regression analysis, data used for the dependent and explanatory variables must be recorded in the same type of geographic unit (e.g., grid cells or census areas). Most of the explanatory variables were from the UK Census, so they already were aggregated to census areas. In most cases, the data was available in the form of small census areas (known as "output areas" in the UK), but some others were available only at the next geographic level of scale, lower super output areas (LSOAs). Data for output areas had to be aggregated to the same geographic unit of other data, recorded in LSOAs. Data that was originally in point object format (e.g., crime data) had to be aggregated as crime counts to the same LSOAs.

Figure 8.4 shows the distribution of residential burglaries for the district of Newcastle for a one-year period. The residential burglary count data was used to calculate burglary crime rates for each census area (using the census data on the number of households and calculated as a rate per 1,000 households). The residential burglary rate was used as the dependent variable in the regression analysis. The initial step in a spatial lag regression analysis involves conducting an OLS regression analysis to explore relationships in the data and determine whether spatial effects are present. Many statistics are generated from regression and spatial regression analysis processes. It is easy, particularly for those unfamiliar with spatial regression, to be a little overwhelmed with all these statistics and what they mean. In the sections that follow, we introduce you to the most important statistics that bear consideration but do so in stages rather than overwhelming you in one take.

In addition to generating regression coefficients, standard errors, p-values, and an R^2 value, the output from the GeoDa software (which is similar to output from other software for spatial regression analysis) generates two other useful statistics: a *log likelihood* value and the *Akaike Information Criterion* (AIC). Both statistics are measures of model performance. The higher the log likelihood value, the better the model. The lower the AIC, the better the model. We can use both the log likelihood and AIC to compare the performance of different regression models—especially the results from OLS regression to the results from spatial regression.

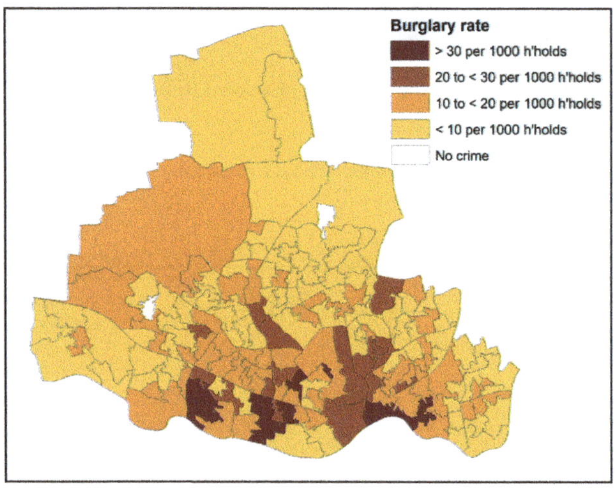

Figure 8.4. Distribution of residential burglary in Newcastle, England.

We begin with a simple OLS regression model that includes six explanatory variables: population density, student population, income deprivation, education and skills deprivation, living conditions deprivation, and owned housing (see table 8.2 for more details on each variable. The table lists several other variables that will be used in subsequent sections). We will call this Model 1 OLS Burglary. These six variables were chosen as it was thought they might, in some way, influence the variation in residential burglary rates. In a subsequent section, we give the selection of variables more considered thought. To run the OLS regression in GeoDa (and most other software) is straightforward. Once the data is arranged into a single file (e.g., a shapefile or ArcGIS feature class), it simply requires specifying the data that is the dependent variable and the data making up the explanatory variables.

Table 8.2. Explanatory variables used in the regression analysis of residential burglary and assaults in Newcastle, England

Data variable	Source	Description	Original spatial scale
Household density	Census 2011	Households per km^2.	Output areas (OAs)
Population density	Census 2011	Population per km^2.	OAs
Low level of education	Census 2011	Percentage of population with no qualifications or having achieved only Level 1 status (1+'O' level passes, 1+CSE/GCSE any grades, NVQ level 1, Foundation GNVQ).	OAs
Asian population	Census 2011	Percentage of population that is Asian.	OAs
Black population	Census 2011	Percentage of population that is Black.	OAs
Chinese population	Census 2011	Percentage of population that is Chinese.	OAs
Student population	Census 2011	Percentage of the normally resident population that are students (attending a higher-education establishment).	OAs
Owned housing	Census 2011	Percentage of households in which the household head owns the property (with or without a mortgage).	OAs
Socially rented housing	Census 2011	Percentage of households rented from the local council or housing association.	OAs
Private rented housing	Census 2011	Percentage of households privately rented.	OAs
Population age 20–29	Census 2011	Percentage of the normally resident population age 20 to 29.	OAs
Population age 30–39	Census 2011	Percentage of the normally resident population age 30 to 39.	OAs
Population age 40–49	Census 2011	Percentage of the normally resident population age 40 to 49.	OAs
Population age 50–59	Census 2011	Percentage of the normally resident population age 50 to 59.	OAs
Population age 60–69	Census 2011	Percentage of the normally resident population age 60 to 69.	OAs
Population age 70 and over	Census 2011	Percentage of the normally resident population age 70 and over.	OAs
Population born in the UK	Census 2011	Percentage of the population that was born in the UK.	OAs
Population that arrived from outside the UK 2001–03	Census 2011	Percentage of the population that arrived in the UK between 2001 and 2003.	OAs
Population that arrived from outside the UK 2004–06	Census 2011	Percentage of the population that arrived in the UK between 2004 and 2006.	OAs
Population that arrived from outside the UK 2007–09	Census 2011	Percentage of the population that arrived in the UK between 2007 and 2009.	OAs

(continued)

Table 8.2. (continued)

Data variable	Source	Description	Original spatial scale
Serious Acquisitive Crime (SAC) offenders	Northumbria Police (2010)	The number of people arrested for SAC offenses per 1,000 population. SAC offenses include residential burglary, theft of vehicles, theft from vehicles, and robbery (from people and businesses).	Home address
Income Deprivation	Index of Deprivation (2010)	The percentage of children who live in families that are income deprived (i.e., in receipt of income support, income-based jobseeker's allowance, working families' tax credit, or disabled person's tax credit below a given threshold). The higher the percentage, the greater the deprivation.	Lower super output areas (LSOAs)
Education and skills deprivation	Index of Deprivation (2010)	A score relating to the education and skills attained by children and young people. The higher the score, the greater the deprivation.	LSOAs
Living environment deprivation	Index of Deprivation (2010)	A score relating to quality of the indoor living environment of housing. The higher the score, the greater the deprivation.	LSOAs
Licensed premises	Newcastle City Council (2010)	Registered licensed premises, including nightclubs, bars, pubs, and off-licenses.	Address of premise

OLS regression results and spatial autocorrelation

A visual inspection of figure 8.4 indicates clustering in the distribution of residential burglaries in Newcastle. An analysis using Moran's I confirmed this clustering: Moran's I = 0.39 ($p < .001$). The OLS regression was then run. The R^2 value (.209) and *adjusted* R^2 value (.181) for this first OLS regression model (Model 1 OLS Burglary) were weak for the model (table 8.3). Inspection of the results for each explanatory variable showed only that student population, education and skills deprivation, and living conditions deprivation were significant, and that the other three variables (population density, income deprivation, and owned housing) were not significant. At this stage, the log likelihood and AIC meant little until other model results were generated for comparison.

The regression diagnostic results from an OLS regression also provide three important measures (see also table 8.3):

- *Multicollinearity condition number*: for the OLS regression model that was specified (Model 1 OLS Burglary), as this number was greater than 30, it suggests there were problems with the stability of the results because of *multicollinearity*. That is, the explanatory variables were too correlated with each other.
- *Jarque-Bera test* on the normality of the errors: suggests there was a significant problem with the nonnormality of the errors, making the inference of the results unreliable. The model has significant bias.
- Tests for *heteroscedasticity*: at least one of the *Breusch-Pagan* and *Koenker-Bassett tests* was significant. Significance in a heteroscedasticity test of the error terms means that the relationship between the variables varies significantly across the study area. This is

not usually a good thing to have in a model, but for us, it may demonstrate that an OLS regression analysis is insufficient because the relationship between variables varies spatially, so the model potentially can be improved using spatial regression analysis. What we want to achieve is for heteroscedasticity not to be present once we effectively incorporate the spatial effects in our data to a spatial regression model.

Table 8.3. Results from a simple OLS regression analysis (Model 1 OLS Burglary) of residential burglary and several explanatory variables in Newcastle, England

				Regression diagnostics			
						Heteroscedasticity tests:	
Explanatory variables: coefficient (standard error)	**R^2; adjusted R^2**	**Log likelihood**	**AIC**	**Multicollinearity condition number**	**Normality test: Jarque-Bera**	**Breusch-Pagan test**	**Koenker-Bassett test**
Population density 0.00002 (0.0002) Student population 0.305** (0.095) Income deprivation −5.862 (15.434) Skills and education deprivation 0.150* (0.062) Living environment deprivation 0.177* (0.086) Owned housing 6.336 (6.892)	.209; .181	−604.73	1.223.48	31.72	213.57***	27.61***	8.79

* p <.05; ** p <.01; *** p <.001

OLS regression results using a spatial weights matrix

Next, the model was run again (in GeoDa), this time applying the queen's case contiguity weights matrix within the OLS model (Model 2 OLS Burglary) to generate diagnostic statistics that can tell us something about the spatial effects in the model. The results from this model generated the same R^2 and coefficient values (the specification of the model still follows a classic OLS approach rather than introducing any spatial effects at this stage) but in addition generated statistics that identified whether a spatial regression analysis could be of potential benefit. This model includes information about spatial autocorrelation, as well as statistics that help us determine whether we should progress with a spatial lag model or a spatial error model.

The Moran's I result from the OLS regression, applied with the queen's case contiguity weights matrix (Model 2 OLS Burglary), was positive and significant, suggesting there was a problem with spatial autocorrelation in the model (table 8.4). OLS regression assumes the error terms from a model are independent of each other. The Moran's I result stated that we are in violation of this assumption, meaning that our results can be misleading and that our regression analysis of this spatial data requires special *spatial* treatment (i.e., requires us to perform some form of spatial regression analysis to generate a reliable model). A set of *Lagrange multiplier* results is generated when the OLS regression includes a spatial weights matrix. The Lagrange multiplier results help us determine whether a spatial lag model is more appropriate or if the application of a spatial error model is how we should proceed. The rules are as follows:

- If neither the Lagrange multiplier (lag) nor Lagrange multiplier (error) test results are significant, stick with the OLS results. But if there are problems with the OLS results that are indicated in the regression diagnostic statistics (such as heteroscedasticity and spatial autocorrelation being present and the Jarque-Bera test being significant), the OLS model will need respecification (i.e., the content of the model must be changed in some way, such as changing the explanatory variables used) to address these issues.
- If the Lagrange multiplier (lag) is significant and the Lagrange multiplier (error) is not significant, proceed with a spatial lag model.
- If the Lagrange multiplier (error) is significant and the Lagrange multiplier (lag) is not significant, proceed with a spatial error model.
- If both the Lagrange multiplier (lag) and Lagrange multiplier (error) are significant, refer to the robust Lagrange multiplier results.
- If the robust Lagrange multiplier (lag) is significant and the robust Lagrange multiplier (error) is not significant, proceed with a spatial lag model; if the robust Lagrange multiplier (error) is significant and the robust Lagrange multiplier (lag) is not significant, proceed with a spatial error model.
- If both robust Lagrange multiplier (lag) and robust Lagrange multiplier (error) are significant, proceed with a spatial regression model for the Lagrange multiplier test that is more significant (e.g., if the robust Lagrange multiplier (lag) is more significant, proceed with a spatial lag model) (Anselin 2005).

The OLS regression model results, applied with the queen's case contiguity weights matrix (Model 2 OLS Burglary), showed that both the Lagrange multiplier (lag) and Lagrange multiplier (error) statistics were significant (table 8.4). This significance meant the robust Lagrange multiplier statistics had to be checked. Only the robust Lagrange multiplier lag statistic was significant, suggesting a spatial lag model rather than a spatial error model should be used.

Table 8.4. Regression diagnostic results for the OLS regression analysis for residential burglary and several explanatory variables in Newcastle, England

	Moran's I (residuals)	LM (lag)	Robust LM (lag)	LM (error)	Robust LM (error)
Model 2 OLS Burglary	0.3106***	51.36***	6.83**	44.54***	0.0024

Note: LM = Lagrange multiplier. The model uses a queen's case contiguity weights matrix.
* $p < .05$; ** $p < .01$; *** $p < .001$

Spatial lag regression results

The next step involved running a spatial lag (SL) model (Model 3 SL Burglary), using the same variables and same weights matrix used in Model 2 OLS Burglary. Table 8.5 lists the results for the spatial lag model. Although a spatial lag model generates an R^2 value, this is a *pseudo* R^2 value and is not directly comparable to the R^2 result from the OLS model. The appropriate measures to use for comparing the OLS and spatial lag models are the log likelihood and AIC values. The log likelihood for the spatial lag model in comparison with the OLS model increased and the AIC decreased, both suggesting the spatial lag model provided an improvement in the fit of the relationships between residential burglary and the explanatory variables. That is, by considering spatial effects, we have a better model of these relationships.

The spatial lag model (using GeoDa) does not generate a multicollinearity condition number or a Jarque-Bera statistic result but generates an additional indicator that relates to the spatial lag effect of the dependent variable (in this example, for residential burglary). It appears as the coefficient value for this variable (the weighted residential burglary variable in table 8.5) and as the lag coefficient (Rho) in the model results. This coefficient value (often referred to as the "spatial autoregressive coefficient") reflects the degree of spatial dependence inherent in the dependent variable. The spatial autoregressive coefficient for residential burglary of .554 was highly significant, suggesting there was a great degree of positive influence on residential burglary in each area from its neighboring areas (i.e., there is a significant spatial lag effect from the dependent variable).

Overall, the significance of the coefficients for each explanatory variable from the spatial lag model were like those from Model 2 OLS Burglary: student population, and education and skills deprivation remained significant; population density, income deprivation, and owned housing remained nonsignificant. The main change was that living conditions deprivation was no longer significant in the spatial lag model, and the magnitude of the coefficients that were significant all showed a decrease in value. The latter is because of the spatial lag term in the model (represented by the weighted residential burglary variable) taking some of the strength away from the relationships between burglary and these other statistically significant explanatory variables and appropriately indicating the influence of the spatial lag effect.

Table 8.5. Results from a spatial lag regression analysis (Model 3 SL Burglary) of residential burglary and several explanatory variables in Newcastle, England

Coefficient (standard error)	R^2; adjusted R^2	Log likelihood	AIC	Spatial autoregressive fit - Lag coefficient, Rho (z-score)	Regression diagnostics			
					Multicollinearity condition number	Jarque-Bera	Breusch-Pagan test (heteroscedasticity)	Likelihood ratio test (spatial dependence)
Weighted residential burglary 0.5541*** (0.0791) Population density 0.00004 (0.0002) Student population 0.1918* (0.0814) Income deprivation -4.6844 (13.0582) Skills and education deprivation 0.1177* (0.0524) Living environment deprivation 0.1008 (0.0732) Owned housing 3.9629 (5.8257)	.41; n/a	−584.87	1185.74	0.5541*** (7.001)	n/a	n/a	20.6719**	39.7377***

* p < .05; ** p < .01; *** p < .001

The spatial lag regression model diagnostic results still show, however, that heteroscedasticity remains a serious problem (see table 8.5), even after introducing the spatial lag term in the model (i.e., we must be careful in the trust we place in our results, especially as to whether the explanatory variables that appear as significant are indeed significant). The likelihood ratio test result that is also reported in the results provides confirmation of the significance of the spatial autoregressive coefficient. The likelihood ratio test compares the classic OLS model specification to the spatial lag model, and as it is significant, it confirms the importance of considering the spatial dependence specified in the model.

An accurate and reliable spatial lag model stipulates there is an influence on the dependent variable from neighboring values (the lagged dependent variable). The spatial lag model works by filtering out the confounding effect of spatial autocorrelation in the dependent variable, and by doing so generates a more accurate inference (in the coefficients generated) from the explanatory variables in the model. If the spatial lag model is specified correctly, we should have evidence that consideration of spatial dependence has improved the model, and that heteroscedasticity is not present. The last model (Model 3 SL) did not pass the test of heteroscedasticity.

Interpretation and regression model respecification

So, what do we do next? Generally, these results tell us that the pattern of residential burglaries in our study area appears potentially to be related to some explanatory variables (e.g., where there is a high proportion of students living). In this example, we included only a limited number of explanatory variables to introduce spatial regression analysis, the processes involved, statistical terms, and how to interpret the results. There is a lot to take in when doing spatial regression analysis. Our model is not perfect, it is not reliable, and currently, we cannot be confident in any inferences we want to make from these results because of problems of heteroscedasticity remaining, even after applying special spatial treatment in the regression analysis. What we can be clear about, at this stage, is that the relationship between the distribution of residential burglary and the variables that we included cannot be explained simply as linear relationships. Instead, the results show that by considering the spatial distribution of the dependent variables (in this case, the influence of residential burglaries in neighboring areas), the model improved, suggesting that a spatial consideration of the distribution of residential burglary and other variables will be more accurate and reliable if spatial effects are considered.

The next step would involve respecifying our spatial lag regression model. This could involve removing the variables that were not significant and running the model with only those that were significant, respecifying our weights matrix (e.g., using distance rather than contiguity—census areas vary in size, so in some cases, our model includes neighbors that were very close and neighbors that were relatively far away), or adding other variables that we believe better explain the spatial distribution of the dependent variable. We will perform these tasks as we progress through the next sections rather than examining several additional model specifications only to satisfy the spatial lag model diagnostic requirements at this stage. Once we have a well-specified, accurate, and reliable model, we can map the parameter estimates of the dependent variable for each geographic unit from our spatial lag model results, as well as their residuals to show the areas in which the explanatory variables perform best and in which they do not perform as well.

We continue by examining other spatial regression techniques, so we can learn how other techniques operate and describe additional processes involved in spatial regression analysis. The goal is to produce a well-specified model that is accurate and reliable using the variables we have begun to consider. At the end of the chapter, we return to the spatial lag model, map the outputs it generates, and illustrate its application once we have a model that meets all the requirements for accuracy and reliability.

Spatial error regression

Definition and use

By now, you should be comfortable with the term *spatial dependence*, and its importance in conducting regression analysis of spatial data. Spatial data may show spatial dependence in the variables being analyzed, and in the residuals of these variables. If residuals are correlated, the estimates the model produces are inefficient, and indicative that one or more variables are currently missing from the model. An examination of spatial error can indicate that a key variable or variables have been omitted from the model, and if left unattended would affect the inference we can make from the model.

An effective spatial error model indicates there is no distinctive influence from neighboring areas on the dependent variable. The spatial patterns observed in the dependent variable, such as clustering of crime, are accounted for simply by the spatial patterning of explanatory variables, and not influenced by the values of the dependent variable in neighboring areas. When clustering is present in the dependent variable, the spatial error model provides an indication that it is influenced by unmeasured variables. Similar to the spatial lag model, the outputs consist of a coefficient value that relates to spatial error, plus coefficient values that describe the relationship between the dependent variable and each explanatory variable included in the model once the spatial error term is taken into consideration. Again, it is important to recognize that the coefficients from a spatial error model refer to the relationship with the dependent variable across the whole study area rather than different coefficient values being produced for each observation. Parameter estimates are produced from the spatial error model for each observation, which can be mapped to show how they vary across space. Coefficient values for each observation are an output from geographically weighted regression, the technique we examine in the next section.

Data and spatial weights matrix

To illustrate the application of a spatial error model, we use data again from Newcastle, England, and perform the initial OLS regression analysis with a spatial weights matrix to generate diagnostic results that indicate whether a spatial lag or spatial error model is the way to proceed. For this section, we hope it will be the latter. In this example, the type of crime that is analyzed is violent assaults, but we start by using several of the same explanatory variables used in the previous OLS and spatial lag models: student population, income deprivation, education and skills deprivation, and living conditions deprivation. The assaults data is counts of crime. Like the previous residential burglary example, all data is aggregated to lower super output areas. GeoDa was used to perform the spatial regression tasks, and the same queen's case contiguity weights matrix used for the previous OLS and spatial lag examples was used for the spatial error model.

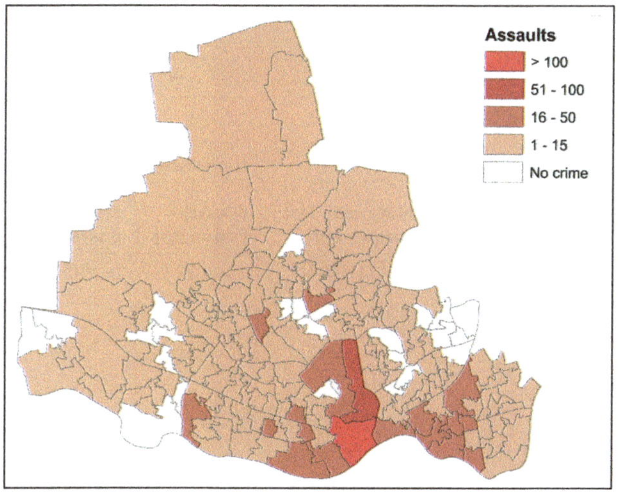

Figure 8.5. Distribution of assaults in Newcastle, England.

OLS regression results using a spatial weights matrix

Figure 8.5 shows that assaults appear to be clustered and were particularly high in one area of Newcastle. The Moran's I result of 0.076 ($p < .001$) suggests significant evidence of clustering, but it is noted that this result is close to zero. The first OLS model (Model 1 OLS Assaults) produced R^2 (.106) and adjusted R^2 values (.085) that are weak (table 8.6). Inspection of the results for each explanatory variable shows that income deprivation, education and skills deprivation, and living conditions deprivation are significant, and student population is not significantly correlated to the distribution of assaults. The log likelihood and AIC are noted in the table for subsequent comparison. The multicollinearity condition number of 11.29 suggests there were no problems with the stability of the regression results because of multicollinearity. The Jarque-Bera test result suggests there was a significant problem with the nonnormality of the errors, and both tests for heteroscedasticity were significant.

Table 8.6. Results from a simple OLS regression analysis (Model 1 OLS Assaults) of assaults and several explanatory variables in Newcastle, England

				Regression diagnostics			
				Multicollinearity condition number		Heteroscedasticity tests:	
Coefficient (standard error)	R^2; adjusted R^2	Log likelihood	AIC		Normality test: Jarque-Bera	Breusch-Pagan test	Koenker-Bassett test
Student population 0.184 (0.327) Income deprivation −140.809* (62.420) Skills and education deprivation 0.9154** (0.347) Living environment deprivation 1.3864** (0.447)	.106; .085	−904.47	1,818.94	11.29	127,341.68***	1,195.29***	0.0013**

* p <.05; ** p <.01; *** p <.001

The Lagrange multiplier results showed that only the robust Lagrange multiplier (error) test result was significant (table 8.7), suggesting the way to proceed should be to use a spatial error model.

Table 8.7. Regression diagnostic results for the OLS regression analysis for assaults and several explanatory variables in Newcastle, England

	Moran's I (residuals)	LM (lag)	Robust LM (lag)	LM (error)	Robust LM (error)
Model 1 OLS Assaults	0.0786*	1.37	2.69	2.85	4.18*

Note: The model uses a queen's case contiguity weights matrix.
* p <.05

Spatial error regression results—first attempt

Table 8.8 lists the results for the first spatial error (SE) model (Model 2 SE Assaults), using the same variables from Model 1 OLS Assaults. The R^2 value (.13) is a pseudo R^2 value and is not directly comparable to the R^2 result from the OLS model. The main thing to note, however, is the R^2 value from the spatial error model was still weak. The log likelihood for the spatial error model in comparison with the OLS model increased, and the AIC decreased, both suggesting the spatial error model provided an improvement in the fit of the relationships between assaults and the explanatory variables. The spatial error model generates an additional variable for the spatial autoregressive term that relates to the spatial correlation of the errors, Lambda. The Lambda value ranges from zero to one. Lambda was .2193 and was not significant (but only marginally so, $p = .0516$), suggesting the residuals are not significantly spatially correlated.

Overall, the significance of the coefficients for each explanatory variable is like that for the first OLS model: income deprivation, education and skills deprivation, and living conditions deprivation were significantly correlated to the distribution of assaults, and student population was not significant. The spatial error regression model diagnostic results show, however, that heteroscedasticity was a serious problem (i.e., we must be careful in the trust we place in our results, in particular, as to whether the explanatory variables that appear as significant are, indeed, significant). The likelihood ratio test in this model relates to spatial dependence of the error terms in the model. The test uses the spatial autoregressive coefficient (i.e., Lambda) to compare the spatial error model with the OLS model. As the result was not significant, it suggests no significant presence of spatial dependence, and that the current poor performance of the model is likely to be because variables are missing from the current model.

Table 8.8. Results from a spatial error regression analysis (Model 2 SE Assaults) of assaults and several explanatory variables in Newcastle, England

Coefficient (standard error)	R^2; adjusted R^2	Log likelihood	AIC	Spatial autoregressive fit - error coefficient Lambda (z-score)	Regression diagnostics			
					Multicollinearity condition number	Jarque-Bera	Breusch-Pagan test (heteroscedasticity)	Likelihood ratio test (spatial error dependence)
Lambda 0.219 Student population 0.110 Income deprivation −175.959** Skills and education deprivation 1.078** Living environment deprivation 1.474**	.1294; n/a	−902.93	1,815.86	0.2193 (1.947)	n/a	n/a	1,185.82***	3.08

* p <.05; ** p <.01; *** p <.001

Spatial error (and spatial lag) regression results—second attempt

The process was then repeated but this time adding the variable licensed premises to the model and retaining all the previous explanatory variables. Licensed premises is a key explanatory variable that was missing from the model, and hence, perhaps why a spatial error model was initially suggested from the Lagrange multiplier results as the way to proceed. This process began by conducting an OLS regression analysis (Model 3 OLS Assaults) with the queen's case contiguity weights matrix. After the inclusion of the licensed premises variable, the Lagrange multiplier test results indicated that a spatial lag model was now the most appropriate way to proceed (table 8.9). However, for purposes of experimentation and comparison, both a spatial lag (Model 3 SL Assaults) and spatial error regression analysis (Model 3 SE Assaults) were applied, with the results shown in table 8.10 alongside the OLS regression results.

Table 8.9. Regression diagnostic results for the OLS regression analysis for assaults and several explanatory variables in Newcastle, England

	Moran's I (residuals)	LM (lag)	Robust LM (lag)	LM (error)	Robust LM (error)
Model 3 OLS Assaults	-0.03	10.86***	10.66**	0.32	0.12

Note: Model uses a queen's case contiguity weights matrix.
* p < .05; ** p < .01; *** p < .001

Table 8.10. Results from an OLS, spatial lag, and spatial error regression analysis of assaults and several explanatory variables in Newcastle, England: (a) regression results and (b) coefficients

a

					Regression diagnostics			
Model	R^2; adjusted R^2	Log likelihood	AIC	Spatial autoregressive fit—error coefficient Lambda (z-value)	Multicollinearity condition number	Jarque-Bera	Breusch-Pagan test (heteroscedasticity)	Likelihood ratio test
Model 3 OLS Assaults	.97; .96	-624.08	1260.16	n/a	11.68	902.70***	166.14***	n/a
Model 3 SL Assaults	.97; n/a	−618.38	1250.75	−0.114*** (−3.41)	n/a	n/a	101.57***	11.40***
Model 3 SE Assaults	.97; n/a	−623.86	1,259.71	−0.098 (−0.769)	n/a	n/a	178.11***	0.45

* p < .05; ** p < .01; *** p < .001

b

Coefficients	Model 3 OLS	Model 3 SL	Model 3 SE
Spatial autoregressive coefficient	n/a	−0.114***	-0.098
Student population	−0.169*	−0.134*	−0.157*
Income deprivation	9.038	20.037	12.180
Skills and education deprivation	0.084	0.046	0.069
Living environment deprivation	−0.102	−0.090	−0.106
Licensed premises	1.781***	1.798***	1.775***

* p < .05; ** p < .01; *** p < .001

In this third set of models, the spatial lag model was found to perform better—all models had the same R^2, but the spatial lag model had the highest log likelihood value and the lowest AIC. However, heteroscedasticity was still present, suggesting we must be careful in the trust we place in our results, especially as to whether all the explanatory variables that appear as significant are, indeed, significant.

Introducing another test for model performance and model specification

A final test to introduce in this section is another model performance measure that helps confirm whether the better performing model used to this point has been specified effectively. This test involves examining the order of three statistics: the Wald statistic value for our model should be greater than the likelihood ratio for the model, and the likelihood ratio should be greater than the relevant Lagrange multiplier value. The Wald statistic is simply the z-value of the spatial autoregressive coefficient squared. This z-value from the spatial lag model was −3.41 (table 8.10). So, for the model to be effectively specified, we should find the Wald statistic ($-3.41^2 = 11.63$) > log likelihood (11.40) > Lagrange multiplier (10.86). This is the case for the spatial lag model, suggesting we have a well-specified model, but we still must perform some slight respecification to address heteroscedasticity (e.g., examine whether the nonnormality, as indicted by the Jarque-Bera statistic in the OLS model, is creating a problem). We will not do further model specifications on the spatial lag (or error) model yet. We will continue by examining our final spatial regression technique, use it to learn more about model specification, and return to the spatial lag and error models with these examples at the end of the chapter.

Geographically weighted regression

Definition and use

In this section, we describe the use of GWR. Although each of the spatial regression techniques described so far are different, they each follow the same general process. The same is the case with GWR: decide on dependent and explanatory variables, run an OLS regression analysis, review the results, and then (if suitable), progress with the spatial regression model. In the processes that follow, we spend more time examining and comparing model specifications than we have done previously, until we reach a point at which we consider the model to be satisfactory (i.e., a good fit, not biased—and a good candidate for spatial regression analysis).

GWR involves generating regression equations for explaining the spatially varying relationship between the dependent and explanatory variable or variables for every area in a study area (Fotheringham, Brunsdon, and Charlton 2002). In several tests that have compared spatial regression techniques, GWR was found to be better in modeling performance (Waller et al. 2007; Wang, Ni, and

Tenhunen 2005; Zhang, Gove, and Heath 2005). However, GWR is not without its critics, including some commentators questioning its use for determining inference (Griffith 2008; Wheeler and Tiefelsdorf 2005; Wheeler and Waller 2009). We will review the issue of its use for inference later in this section.

GWR produces many outputs that can be mapped. These include the parameter estimates (i.e., the coefficient describing the relationship between the explanatory variable or variables and the dependent variable for each geographic unit), t-values (indicating the statistical significance of the parameter estimates), and parameter residuals (the level of error or unexplained component of the relationship between the explanatory variable or variables and the dependent variable in the model). It is important to consider all these outputs (and others that we will illustrate) to interpret the results rather than simply relying on generating a map of parameter estimates.

Data and regression approaches

The application of GWR is illustrated using the data on crime for Newcastle, England, that we already have used in this chapter—residential burglary and assaults. In this section, we also illustrate the two approaches to regression analysis: the hypothesis-driven approach and the exploratory approach. We begin with an exploratory approach of residential burglary in Newcastle. This includes examining a large range of explanatory variables, and then testing different model specifications to produce a model that performs well, is statistically significant, and is not biased.

OLS and GWR diagnostic tests and model respecification process

Like spatial lag and spatial error regression, the first stage of GWR modeling requires an OLS regression analysis to test whether the explanatory variables included provide a suitable model for explaining the presence of the dependent variable. Where explanatory variables are not significant, they are excluded from the model until only those variables that are significant remain. Another part of this initial process involves performing several statistical diagnostic tests to determine whether the model is suitable for GWR, reviewing the explanatory variables included, and examining the residuals from each model. This may involve experimenting with many models that include or exclude explanatory variables that were previously included. In this chapter, we already have described several diagnostic tests for our regression models. Here, we expand on the diagnostic tests previously described, organizing them into three categories: model performance, model significance, and model bias.

Model performance

Tests for model performance determine whether there is, in a statistical sense, a relationship between the dependent variable and the explanatory variables. The test results of interest are the adjusted R^2 value, the coefficient of each explanatory variable and their significance, the AIC, the *variance inflation factor* (VIF), and the *Koenker statistic*. By now (if you have read the previous sections of this chapter), you should be familiar with adjusted R^2, the coefficients, p-values, and AIC. The VIF quantifies the severity of multicollinearity in the model. If the modeling begins by including many explanatory variables, it is likely our results will show that several of the explanatory variables were not statistically significant. Consideration then should be given to removing these variables from the OLS model, as they do not appear to be contributing in any way to explaining the variation in the dependent variable. To help with this process of selecting which variables to remove from the model, the VIF for each explanatory variable indicates if it is not contributing to the model. If the VIF value for a variable is greater than 7.5, it suggests the variable is redundant in the model and should be removed. The same removal procedure applies when an explanatory variable is significant but has a VIF value greater than 7.5. This situation is likely because this variable offers a similar explanation to another statistically significant variable currently present in the model. The rule to follow is to remove the explanatory variable with the highest VIF, although experimentation that involves interchanging these collinear variables also is recommended. The *Koenker statistic* primarily is used as a model significance measure (explained in the following section), but its use for model performance is to determine whether p-values or robust p-values should be used. If the Koenker statistic is significant, robust p-values should be used rather than standard p-values.

Understanding multicollinearity: An analogy

An analogy that I think is useful for understanding multicollinearity is to think of a discussion with a class of students, all speaking at the same time to you, the teacher. Some of the students may be saying the same thing, whereas others will be offering different points of view. If the students saying the same thing speak in unison, their opinion limits your ability to hear other opinions. In regression terms, when you have multiple explanatory variables saying the same thing, multicollinearity exists. The inclusion of explanatory variables saying the same thing in your model, contributing with the same opinion at the same time, basically means this single point of view is being shouted in your model. With this inclusion, you are introducing bias into the model because more attention is being drawn to the single point of view that more than one explanatory variable is providing, which in turn undermines the other points of view (i.e., other explanatory variables that may be equally, if not more, important to the model but that are not being effectively heard). Removing the variables that give the same message, and leaving only one (e.g., hearing from only one spokesperson for the students who have the same point of view), means that all messages (i.e., all explanatory variables that are useful to the model) can be heard. In regression analysis, you then can determine which points of view, in a statistical sense, are worth listening to (i.e., are significant), and which are not.

Model significance

Tests for model significance help determine whether the model is a good candidate for GWR. Results from three tests require consideration: the Koenker statistic, the *joint F statistic*, and the *joint Wald statistic*. If the Koenker statistic from an OLS regression is significant, it suggests there is heteroscedasticity (or *nonstationarity*) in the model, meaning the relationships between the dependent and explanatory variables vary across space. Heteroscedasticity in an OLS model is a good sign (thinking in spatial terms), because it suggests the model is a good candidate for GWR. The joint F and joint Wald statistics are measures of overall model statistical significance. If the Koenker statistic for the regression model is significant, the joint F statistic is unreliable, and therefore, the joint Wald statistic should be used instead to determine whether the regression model is significant.

Model bias

The third group of tests is for model bias. Testing for bias requires an analysis of the residuals from the regression model to determine whether they are normally distributed. To do so, we use the Jarque-Bera statistic: if this statistic is significant, it suggests the model is biased. If this is the case, we have three options for attempting to solve the bias in the model (or a combination of the three): we include other explanatory variables, we remove influential outliers (Rosenshein, Scott, and Pratt 2011), or we perform a transformation on the variables to address skewness in their distribution (e.g., a *natural log transformation*). Application of these options is described in the examples that follow.

Spatial clustering of residuals

If we have a good performing model, the model is significant (and appears to be a good candidate for GWR), and we have passed the Jarque-Bera statistical test (the model is not biased), the model's residuals then are tested to determine whether they are spatially clustered, using Moran's I. If Moran's I is positive and significant, it suggests a key variable is missing from the model (to explain the spatial patterns in the dependent variable). This means we must continue respecifying our model and repeat all the tests described above until we have a model that passes these tests and passes the Moran's I test on the residuals. When the Moran's I result suggests evidence of clustering in the residuals, mapping the residuals thematically may help identify the explanatory variable that is missing. That is, where residual values are high and clustered, we are required to consider what else is present in this area that could help explain the spatial patterns in the dependent variable, in addition to, or in replacement of, the existing explanatory variables.

Coping with outliers

In regression analysis, one way to address issues associated with outliers, such as the influence they may have in biasing the regression model, is to remove them. Removing influential outliers in spatial regression analysis is just such a practice several researchers have suggested (e.g., Rosenshein, Scott, and Pratt 2011). The spatial patterning of crime into hot spots naturally results in producing statistical outliers—the hot spots. So, do we remove the areas we are most interested in to satisfy the regression model diagnostic tests we are required to pass? My suggestion is no. Rather than removing outliers, embrace them. Our hot spots are important to us. This means we must find some other way to address bias issues in the regression model caused by outliers.

Is it worth it?

There is a lot to take in, but by progressing through the chapter in the manner we have, hopefully we have built up confidence in conducting a full spatial regression analysis. All the statistics described in this section are generated automatically by the software used for GWR. ArcGIS includes tools to perform the initial OLS regression analysis and perform a GWR analysis. Dedicated GWR software also is available from the technique's creators (see the "Further Resources" section at the end of this chapter), as well as in several other software solutions, such as R. The examples that follow were all generated using ArcGIS tools. Spatial regression analysis is a recursive process. It is important to think of regression analysis as involving a set of statistical processes from which the results are used collectively to interpret the relationships that have been tested, rather than it being a single process generating a single result. Effective spatial regression analysis takes time but is valuable, as the results from the following examples will show.

OLS model specification in preparation for GWR analysis

The initial OLS regression and statistical diagnostic test results identify whether the model is suitable for GWR. If not—for example, if the Jarque-Bera statistic is significant—a respecification of the model is required before progressing. As has been seen in the previous sections, this respecification process is not exclusive to GWR but is expected in all types of spatial regression analysis. Once a suitable OLS model is identified, the explanatory variables from this model then can be applied to a GWR model. This application involves a straightforward process of including the explanatory variables from the final OLS regression model in the GWR model. Recording the AIC value calculated as part of this initial process for each OLS regression model is important, as it can be used later to determine whether the GWR model performs better than the OLS regression model.

Example: GWR regression of residential burglary

OLS model specification and results

The analysis of the spatial distribution of residential burglary in Newcastle (the dependent variable) followed an exploratory approach. It initially involved including all the 25 explanatory variables listed in table 8.2 in a single preliminary exploratory OLS regression model. Only four explanatory variables were correlated significantly to residential burglary: serious acquisitive crime (SAC) offenders, deprivation of education and skills, the 50–59-year-old age group, and residents born in the UK. These variables were selected for Model 1 OLS Burglary, and the OLS regression and diagnostic analysis process was repeated. The analysis processes then would determine whether, together, these variables were significant, and whether the model performed well, was significant, and unbiased. The OLS regression analysis process was repeated several times (a reflection of the recursive nature of regression analysis) using different combinations of explanatory variables (the logic to these different combinations is described below) until the model passed the necessary tests for a GWR analysis.

Because of the large number of variables included in the preliminary model, a high degree of multicollinearity was expected to be present in this first model. This was the case, with many variables identified as not significant, having VIF values greater than 7.5. For each model that followed, the VIF for each explanatory variable was less than 7.5, indicating the issue of multicollinearity had been addressed. The decision-making behind the choice of explanatory variables was as follows (the key results from each model are listed in table 8.11):

- Model 1 OLS Burglary: This model included only those explanatory variables (from all the explanatory variables listed in table 8.2) identified from the preliminary OLS model as significantly related to residential burglary: SAC offenders, deprivation of education and skills, 50–59-year-old age group, and residents born in the UK. It was noted from the preliminary OLS model that the VIF value for SAC offenders was less than 7.5, but the VIF values for the three other variables were greater than 7.5. It was anticipated that the removal of most of the explanatory variables from the preliminary OLS model for the creation of model 1 would result in each of the four explanatory variables being significant, with VIF values that were less than 7.5. However, although deprivation of education and skills, the 50–59-year-old age group, and residents born in the UK were significant variables in the preliminary OLS model, warranting their inclusion in Model 1 OLS Burglary, these may not be the correct explanatory variables to include (because they were correlated with other explanatory variables that instead may be more relevant to the model). From this, it was anticipated that some further experimentation in model respecification would be necessary after results from each OLS model were examined. Following the respecification of the preliminary model using only these four variables, all variables were found to be significant in this model. SAC offenders were positively associated with residential burglary, and the deprivation of education and skills, the 50–59-year-old age group, and UK-born residents were negatively associated with residential burglary. The model generated an adjusted R^2 value of .42 and was found to be statistically significant (based on the joint Wald statistic result). The Koenker statistic was significant, suggesting the model was a good candidate for GWR (i.e., heteroscedasticity or nonstationarity was present in the OLS model), but the Jarque-Bera statistic was significant, suggesting the model was biased because the residuals were not distributed normally. A different model specification, therefore, was required.

- Model 2 OLS Burglary: OLS regression does not require variables to be normally distributed. However, if problems exist in the specification of the model (in particular, if associated with the nonnormality of residuals), one way to attempt to address it is to transform any strongly skewed variables and see if it corrects the issues. The residential burglary rate, SAC offenders, deprivation of education and skills, and percentage born in the UK all were skewed in their distribution, so to correct this skewness, a natural log transformation was performed on these variables. The natural log transformation versions of these variables were included in model 2. SAC offenders, the 50–59-year-old age group, and UK-born residents were variables that remained significant, but deprivation of education and skills no longer was significant. Heteroscedasticity now had been removed, suggesting this model was not a good candidate for GWR. The Jarque-Bera statistic remained significant, suggesting the model remained biased. Model respecification was required.
- Model 3 OLS Burglary: Mapping a model's residuals thematically can help identify explanatory variables that could contribute to the model. An analysis of the distribution of Model 1 OLS Burglary residuals was conducted (by mapping them thematically), followed by a thematic spatial analysis of each of the 21 explanatory variables that had not been included in model 1. The thematic spatial analysis of these 21 explanatory variables suggested that the proportion of the population that was Asian visually matched in some areas with the mapped distribution of the model 1 residuals and, therefore, was considered to be an explanatory variable that could contribute to the explanatory variables used in models 1 and 2. All variables in model 3 used original values rather than transformed values. The results from this model found SAC offenders, deprivation of education and skills, the 50–59-year-old age group, and residents born in the UK to be significantly correlated to residential burglary, but the Asian population was not correlated significantly to residential burglary. Heteroscedasticity was present, but issues with model bias remained.

- Model 4 OLS Burglary: An analysis of the Model 3 residuals, and further analysis of the spatial distribution of all the other 20 explanatory variables, indicated the proportion of the population that was students could help explain the spatial distribution of residential burglary. Therefore, the student population was included in model 4 alongside the four variables included in models 1 and 2. Also, the distribution of the student population was found to be significant in the residential burglary spatial lag models described previously in this chapter, further justifying its possible inclusion in explaining the distribution of residential burglary. The results from model 4 revealed the student population to be significantly positively related to residential burglary, but deprivation of education and skills and the 50–59 age group were no longer significant. Model 4 led to an improvement in the model, as shown by a small increase in the adjusted R^2 value (from .38 in model 1 to .40 in model 4) and a small reduction in the AIC (from 1,174 in model 1 to 1,172 in model 4). The model was significant, and the Koenker statistic was significant (i.e., the model was a good candidate for GWR), but model bias remained.
- Model 5 OLS Burglary: This model used only those variables that were significant in model 4: SAC offenders, UK-born population, and student population. All explanatory variables were significant, there was a further improvement in model performance (the AIC decreased to 1,169), and the Koenker statistic was significant, but the model was still biased.
- Model 6 OLS Burglary: Following an analysis of the distribution of model 5 residuals and a thematic spatial analysis of all 25 explanatory variables, it was decided to reintroduce the Asian population variable into model 6 alongside the three variables from model 5. The distribution of the residuals matched in some areas with the distribution of the Asian population. Model 6 resulted in SAC offenders and the student population being significant and the Asian population and the UK-born population not being significantly correlated to residential burglary. The Koenker statistic was significant, but the Jarque-Bera statistic remained significant, indicating the model was still biased.
- Model 7 OLS Burglary: Model 7 involved experimenting with the inclusion of the Asian population rather than the UK-born population variable. Hence, model 7 included the Asian population, SAC offenders, and the student population as the explanatory variables. This combination resulted in all three being significant and positively associated to residential burglary, with the Koenker statistic being significant but the Jarque-Bera statistic remained significant.
- Model 8 OLS Burglary: The next stage involved experimenting with transforming variables to reduce skewness in their distribution. Model 8 involved performing a natural log transformation on the burglary data but retaining the original values for the three variables from model 7: SAC offenders, the student population, and the Asian population.

All three variables were significant and positively associated with residential burglary; the model was significant; heteroscedasticity was present, suggesting the model was a good candidate for GWR; and the Jarque-Bera statistic was not significant. An adjusted R^2 value (and AIC) for a model that includes log-transformed data is not comparable to the adjusted R^2 value (or AIC) for a model with data that has not been log transformed. The comparison that could be made was to model 2, in which log transformations were used. The adjusted R^2 value from model 8 was better than from model 2. Model 8 also generated a lower AIC value than from model 2. A Moran's I test was performed on the residuals of model 8, and these were not clustered.

Table 8.11. The key results from OLS models, identifying explanatory variables that were significant, model performance, model significance, heteroscedasticity (or nonstationarity), and bias

Residential burglary models	Model variables and coefficients	Adjusted R^2	AIC	Koenker statistic	Joint F or joint Wald statistic	Jarque-Bera
Model 1 OLS	SAC offenders (2.13***) Deprivation of education and skills (−0.06*) 50 to 59 age group (−49.07*) Residents born in UK (−26.03***)	.38	1174.4	25.28***	73.46***	21.63***
Model 2 OLS (NL of burglary)	NL SAC offenders (0.30**) NL Deprivation of education and skills (−0.006) NL 50 to 59 age group (−7.30***) NL Residents born in UK (−1.44**)	.26	379.7	3.58	15.97***	10.13**
Model 3 OLS	SAC offenders (2.13***) Deprivation of education and skills (−0.06*) 50 to 59 age group (−49.24*) Residents born in UK (−25.62**) Asian population (0.004)	.38	1176.6	25.45***	73.40***	21.75***
Model 4 OLS	SAC offenders (2.10***) Deprivation of education and skills (−0.03) 50 to 59 age group (−14.44) Residents born in UK (−23.74**) Student population (0.14*)	.40	1172.2	27.88***	81.12***	23.78***
Model 5 OLS	SAC offenders (1.83***) Residents born in UK (−24.79***) Student population (0.17**)	.40	1169.1	29.53***	63.28***	24.37***
Model 6 OLS	SAC offenders (1.92***) Residents born in UK (−17.66) Student population (0.19***) Asian population (0.07)	.40	1,170	30.56***	72.70***	28.89***

(continued)

Table 8.11. (continued)

Residential burglary models	Model variables and coefficients	Adjusted R^2	AIC	Koenker statistic	Joint F or joint Wald statistic	Jarque-Bera
Model 7 OLS	SAC offenders rate (2.17***) Student population (0.25***) Asian population (0.16**)	.40	1,172	25.46***	80.58***	42.05***
Model 8 OLS (NL of burglary)	SAC offenders (0.156***) Student population (0.024***) Asian population (0.014**)	.30	368.7	10.43*	114.18***	5.84

Note: NL = natural log transformation.
* $p < .05$; ** $p < .01$; *** $p < .001$

Several additional exploratory regression models were calibrated to improve model 8. This attempt involved adding the following variables in turn to the three explanatory variables from model 8: 30–39 age group population, 40–49 age group population, 50–59 age group population, 60–69 age group population, 70 years and older age group population, population that arrived and became resident in 2007–09, population that arrived and became resident in 2004–06, social rented housing, owned housing, living conditions deprivation, income deprivation, and the deprivation of education and skills. None of these other combinations improved model 8.

GWR candidate model

Model 8, consisting of SAC offenders, the student population, and the Asian population, offered the best combination of explanatory variables to produce a model with the highest level of performance, significance, and nonbias, and was a suitable candidate for GWR. Figure 8.6 shows the distribution of each of these explanatory variables across Newcastle. The maps show the SAC offenders (figure 8.6a) to be highest across the southern border of the district and in the central district area, the student population (figure 8.6b) to concentrate around the southern central area of the district, and the Asian population (figure 8.6c) to be greatest in the southern central and northeastern areas of the district. Figure 8.6d shows the residential burglary rate to be highest in the southern and central regions of the district of Newcastle. Each of the three variables positively correlated with residential burglary. An interpretation of the coefficients is provided later with the GWR results.

The regression modeling approach applied to residential burglary was exploratory. This approach requires you to identify explanatory variables that result in a good OLS regression model, and a model that is a good candidate for GWR (i.e., heteroscedasticity or nonstationarity is present). The onus is then on you to interpret the results on the basis of plausible logic or some theoretical basis. An interpretation of the relationship between the distribution of residential burglary, and SAC offenders, the student population, and the Asian population follows the GWR analysis of these variables later in this chapter.

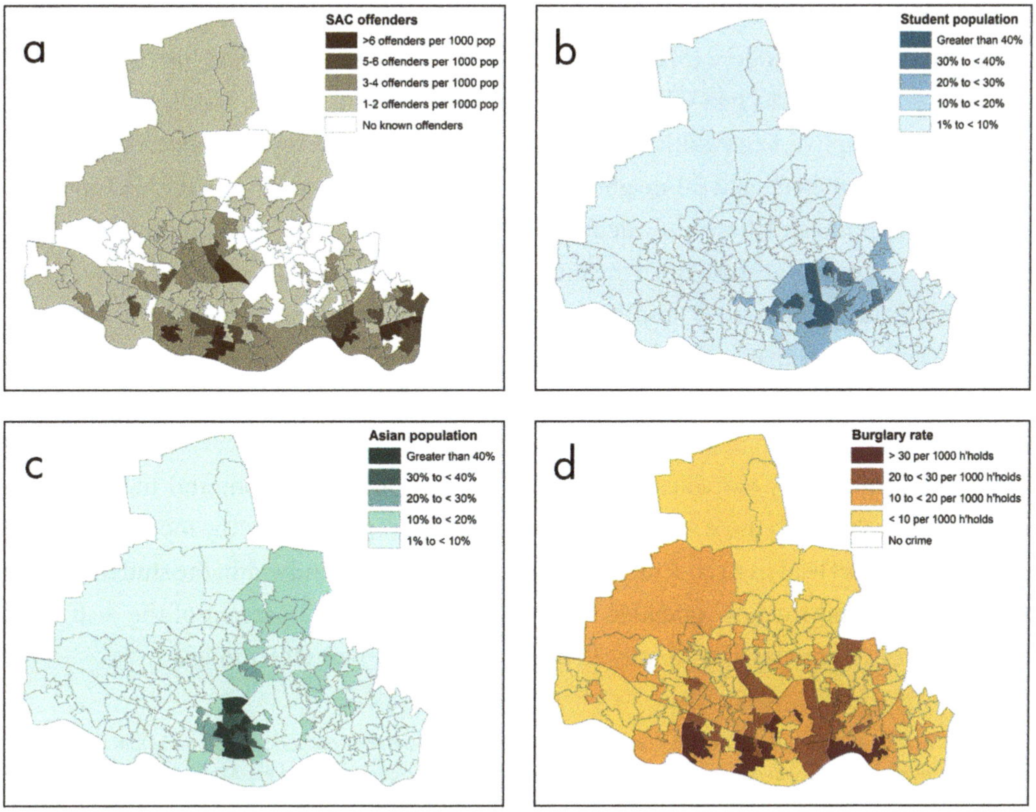

Figure 8.6. The distribution of (a) SAC offenders, (b) the student population, (c) the Asian population, and (d) the residential burglary rate in the district of Newcastle, England.

GWR modeling and spatial bandwidths

The GWR model requires the specification of a spatial bandwidth that defines the size of the neighborhood over which relationships between the dependent variable and the explanatory variables are examined. The AIC is a useful bandwidth optimization method (built into the GWR software options) (Fotheringham, Brunsdon, and Charlton 2002). In most software applications, the GWR modeling process results in the automated calculation of the bandwidth size, determined by minimizing the AIC value for that model. An alternative approach is for the user to specify a bandwidth size, with the user then evaluating the performance of this GWR model by comparing AIC values.

A fixed or adaptive bandwidth can be applied. The fixed approach uses a single metric bandwidth size for examining the relationship between variables across the whole study area. The adaptive approach uses the number of neighbors to include, therefore allowing the metric size of the bandwidth to vary across space (i.e., smaller in metric size for areas with small geographic units of analysis and larger in metric size for areas with larger geographic units of analysis).

GWR analysis

The three variables from model 8 (SAC offenders, the student population, and the Asian population) then were used as inputs to two GWR models: one that applied a fixed bandwidth and a second that applied an adaptive bandwidth to the spatial regression. The optimal fixed bandwidth was calculated as 2,422 meters, and 30 neighboring LSOAs were determined as optimal for the adaptive-bandwidth approach (these are calculated when running GWR). Calibrating and running a GWR analysis is straightforward, once the model is correctly specified.

Table 8.12 lists the adjusted R^2 and AIC values for the OLS model (model 8) and the two GWR models. The table shows that both GWR models improved on the OLS model and that the adaptive GWR model performed the best of all three—with an adjusted R^2 value of .42 and an AIC of 349. These results show that rather than a stationary relationship existing between residential burglary and the three explanatory variables (SAC offenders, student population, and the Asian population), the relationship with these variables varied spatially, as indicated by the model improvements using GWR. This can lead us to conclude that, using an adaptive bandwidth, the student population, Asian population, and the SAC offender variables could explain 42 percent of the distribution of residential burglary, compared with only 30 percent using the best OLS model.

Table 8.12. Adjusted R^2 and AIC results comparing the OLS model to fixed- and adaptive-bandwidth GWR models for residential burglary in relation to SAC offenders, the student population, and the Asian population

Type of residential burglary regression model	Adjusted R^2	AIC
OLS model (Model 8)	.30	369
GWR fixed-bandwidth model	.39	353
GWR adaptive-bandwidth model	.42	349

The GWR results from the adaptive-bandwidth model produced the best model performance results, so this model then was chosen for further analysis. Several outputs are provided from a GWR model for each geographic unit:

- Multicollinearity condition number: Recall from earlier in this chapter that if this value is greater than 30, it suggests strong multicollinearity, and the results can be unreliable. None of the geographic units recorded a condition number greater than 30. The highest was 8.3, indicating that multicollinearity between variables for each geographic unit was not a problem in the results from the GWR model.
- Local R^2 values: These values describe how well the model fits the observed values for each geographic unit. The values range from zero to one, and are reflective of the model's global R^2 value. Low values indicate where the model does not perform well, whereas higher values show where it performs best. Figure 8.7a shows the mapped local R^2 values

for the GWR adaptive model. This local R^2 map shows that in the southern and south-eastern parts of Newcastle (the more urban parts of the district), the model performed best, whereas in the more suburban and rural areas of the district, and in the central business district of Newcastle (in the central southeastern area of the district), the model did not perform as well. These results suggest that certain conditions relating to residential burglary, and its relationship with the distribution of SAC offenders, the student population, and the Asian population, differed between urban residential areas, rural parts of the district of Newcastle, and the Newcastle central business district. It also suggests, because the R^2 value was not particularly strong, a large proportion of the variance in residential burglary is not explained by the explanatory variables included in the model.

- Standardized residuals: The standardized residuals have a mean of zero and a standard deviation of one. Thematically mapping the distribution of the standardized residuals shows where the estimated value of the dependent variable (based on the relationship with the explanatory variables) has been overestimated or underestimated, relative to the standard deviation of the residuals (figure 8.7b). Mapping these residuals also can help identify other variables that could help improve where the model is currently underperforming and, by their inclusion, improve the performance of the model (i.e., identify which features are present in the areas in which the standardized residuals are greater than one standard deviation from the mean). Several other results from the GWR outputs also can be mapped. These are illustrated in the second example.

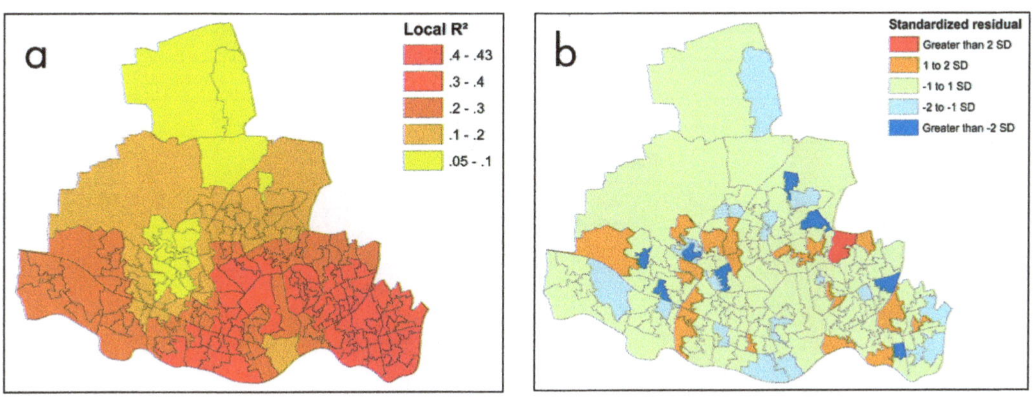

Figure 8.7. (a) Local R^2 and (b) standardized residual maps of the GWR model for residential burglary related to SAC offenders, the student population, and the Asian population in Newcastle, England.

Figure 8.8 shows the mapped coefficient values and mapped standard errors of each of the three explanatory variables. Each coefficient illustrates the spatially varying relationship between the variable and residential burglary. A first point to note from these maps is that, in some areas for each explanatory variable, the coefficients between residential burglary and the explanatory variable were negative. In each case, the OLS regression model results suggested these relationships were always positive. An example of the spatial variation in relationships is illustrated in figure 8.8c, showing a stronger positive relationship between residential burglary and the student population in the central and eastern regions of the district and that in the western region, this relationship was negative. Similarly, the maps showing the relationship between the Asian population and residential burglary (figure 8.8e), and SAC offenders and residential burglary (figure 8.8a) varied spatially. These results illustrate the value of analyzing and determining whether the relationships between the dependent variable and explanatory variables vary across space rather than assuming the relationship to be stationary. The maps of standard errors (figures 8.8b, 8.8d, and 8.8f) show the areas in which there was a greater level of reliability in the results for each explanatory variable—that where these errors were small, more confidence can be placed in the relationship between the dependent variable (residential burglary) and each of the explanatory variables.

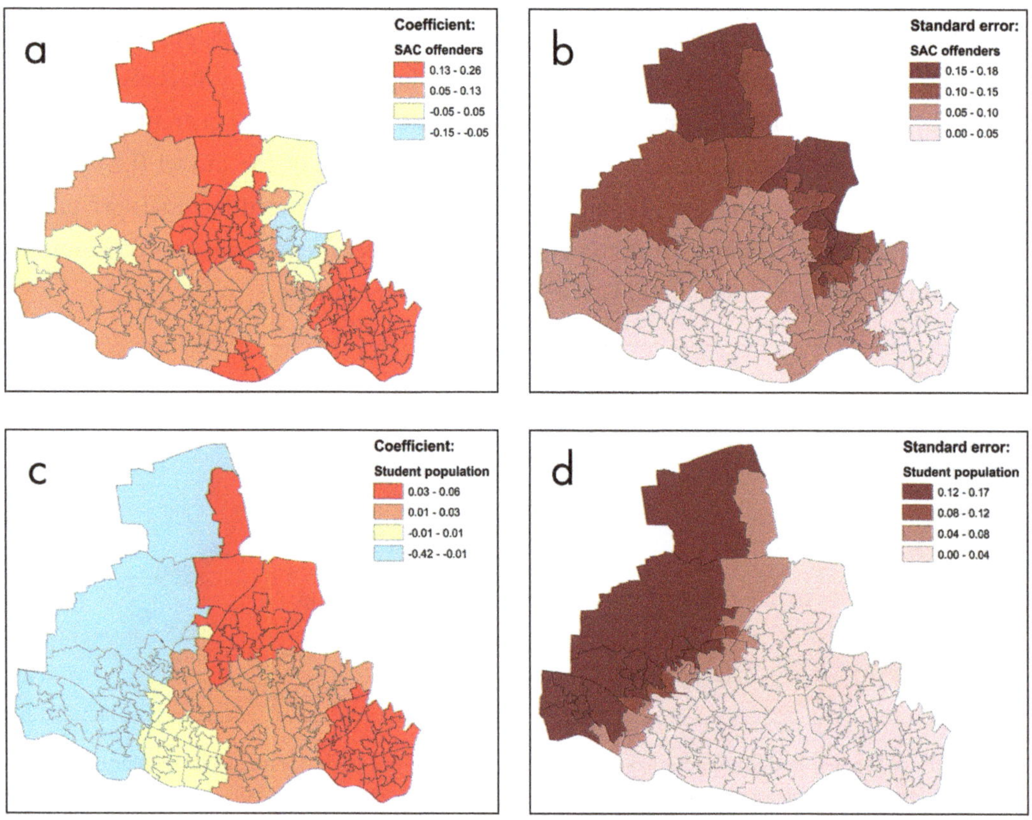

(continued)

Figure 8.8. GWR model outputs of the SAC offender rate, the student population, and the Asian population in relation to residential burglary in Newcastle, England: (a) coefficient and (b) standard error of SAC offender rate; (c) coefficient and (d) standard error of student population; (e) coefficient and (f) standard error of Asian population.

Figure 8.8. (continued)

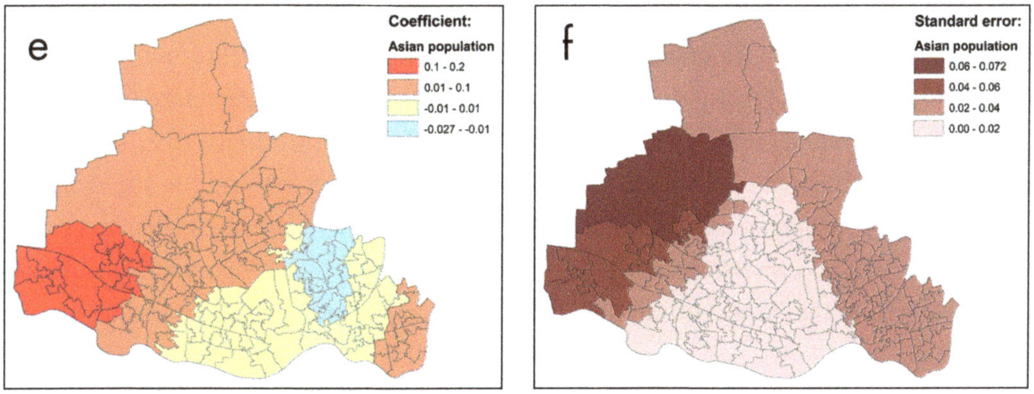

Interpreting the results

In this example of a spatial regression analysis of residential burglary, an exploratory approach identified three variables that were significantly correlated with the distribution of burglary, and that the relationships between these explanatory variables and burglary varied spatially. The exploratory approach to regression analysis then requires an attempt to interpret the results on the basis of plausible logic or some theoretical basis. Recall that, in the introduction of this chapter, to understand the underlying causes of crime requires knowledge of appropriate theory and an iterative interchange between analysis and theory to help make judgments on why a crime problem exists. What can help this process is to draw from previous studies that have examined the relationships between dependent variables and explanatory variables the findings that the exploratory regression analysis has exposed.

The positive relationship between SAC offenders and the residential burglary rate is supported by research into offender journey to crime patterns—offenders who commit burglary are likely not to travel far and to live locally to the locations in which they commit these offenses (Rossmo 2000 contains an extensive review of the journey to crime literature; journey to crime also is discussed in chapter 2). The positive relationship between the distribution of SAC offenders and the burglary rate, though, did vary, with the relationship being stronger in some areas than it was in others. In one area, the relationship between SAC offenders and burglary levels was negative (in the central eastern part of the district). This finding could be a result of local offenders who reside in the district of Newcastle not offending in this area or could be associated with offenders who live close to this area but over the border in the neighboring district of North Tyneside being responsible for the levels of burglary in this area of Newcastle. Or the explanation for the level of burglary in this area could be for reasons not identified from this analysis (recall, the best-performing model explained 42 percent of the distribution of residential burglary) and remains unexplained.

The positive relationship between residential burglary and both the student population and the Asian population most likely can be explained by the higher level of vulnerability to burglary these two demographic groups often experience. For example, with students, multiple-occupancy living and the ownership of many electronic items, combined with a general relaxed attitude toward personal security, presents them as attractive targets to offenders (McCreith and Parkinson 2004). Similarly, Asian households (and note, in England, we mainly are referring to the demographic group consisting of people of Indian ethnicity) tend to be attractive targets to offenders because of the higher level of gold ownership associated with this demographic group and the likelihood that this gold is stored in the house with limited built-in security to protect these valuable possessions (Gray 2000).

These interpretations of the results from the GWR analysis then could inform policing and public safety initiatives that help reduce the positive and significant relationships between SAC offenders, the student population, and the Asian population relating to residential burglary in Newcastle. Additionally, the GWR analysis has identified variables that appear to have a spatially varying influence on residential burglary levels across Newcastle. These findings suggest the results from a GWR analysis could inform where to specifically target crime prevention interventions. We will consider the practical use of these results after considering a final example.

Example: GWR regression of violent assaults

OLS model specification and results

In this example, we illustrate the use of GWR on a highly spatially clustered variable—violent assaults in the district of Newcastle. The use of GWR for modeling assaults followed a hypothesis-driven approach. It involved selecting explanatory variables determined to explain theoretically the variation in assault levels. Much evidence suggests that the distribution of assaults is influenced heavily by the nighttime economy (NTE) (Maguire and Hopkins 2003; Babor et al. 2003; Graham and Homel 2008), with which pubs, bars, and nightclubs are highly associated. Figure 8.9a shows the distribution of assaults in Newcastle (using KDE), and figure 8.9b shows the distribution of licensed premises (places with a license to sell alcohol), with both maps showing a high degree of clustering of these two variables in Newcastle city center. In most situations in which we want to identify factors that explain the presence of a phenomenon, we would consider using more than one variable. In the first instance, only licensed-premises data was included in the OLS regression model as the single explanatory variable to examine in relation to assaults.

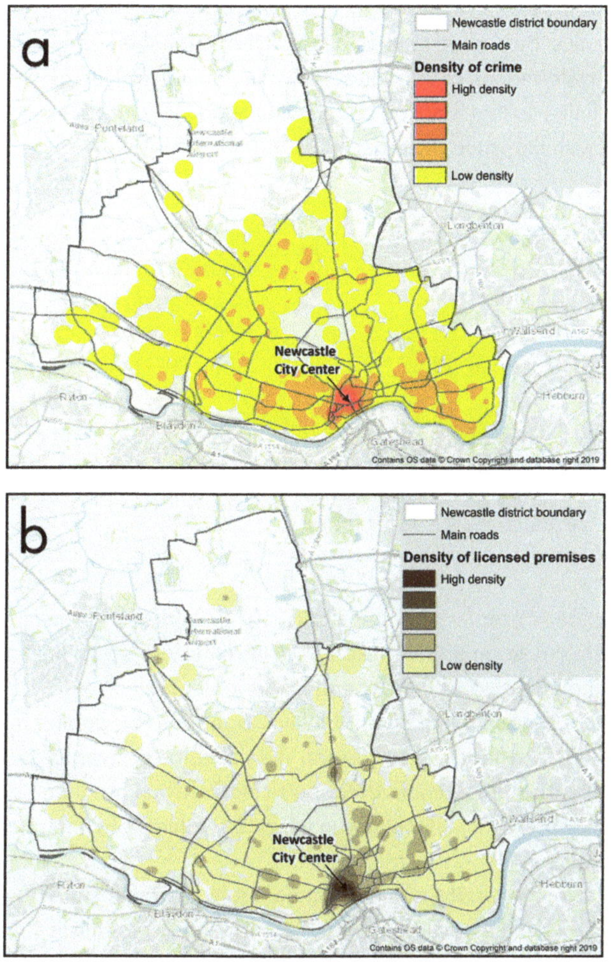

Figure 8.9. KDE maps showing (a) hot spots of assaults, and (b) the distribution of licensed premises in Newcastle, England.

Poisson distribution, overdispersion, and Poisson regression analysis

Crime data, and data for many other variables, that we want to include in spatial regression analysis may exist originally as point data. To conduct spatial regression analysis, we must aggregate this point data to geographic units (e.g., calculate a count of crimes for each census area). When analyzing patterns of crime, we often want to be spatially precise, and hence will err toward using small geographic units rather than large ones. What often happens as a result of using small geographic units is that our data contains many geographic units with zero counts (i.e., in which no crime has occurred), and relatively few geographic units in which crime has occurred. This then means that our data is likely to follow a Poisson distribution rather than a normal distribution (figure 8.10). Although OLS regression analysis does not require our data to be normally distributed, data that is Poisson distributed can create challenges in regression analysis that may need to be addressed. If the data we use is highly skewed, it may result in the residuals not being normally distributed and being spatially autocorrelated (two tests that must be passed to have a model that is reliable). One option you might consider is the log transformation of variables. However, the natural log of zero cannot be defined, or rather, is infinity, so that is not an option when working with count data that contains many observations with the value of zero.

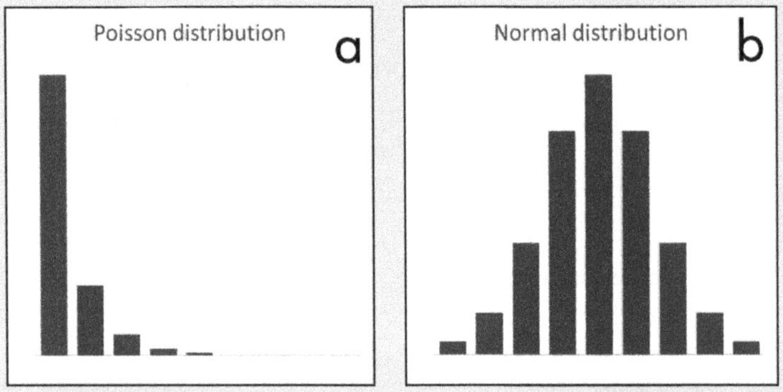

Figure 8.10. (a) A Poisson distribution and (b) a normal distribution.

In a distribution in which the variance is greater than the mean, overdispersion is present. The presence of overdispersion can cause an underestimation of the standard errors in the parameter estimates (i.e., the estimated value of the dependent variable, based on its relationships with explanatory variables), meaning that a variable may appear to be significantly correlated to the dependent variable when, in fact, it is not. The presence of overdispersion is a common feature in spatially distributed crime data, particularly for crimes that are highly clustered, because the data displays some locations in which the volume of crime is high and many places in which it is zero.

To limit overdispersion, one option is to convert the crime count data to some other statistic, for which fewer geographic units have values equal to zero. A good example of this is using the Gi* statistic (see chapter 3), which populates each cell with a positive or negative value representing the spatial distribution of crime, and hence avoiding the issue of many cells having zero counts. A Poisson distribution remains, but overdispersion may have been addressed. An option then is to use a Poisson spatial regression analysis technique. There is a problem, though. Poisson regression analysis cannot be applied to variables that have negative values. A log transformation of Gi* statistic values would not work either because log transformation of negative values is not possible. Another option would be to add a constant to each value before applying the log transformation (e.g., the value of 1) to change the negative values to positive values. This is referred to as a "zero-inflated model." However, for the analysis of crime data, it is important to distinguish areas in which there has been no crime from those in which there has been crime, particularly for rare crimes (e.g., sexual assaults). Some zero-inflated models specify separating the analysis of the cells that originally contained a value of zero from those that contained a value greater than zero. However, the analysis and interpretation of spatial data that is split in this way can be cumbersome without specialized software for treating zero-inflated models.

Another option is to aggregate crime data (and any other data in which overdispersion is present) to larger geographic units, and hence minimize the chance that many of these units will contain a zero value. This solution, however, goes against our desire to examine spatially precise relationships between crime and other variables.

If we managed to sort our data, but the data still followed a Poisson distribution, we then might consider again the option of using a Poisson regression analysis technique such as Poisson GWR, which is available in some specialist GWR software. But there is also a problem that particularly relates to crime data. Poisson regression assumes that the occurrence of an event (e.g., crime) will not make another event more or less likely. However, codependence among events is a common observation with crime. We have seen this pattern already in previous chapters, evident in repeat victimization, risky facilities, and near repeat victimization, and with the codependence among events explained explicitly in relation to boost account and optimal foraging theories.

Although Poisson-based models have been developed (with examples of its use including disease mapping—see Nakaya et al. 2005), to date, little research has been conducted that identifies the specific treatment required for an effective Poisson spatial regression analysis of crime data. Poisson spatial regression is challenging. If you think the issues it presents, and the assumptions it makes, pose challenges in the use of your data, I would recommend avoiding Poisson spatial regression techniques until more is published on their practical use with crime data. If you have a serious problem with overdispersion, and a natural log transformation does not address the issues that skewness presents in your regression models, often the only option is to use larger geographic units for your analysis.

The first model (Model 1 OLS Assaults) that used only licensed premises as an explanatory variable found this variable to be positively and significantly associated with assaults (table 8.13). The model, albeit simple, generated a strong adjusted R^2 value of .956 and was statistically significant (based on the joint Wald statistic result). The Koenker statistic of heteroscedasticity or nonstationarity was significant, suggesting the model was a good candidate for GWR, but the Jarque-Bera statistic was also significant, suggesting the model was biased.

A second model was calibrated (Model 2 OLS Assaults) to include the student population, income deprivation, and living conditions deprivation, alongside licensed premises, as explanatory variables on the hypothesis that each of these three new variables contributed to patterns of assaults across Newcastle—students, through their active engagement with the NTE, income deprivation because several studies have identified a link between this variable and assaults (e.g., Ching-Chi and Pugh 1993), and living conditions because of research findings linking this variable to assaults (e.g., Roncek 1981). Licensed premises remained statistically significant, and of the additional variables, only income deprivation was found to be significant (and positively associated to assaults). The VIF for each was lower than 7.5. Model 2 produced a higher adjusted R^2, and a lower AIC, suggesting it was an improvement on the first model. Heteroscedasticity was present, but the model remained biased.

A third model (Model 3 OLS Assaults) included only the significant variables from model 2—licensed premises and income deprivation. Both variables were significant, with the VIF for each variable of 1.0, indicating that both variables contributed to the model, but there was no overall improvement in model performance in comparison with model 2. Heteroscedasticity or nonstationarity was present, but the model remained biased.

Model 4 (Model 4 OLS Assaults) involved repeating model 3 but using a log transformation of the assault data. Licensed premises and income deprivation remained significant, heteroscedasticity or nonstationarity was present, and the Jarque-Bera statistic was not significant. The adjusted R^2 decreased to .52, but it was because a log-transformed version of the assaults data was used and is not comparable to the adjusted R^2 in previous models when the dependent variable was not transformed. The AIC decreased from 1,272 in model 3 to 370.6 in model 4, but again, these are not comparable because of the use of the log-transformed dependent variable. An analysis of the residuals from model 4 using Moran's I determined they were not clustered.

Table 8.13. Main results from OLS models of assaults in Newcastle, identifying explanatory variables that were significant, model performance, significance, heteroscedasticity, and bias

Assault models	Model variables and coefficients	Adjusted R^2	AIC	Koenker statistic	Joint F or joint Wald statistic	Jarque-Bera
Model 1 OLS	Licensed premises (1.76***)	.956	1,297.8	33.13***	327.90***	1143.0***
Model 2 OLS	Licensed premises (1.79***) Student population (−0.18) Income deprivation (22.75***) Living conditions deprivation (−0.11))	.964	1,262.1	38.78***	580.66***	904.43***
Model 3 OLS	Licensed premises (1.76***) Income deprivation (26.67***)	.961	1,272.0	35.39***	456.55***	1439.90***
Model 4 OLS (NL of assaults)	Licensed premises (0.018***) Income deprivation (3.90***)	.521	370.6	9.56**	212.18***	1.61
Model 5 OLS (NL of assaults)	NTE licensed premises (0.043***) Income deprivation (3.86***)	.489	381.7	7.63*	239.36***	6.28*
Model 6 OLS (NL of assaults)	NL of NTE licensed premises (0.81***) Income deprivation (3.80***)	.514	373.2	8.80*	228.51***	0.08

Note: NL = natural log transformation.
* *p* < .05; ** *p* < 0.01; *** *p* < 0.001

GWR candidate model

Even though model 4 was now a suitable candidate for GWR, another model was calibrated, including income deprivation and only those licensed premises that were associated with the NTE (Model 5 OLS Assaults). This subset of licensed premises included only bars, pubs, and nightclubs rather than shops and off-licenses (i.e., equivalent to liquor stores) that sold alcohol because it was hypothesized that this subset was more likely to be associated with assaults in Newcastle. From model 5, the two explanatory variables were significant. However, the model did not improve in its performance, and bias was reintroduced into the model. Model 5 was repeated by performing a natural log transformation on the highly skewed NTE licensed premises data (Model 6 OLS Assaults). The explanatory variables remained significant, heteroscedasticity or nonstationarity was present, and the model was not biased. However, the adjusted R^2 from model 6 decreased slightly and the AIC increased, so it was decided to use model 4 for GWR analysis.

GWR analysis

Table 8.14 lists the GWR model performance results (using both a fixed and adaptive bandwidth) of assaults in relation to licensed premises and income deprivation. A bandwidth of 3,242 meters was calculated as optimal for the fixed GWR model, and a bandwidth of 55 neighbors was calculated as optimal for the adaptive GWR model. The results show that both GWR models improved the performance of the OLS model, but the adaptive GWR model performed marginally better with an adjusted R^2 of .57 and a lower AIC of 363.

Table 8.14. Adjusted R^2 and AIC results comparing the OLS model for assaults against licensed premises and income deprivation, in relation to fixed- and adaptive-bandwidth GWR models

OLS adjusted R^2	OLS AIC	GWR fixed adjusted R^2	GWR fixed AIC	GWR adaptive adjusted R^2	GWR adaptive AIC
.52	371	.55	364	.57	363

Figure 8.11 shows outputs from the adaptive GWR model. Figure 8.11a shows the relationship between assaults, licensed premises, and income deprivation varied spatially, with this relationship strongest in and around the city center. The standardized-residuals map (figure 8.11b) shows where the model performed best and where it did not perform as well. The coefficient map of licensed premises (figure 8.11c) shows how the relationship with assaults varied across Newcastle, with this relationship being highly positive (i.e., as the number of licensed premises increased, so too did assaults). However, in one area, to the west of Lemington, this relationship was negative. The supporting standard error map of the licensed-premises coefficients (figure 8.11d) shows the results were most reliable in the south central and southeastern areas of the district. Figure 8.11e shows the relationship between assaults and income deprivation also varied spatially, with all areas exhibiting a positive relationship between these two variables (i.e., as income deprivation increased, so too did assaults), with the relationship strongest in the Lemington neighborhood and south of Gosforth. The supporting standard error map of the income deprivation coefficients (figure 8.11f) shows the results were most reliable in the south central and southeastern areas of the district. Other outputs that some software applications of GWR produce include t-statistics that can indicate spatial differences in the statistical significance of each relationship.

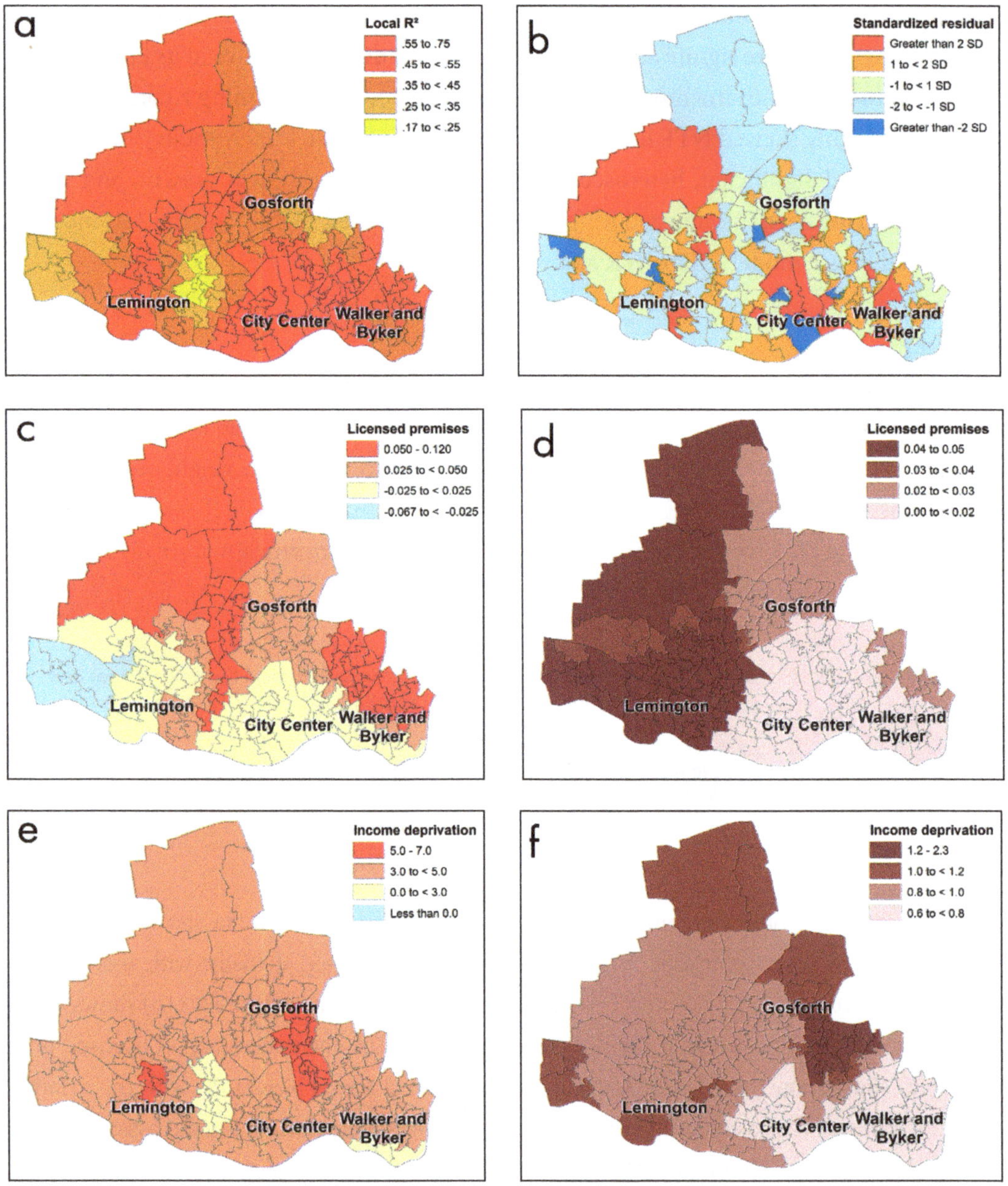

Figure 8.11. GWR model outputs of (a) local R^2 and (b) standardized residuals for assaults related to licensed premises and income deprivation in Newcastle, England; (c) coefficient and (d) standard error values of licensed premises; and (e) coefficient and (f) standard error values of income deprivation.

The results of the OLS regression analyses and GWR modeling for assaults illustrate the value in examining whether relationships between dependent and explanatory variables vary across space. In this example, a hypothesis-driven approach was followed, using an iterative process to include explanatory variables believed to be related to violent assaults, on the basis of sound theory and empirical evidence. Importantly, GWR results show that the relationships between assaults, licensed premises, and income deprivation are not the same in all areas, and instead have a spatially varying influence on assault levels across Newcastle. How to further interpret these results and how they can be used in practice are described in the following section.

Spatial regression analysis applied to spatially precise hot spots

Figure 8.12 shows several crime hot spots and their respective census areas. The figure illustrates that many of the hot spots, and the individual grid cells used for representing the distribution of crime, are much smaller than the geographic units for which most explanatory variables are available. Figure 8.12 also shows that many of the hot spots have a spatial coverage that overlaps several census areas, meaning that any analysis that examined the relationship between the crime hot spots and the explanatory variables would relate to data for the whole of each census area that overlapped the hot spots rather than the specific areas identified as hot spots. A comparison among the specific causes of these spatially precise crime hot spots would not be possible because of this mismatch between the spatial scales of the geographic units for which the dependent variable (e.g., crime) and the explanatory variables relate.

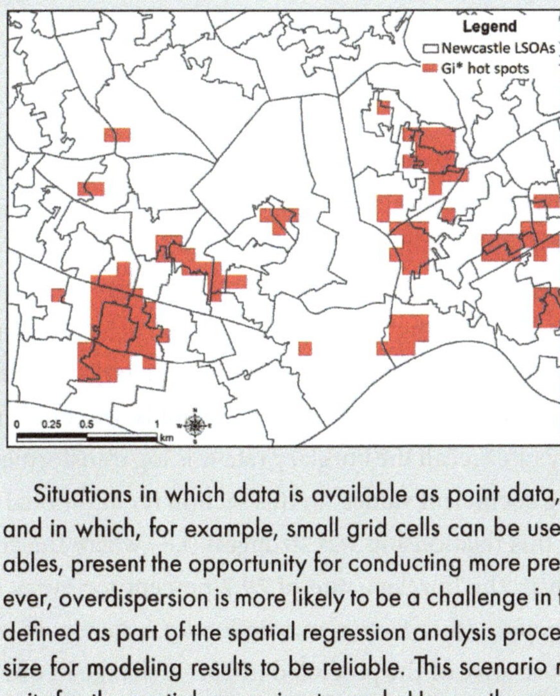

Figure 8.12. Gi* crime hot spots compared with their respective census areas.

Situations in which data is available as point data, as is often the case with crime data, and in which, for example, small grid cells can be used as the aggregating unit for all variables, present the opportunity for conducting more precise spatial regression analysis. However, overdispersion is more likely to be a challenge in the analysis. The neighborhood areas defined as part of the spatial regression analysis process also would need to be of sufficient size for modeling results to be reliable. This scenario may require using larger geographic units for the spatial regression to work. Hence, the spatial regression analysis results would offer a less spatially specific interpretation of the relationship between crime patterns and explanatory variables than the spatial regression analysis determining explanatory differences between individual hot spots when these hot spots are spatially precise.

Interpreting and using GWR results

The parameter estimates and coefficients determined from GWR modeling results are designed to indicate the presence of relationships. The coefficients also can illustrate (rather than offer exact forecasts) potential changes in crime related to changes in explanatory variables. Identifying variables that statistically correlate and spatially vary with patterns of crime provides a useful means of considering options for targeting strategic crime reduction activity, as well as predicting the impact of this activity.

Residential burglary example

An interpretation of the GWR results from the residential burglary example in this chapter (albeit with some caution for the exact extent of the statistical inference results) could inform policy and a strategic reduction plan for burglary. The OLS and GWR analysis of residential burglary (model 8, using a log-transformed version of residential burglary) identified SAC offenders (coefficient = .156), the student population (coefficient = .024), and the Asian population (coefficient = .014) as significant explanatory variables. An adaptive GWR model of these variables resulted in an adjusted R^2 value of .42, with figures 8.7 and 8.8 showing how the relationship between residential burglary and these explanatory variables varied across Newcastle. GWR results can help identify where to target crime reduction programs to address the significantly observed (and reliable) relationship between the dependent variable and explanatory variables. For example, targeted activity toward offenders of burglary (and offenders of other serious acquisitive crimes), which would include their intensive supervision, disruption of criminal activity, reduction in repeat offending, and their removal (e.g., through the serving of custodial sentences), potentially could result in burglary reductions, particularly in neighborhoods in which the relationship between burglary and the distribution of SAC offenders was found to be strongest. For example, using the coefficient of .156 determined from the OLS model (model 8) as an indication of the impact of this type of offender-targeted activity, for every one-unit reduction in the number of SAC offenders per 1,000 households, we could anticipate a reduction of 16.9 percent in the burglary rate (recall the burglary rate was log transformed for the OLS and GWR analysis—see "Interpreting coefficient values" in this section for additional information). In the areas of Newcastle in which this relationship was strongest (i.e., where the coefficient was .26), we could anticipate a reduction in the burglary rate of 29.7 percent for every one-unit reduction in the SAC offenders rate. Although other factors may influence a change in offending behavior, the GWR analysis results are illustrative of the predicted impact a change in one variable may have on changes in crime.

The coefficients for student population and the Asian population can be used in the same way. Using the coefficient of .024 calculated for the student population from the OLS model (model 8), for every one-unit reduction in the proportion of the student population, we could anticipate a reduction of 2.4 percent in the burglary rate. Newcastle has two large universities, and it would be unethical (and outrageous) to think that the local police or local government would want to force a reduction in the number of students living in Newcastle to reduce burglary. The same principles also apply to the Asian population. The results from the GWR analysis showed the relationships between these variables and residential burglary spatially varied. In turn, these results then can help inform the direction of strategic crime reduction policy through the design, development, and spatial targeting of initiatives to reduce the impact these variables have on crime. In statistical terms, this policy would involve aiming to reduce the significant and spatially varying influence these variables have on crime. For example, in the areas in which the relationship between residential burglary and the student population was strongest (i.e., in the areas in which the coefficient values were highest, significant, and reliable), a strategic initiative could be targeted to these areas to reduce this high level of student vulnerability to residential burglary. Additionally, in an evaluation sense, once the initiative was implemented, a way to monitor its impact would be to measure the changes in the coefficient. A reduction in the coefficient value between residential burglary and the student population from an OLS and GWR analysis, both overall but more specifically in the area in which the initiative was targeted, could indicate the initiative had an impact.

Interpreting coefficient values (and when variables have been log transformed)

A coefficient value represents the change in the value of the dependent variable, corresponding to a unit change in the value of the explanatory variable. For example, if we have a coefficient of 1.11 for an explanatory variable, for every one-unit change in the explanatory variable, the dependent variable increases by 1.11. If the coefficient is negative, (e.g., −1.11), for every one-unit change in the explanatory variable, the dependent variable would decrease by the amount equivalent to the coefficient (i.e., decrease by 1.11).

If variables are log transformed, the interpretation of coefficients is different:

- If the dependent variable is transformed using a natural log approach, we perform this calculation: (exp(coefficient value) − 1) × 100. For example, we have a coefficient for an explanatory variable of 1.11, so we exponentiate this value (exp(1.11) = 3.034), subtract 1, and multiply by 100, which equals 203.4. We then can state that for every one-unit increase in the explanatory variable, the dependent variable increases by about 203 percent.
- If an explanatory variable is transformed using a natural log approach, we perform the simple calculation of dividing the coefficient by 100. For example, if we have a coefficient for an explanatory variable of 1.11, we divide it by 100, which equals 0.011. For every 1 percent increase in the explanatory variable, the dependent variable increases by about 0.011; for every 10 percent increase in the explanatory variable, the dependent variable increases by about 0.11.
- If both the dependent and explanatory variables have been log transformed, we perform this calculation: for an x percent increase in the explanatory variable, we

calculate 1.x to the power of the coefficient value, subtract 1, and multiply by 100 to determine the percent increase in the dependent variable. For example, we have a coefficient for an explanatory variable of 1.11. For a 1 percent increase in the explanatory variable, we perform this calculation ((1.01 to the power of 1.11) − 1) × 100 = 1.1 percent; for every 1 percent increase in the explanatory variable, the dependent variable increases by about 1.1 percent. For a 10 percent increase in the explanatory variable, we perform this calculation ((1.10 to the power of 1.11) − 1) × 100 = 11.2 percent; for every 10 percent increase in the explanatory variable, the dependent variable increases by about 11.2 percent.

Violent assaults example

The best-performing model that examined the relationship between assaults and explanatory variables included licensed premises and income deprivation. Recall that the assaults data was log transformed, and the licensed-premises and income deprivation data remained in their original format. Income deprivation index values for Newcastle ranged between 0.01 and 0.61, with higher values representing a higher level of income deprivation. Using the results from the OLS model (Model 4 OLS Assaults), a strategic program aimed at addressing income deprivation that resulted in a reduction in the income deprivation score of 0.1 could yield a reduction in assaults on the order of 4.3 percent (taken from the coefficient of 3.9 in table 8.13). The income deprivation coefficient for this relationship with assaults was highest in the Lemington and Gosforth areas, in which coefficient values in some areas were between 5.0 and 7.0 (figure 8.11e). These results suggest that a reduction in income deprivation in the Lemington and Gosforth areas could have a greater effect on reducing the number of violent assaults. This presents an opportunity, therefore, for a targeted intervention rather than an intervention covering the whole district. Similarly, the results from the OLS regression indicated that a one-unit reduction in the number of licensed premises could yield a 1.8 percent reduction in assaults (taken from the coefficient of 0.018 in table 8.13). However, this relationship with assaults was much higher in certain parts of the district, with coefficient values in some areas of 0.1 on average (figure 8.11c). In these areas, a one-unit reduction in the number of licensed premises could yield a 10.5 percent reduction in assaults. Conversely, if the number of licensed premises in Newcastle increased, we would expect an increase in violent assaults, with this increase in assaults being greatest in the areas in which the coefficient value was highest.

Reducing the number of licensed premises could be problematic, so an alternative crime reduction solution often used instead is to implement stricter licensing conditions on pubs, bars, nightclubs, and off-licenses in crime hot spots. A strategic policy that implemented stricter licensing conditions on licensed premises (such as police and community safety officials working more closely with managers of these premises to reduce violent assaults) could prove fruitful in reducing the strong relationship between assaults and licensed premises. A target for this strategic policy would be to reduce assaults in the areas in which the relationship was observed to be strongest, so these areas did not stand out as being any more problematic than other areas of the district (e.g., reducing the average coefficient value for the areas in which this relationship was strongest from 0.1 to some new, lower target). In practice, this also could involve a more detailed analysis of each licensed premise to identify which ones experienced (or were associated with) the highest number of assaults, with focused policing and community safety activity targeted toward these businesses (i.e., the risky-facilities analysis described in chapter 5). This more detailed analysis of licensed premises also could examine whether the type and size of venue and the volume of customers during open hours also had an influence on the levels of assaults.

GWR results for inference

Critiques of GWR in other fields of science have expressed concerns over the extent of using GWR results for inference. However, there is a danger in statistical modeling of failing to account for the balance required between the technical specification of a technique and its practical application. In practice, the purity of a statistical modeling process must be balanced by specifying the model to suit the data, and the context in which this data is applied. Through the diligence of correctly specifying a model (without compromising the model's technical integrity), balanced with the constraints that data may place on the modeling process, GWR modeling can offer to policing and public safety practice.

The use of GWR for policing and public safety applications may include not only identifying variables that have a spatially varying relationship with crime but also could inform strategic predictions of crime. For example, if the results from a GWR analysis indicate evidence of a relationship between the crime issue being studied (e.g., assaults) and several explanatory variables (e.g., licensed premises and income deprivation), these results could help inform policy direction for long-term crime reduction strategies and offer a predictive indication of how crime may change as a result.

What-if scenarios

The interpretations described help illustrate the what-if scenarios that can be considered from the results of spatial regression analysis. Such as, what if we reduced the number of licensed premises? How would this have an impact on violent crime? And what if we improved our program for supervising offenders released from prison? How might this help reduce burglary? And what if we prioritized these improvements to specific areas? Would this have a better cost-benefit effect?

Correlation and causation

Correlation, of course, does not equal causation. We must ensure that any modeled relationship, and how it spatially varies, can be interpreted on sound theoretical grounds. For example, the journey to crime research literature shows that offenders tend to travel short distances to commit crime, particularly crimes such as burglary. Therefore, a high rate of burglars in an area is likely to result in a high level of burglary in that area. There is often also the need to examine any identified relationships further to interpret accurately what the patterns infer. For example, the presence of a high concentration of licensed premises in an area alone may not necessarily fully explain the high levels of violent assaults in this area. Instead, the high levels of violent assaults could be due to a small number of particularly problematic premises being the main cause of the problem, among a high density of licensed premises. This, therefore, indicates the importance of additional analysis examining and further explaining correlations between explanatory variables and patterns of crime revealed from a GWR analysis before policy decisions are made.

Spatial regression can offer many useful insights that help us better understand crime than is possible from other techniques that examine only crime data, but it is typically a combination of these techniques that must be applied for us to understand crime more fully.

Spatial lag and spatial error models of burglary and assaults

We now will revisit the spatial lag and spatial error models, using the explanatory variables that the recursive OLS modeling processes identified as good candidates for GWR. In each case, each model used the queen's case contiguity weights matrix previously used. For purposes of comparison, spatial lag and spatial error models both were applied to the analysis of residential burglaries and assaults in Newcastle. We also compare the results from these models to the GWR results.

Residential burglary example

The previous OLS regression analysis of residential burglary in Newcastle identified SAC offenders, the student population, and the Asian population as significant explanatory variables (Model 8 OLS Burglary). This model used a natural log transformation of the residential burglary rate. The model was calibrated and run in GeoDa to generate Lagrange multiplier statistics. These results showed that a spatial lag model rather than a spatial error model would be the most suitable spatial regression model of the two to calibrate (table 8.15). Both types of spatial regression models were calibrated for comparison.

Table 8.15. Regression diagnostic results for an OLS regression analysis (Model 8 OLS Burglary) for residential burglary and several explanatory variables in Newcastle, England

	Moran's I (residuals)	LM (lag)	Robust LM (lag)	LM (error)	Robust LM (error)
Model 8 OLS Burglary	0.19***	21.58***	5.14*	16.44***	0.0029

Note: LM = Lagrange Multiplier. Model uses a queen's case contiguity weights matrix.
* $p < .05$; ** $p < .01$; *** $p < .001$

The spatial lag regression model results and the spatial error regression analysis results performed better than the OLS model, but with the spatial lag model, as expected from the Lagrange multiplier results, performing the best of the three (table 8.16). In each case, the explanatory variables were significant and broadly similar in magnitude. The spatial lag model illustrated the significant influence of the spatial autoregressive term of burglaries in the model, in that residential burglaries in each area were related to the burglary rates in their neighboring areas. The spatial lag model did not, however, perform better than the GWR model (indicated by the better R^2 and AIC values in the GWR model), suggesting the GWR model results would be the better ones to use. Figure 8.13 shows the parameter estimates from the spatial lag model and the GWR model, alongside the maps of the burglary rate and, for better comparison, the natural log values of the burglary rate used as the dependent variable in the spatial lag and GWR models. The GWR map of parameter estimates shows the overall better performance in the estimation of parameters and lower levels of residuals in comparison with the spatial lag maps.

Table 8.16. Main results from OLS, GWR, spatial lag, and spatial error models of residential burglaries in Newcastle

Residential burglary models	Model variables and coefficients	R^2; adjusted R^2	AIC	Heteroscedasticity: Koenker; Breusch-Pagan	Nonnormality of errors Jarque-Bera	Likelihood Ratio Test
Model 8 OLS	SAC offenders (0.156***) Student population (0.024***) Asian population (0.014**)	.31; .30	368.7	10.43*; 6.5*	5.84	n/a
Model 8 GWR	SAC offenders (0.156***) Student population (0.024***) Asian population (0.014**)	.51; .42	348.6	Not applicable as specified from OLS model		n/a
Model 8 SL	Weighted burglary rate (0.41***) SAC offenders (0.12***) Student population (0.018***) Asian population (0.0086*)	.40: n/a	349.4	n/a; 7.83	n/a	18.99***
Model 8 SE	Lambda (0.41***) SAC offenders (0.13***) Student population (0.021***) Asian population (0.013**)	.39	352.0	n/a; 7.54	n/a	14.48***

* $p < .05$; ** $p < .01$; *** $p < .001$

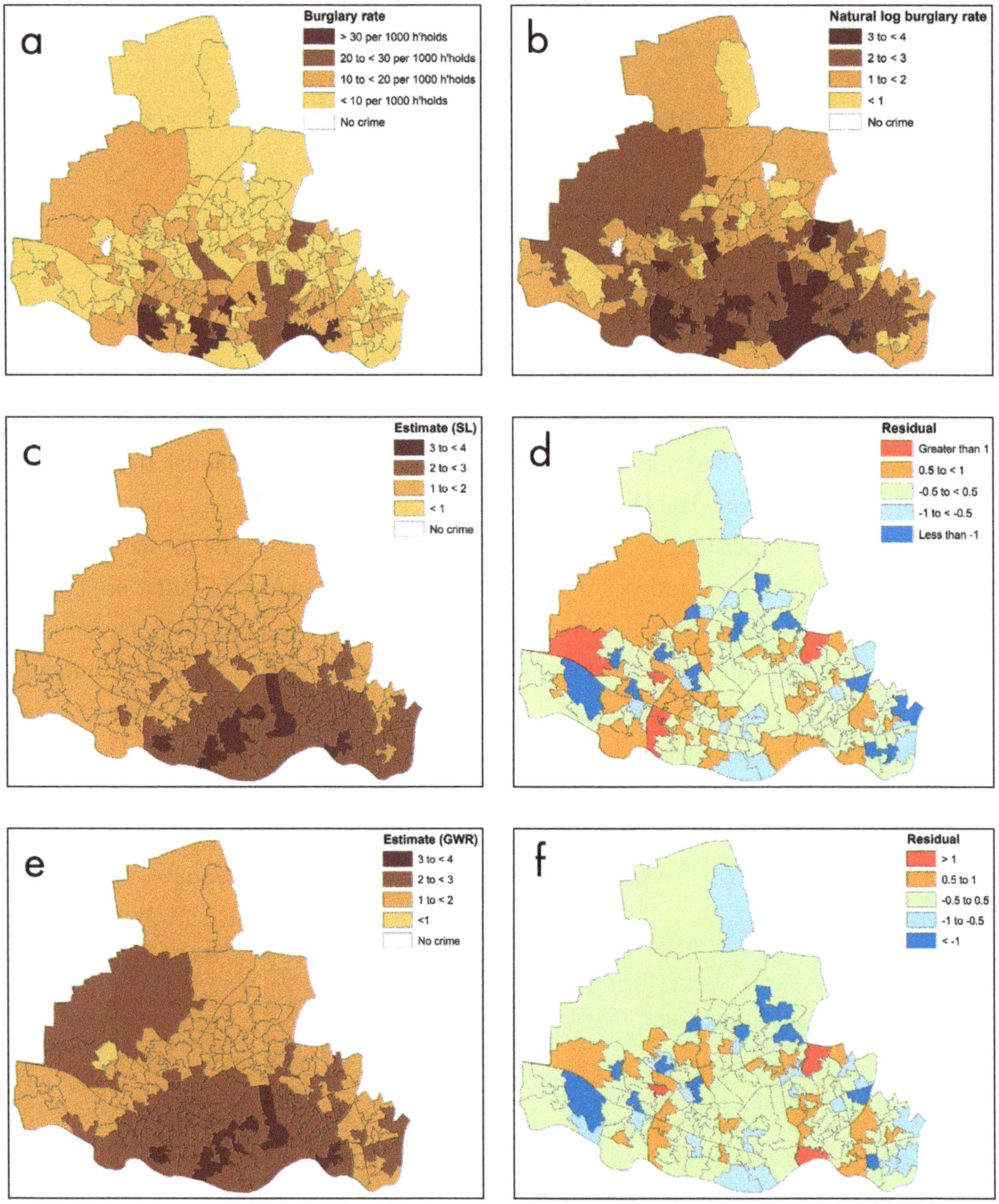

Figure 8.13. (a) Residential burglary rate, (b) natural log transformation of residential burglary rate, (c) spatial lag regression model parameter estimates of residential burglary, (d) residuals of spatial lag regression model, (e) GWR parameter estimates of residential burglary, and (f) residuals of GWR model.

Violent assaults example

The previous OLS regression analysis of assaults identified licensed premises and income deprivation as significant explanatory variables (Model 4 OLS Assaults). This model used a natural log transformation of the original assault variable. The model was calibrated and run in GeoDa to generate Lagrange multiplier statistics. These results showed that a spatial lag model rather than a spatial error model would be the more suitable spatial regression model of the two to calibrate (table 8.17). Both types of spatial regression models were calibrated for comparison.

Table 8.17. Regression diagnostic results for an OLS regression analysis for assaults and several explanatory variables in Newcastle, England

	Moran's I (residuals)	LM (lag)	Robust LM (lag)	LM (error)	Robust LM (error)
Model 4 OLS Assaults	0.12**	19.05***	13.57***	6.87**	1.39

Note: LM = Lagrange multiplier. Model uses a queen's case contiguity weights matrix.
* $p < .05$; ** $p < .01$; *** $p < .001$

The spatial lag regression model results and the spatial error regression analysis results performed better than the OLS model, but with the spatial lag model, as expected from the Lagrange multiplier results, performing better than the spatial error model (table 8.18). The explanatory variables were significant in the spatial regression models and similar in magnitude. The spatial lag model illustrated that the influence of the spatial autoregressive term of assaults in the model was highly significant, in that assaults in each area were related to the level of assaults in neighboring areas. The spatial lag model performed better than the GWR model in terms of the AIC values, but the GWR model generated a better R^2 result. The benefit of the GWR results is that they show how each explanatory variable varies spatially in its relationship with assaults. Figure 8.14 shows the parameter estimates from the spatial lag model and the GWR model, alongside the maps of the assaults and of the natural log values of the assaults used as the dependent variable in the spatial lag and GWR models. The maps of residuals show little difference between the GWR model and the spatial lag model, as both had the same number of areas in the highest thematic classes (positive and negative) of residuals.

Table 8.18. Main results from OLS, GWR, spatial lag, and spatial error models of assaults in Newcastle, England

Assault models	Model variables and coefficients	R^2; adjusted R^2	AIC	Heteroscedasticity: Koenker; Breusch-Pagan	Jarque-Bera	Likelihood ratio test
Model 4 OLS	Licensed premises (0.018***) Income deprivation (3.90***)	.53; .52	370.6	9.56**; 6.92*	1.61	n/a
Model 4 GWR	Licensed premises (0.018***) Income deprivation (3.90***)	.63; .57	363	Not applicable as specified from OLS model		n/a
Model 4 SL	Weighted Assaults (0.358***) Licensed premises (0.016***) Income deprivation (2.95***)	.58; n/a	353.0	n/a; 6.23*	n/a	17.37***
Model 4 SE	Lambda (0.332**) Licensed premises (0.016***) Income deprivation (3.536***)	.55; n/a	361.0	n/a; 5.89	n/a	7.37**

* p < .05; ** p < .01; *** p < .001

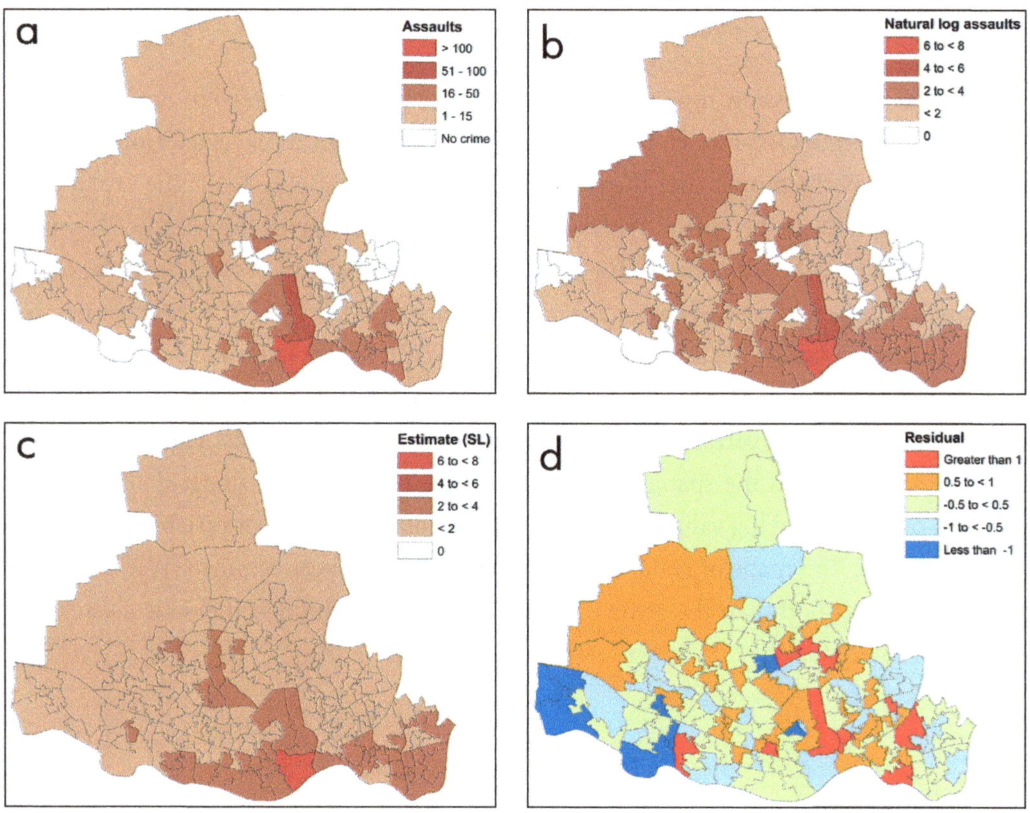

(continued)

Figure 8.14. (a) Assaults, (b) natural log transformation of assaults, (c) spatial lag regression model parameter estimates of assaults, (d) residuals of spatial lag regression model, (e) GWR parameter estimates of assaults, and (f) residuals of GWR model.

Figure 8.14. (continued)

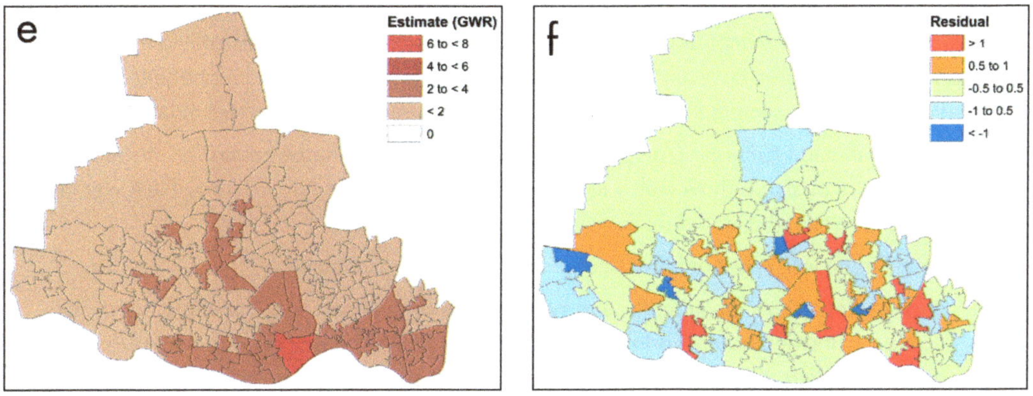

Chapter summary

Spatial regression can be challenging, but with some patience and diligence, it can generate useful results for understanding crime. Spatial regression analysis does not involve a single process, generating a single result. Instead it involves a set of statistical processes from which the results are used collectively to interpret the relationships being examined.

In this chapter, the focus has been on four regression techniques: OLS, as a foundation for exploring relationships between variables, and from which diagnostic statistics can be generated for informing a spatial regression analysis; a spatial lag model, which considers the impact of the dependent variable in neighboring areas in the model alongside the influence of other explanatory variables; a spatial error model, which gives an indication that the clustering observed in the dependent variable reflects the influence of unmeasured explanatory variables; and GWR, which performs local regression equations on each geographic unit in a study area. There are many other spatial regression techniques, some of which include combining the examination of spatial lag and spatial error effects. The descriptions and illustrations of the techniques in this chapter should provide you with a good foundation, and the confidence, to replicate and experiment with the techniques included and to experiment with other spatial regression techniques.

Descriptions of statistical measures, tests, and terms

What follows is an attempt at plain English descriptions of common statistical measures and terms to help interpret the results from regression analysis. They are listed alphabetically. The "Further Resources" section that follows provides links to other useful sources of information.

Akaike Information Criterion (AIC): A measure that compares the performance of statistical models. It evaluates how well a model fits the data it was generated from. The lower the AIC, the better the model (considering model complexity).

Breusch-Pagan test: A test of heteroscedasticity. See *heteroscedasticity*.

coefficient: Represents the strength and type of relationship between the explanatory variable and the dependent variable. It is the change in the value of the dependent variable corresponding to a unit change in the value of the explanatory variable. For instance, for every one-unit change in the explanatory variable, the dependent variable changes (increases if positive, or decreases if negative) by the amount equivalent to the coefficient.

dependent variable: See variable.

error: Refers in statistical terms to the difference between the observed value and its "true" value. Its true value is the mean from the whole population. For example, if we have a sample of 100 census areas and we state that the average is representative of crime levels for each census area, and the average level of crime for this sample is 10 crimes per census area, if a randomly chosen census area experienced 12 crimes, the "error" would be 2.

error terms: See *variance*.

explanatory variable: See variable.

F-test: A measure for comparing the fit of different regression models. The F-test of the overall significance of a model is a specific form of the F-test. It compares a model with no predictors to the model that has been specified. See also *joint F statistic*.

heterogeneity: Some researchers use the word "heterogeneity" when referring to heteroscedasticity or nonstationarity. The latter two are the correct terms to use in regression analysis, but see also *spatial heterogeneity*.

heteroscedasticity (and nonstationarity): OLS regression assumes, in order to trust the results, that the residuals have a constant variance (homoscedasticity) and are uncorrelated. This means it assumes a consistent relationship between explanatory variables and the dependent variable. There are two ways to consider this consistency. First, in data terms, it means that the variation in the relationship between the dependent variable and each explanatory variable does not change with changes in the magnitude of the explanatory variable (i.e., there is no heteroscedasticity in the model). Second, in spatial terms, the spatial processes represented by the explanatory variables behave the same everywhere (i.e., stationarity is present). If the result of a test for heteroscedasticity or nonstationarity is significant, it means the relationship between the variables varies significantly across the study area.

hypothesis test and the *null hypothesis* (see also *p-value*): A hypothesis test in statistics tests the validity of a claim we are making (e.g., that homicides are related to income inequality). The claim that is being tested is called the null hypothesis.

Jarque-Bera test: A test of normality. If this statistic is significant, it suggests the model is biased. This, for example, would be where the model performs well for low values but does not perform well for high values.

joint F statistic and *joint Wald statistic*: Measures of the overall statistical significance of a regression model. They both compare the joint effect of all variables together in the model. If the Koenker statistic for a regression model is statistically significant, the joint F statistic result is unreliable, and, therefore, the joint Wald statistic should be used.

Koenker statistic (*Koenker-Bassett test*): A measure of nonstationarity or heteroscedasticity, or both. The Koenker-Bassett test is a modification of the Koenker test, but Koenker is most used. If either statistic is significant, it suggests heteroscedasticity or nonstationarity, or both, in the model, meaning the relationship between the dependent and explanatory variables varies across space. The Koenker statistic also can determine whether standard p-values or robust p-values generated in the model should be used. If the Koenker statistic is significant, the robust p-values should be used.

Lagrange multiplier: Lagrange multipliers are used in statistics to find the maxima and minima of a function subject to constraints, such as find the quickest route from A to B but via point C, or minimize the distance that must be traveled but delivering items to locations A, B, and C en route. For our purpose, in spatial regression analysis, Lagrange multipliers determine whether we should use a spatial lag model or a spatial error model after preliminary analysis using OLS regression with a spatial weights matrix.

log likelihood: A method *to compare the performance of statistical models. It measures the* goodness of fit *of the coefficient values. The log likelihood is always negative. The higher the log likelihood value (i.e., the closer to zero), the better the model.*

log likelihood ratio: A measure of the goodness of fit of two competing statistical models. For the purposes of spatial regression analysis software, this measure compares the log likelihood measure from the OLS specification of the model to the alternative spatial regression model. Refer to the examples of its use in this chapter for help interpreting this measure.

multicollinearity: A situation in which two or more explanatory variables in a regression model are correlated. Explanatory variables should be independent. If multicollinearity is not addressed in a regression model, the results will be unstable. See also *multicollinearity condition number* and *variance inflation factor*.

multicollinearity condition number: A measure of multicollinearity. If this number is greater than 30, it suggests there are problems with the stability of the regression results because of multicollinearity. That is, the explanatory variables are too correlated with each other. To address this problem, at least one variable should be removed from the regression model. See also *variance inflation factor*.

natural log transformation: A popular way to address skewness in data and to get the data to conform approximately to normality. In regression analysis, on occasion, we may need to remove the skewness in data to address issues in a model's specification.

nonstationarity: See *heteroscedasticity*.

p-value (often referred to as "probability"): A measure of statistical significance. A p-value less than .05 indicates strong evidence for the null hypothesis (see *hypotheses test*), so the null hypothesis can be rejected. A p-value of less than .05 means that we can be at least 95 percent confident in the result generated. A p-value of .001 means we can be at least 99.9 percent confident in the result generated. An explanatory variable's coefficient means little if the relationship is not statistically significant.

parameter estimate (or "estimated value" or "predicted value"): The estimated value of the dependent variable (generated from a regression model), based on its statistical relationship with the explanatory variables.

R^2 and *adjusted* R^2: The R^2 statistic provides a measure of the proportion of the variance in the dependent variable that is explained by the explanatory variable or variables. It provides an indication of the strength in the relationship between these variables. The higher the R^2 value, the stronger the relationships. The adjusted R^2 statistic adjusts for the number of explanatory terms in a model. Unlike R^2, the adjusted R^2 value increases only if an added new variable improves the model more than expected by chance. In the case of using multiple variables to explain the dependent variable, the adjusted R^2 statistic is a better measure to use than R^2.

residual: The difference between the observed value of the dependent variable and the estimated value of the dependent value. In spatial terms, if the residuals for every location in our study area are zero, we have a perfect model. This is unlikely, however. Analysis of the residuals can provide insight into the accuracy of a regression model and also the viability of the regression results.

spatial dependence: The presence of a functional relationship between what happens at one location and what happens elsewhere (Anselin 1988). It often is referred to as "second-order spatial processes," in that it refers to localized spatial structure and processes that operate at short distances. If the presence of an event (the concentration of these events or the presence of relationships with other variables) at one location influences the presence of events at neighboring locations, spatial dependence exists. If spatial dependence is present, the uneven distribution of events (the concentration of these events or the presence of relationships with other variables) is because of some form of spatial interaction between events that is associated with the presence of the small-scale, localized influence from neighbors.

spatial effects: The collective term used for spatial variation, spatial dependence, spatial heteroscedasticity, spatial autocorrelation, and anything else that is spatially of interest in a regression analysis.

spatial heterogeneity: The uneven distribution of events (e.g., crime), concentrations of these events, or the variation in relationships over space. Spatial heterogeneity often is referred to as "first-order spatial effects," in that it is a process that operates across the entire region of study rather than referring to localized spatial structures and spatial processes that operate over short distances (second-order effects—see *spatial dependence*). If the intensity of the occurrence of the event (or concentration of these events or relationships) is equally unevenly distributed across the region (i.e., it is patchy), spatial heterogeneity exists. See also *spatial dependence*.

spatial homogeneity: The even distribution of events or relationships across space.

standard error: A measure of the accuracy of the estimated coefficient, and can be used to create confidence intervals and margins of error. It is calculated by measuring the mean of the sample, calculating how each measurement deviates from the mean, squaring each of the deviations calculated in the prior step, and then summing them.

stationarity: In spatial regression terms, it is when the explanatory variables in the model have a consistent relationship with the dependent variable. Hence, nonstationarity is when the relationships between the dependent variable and the explanatory variables are not consistent over space. See also *heteroscedasticity*.

t-statistic: A simple measure of the significance of each coefficient estimate, calculated by dividing the coefficient estimate by the standard error. We would like this value to be large, and greater than two, as this value indicates the coefficient is more than two standard errors from zero.

variable (*dependent variable* and *explanatory variable*): The technical term for the phenomenon, factors, features, or characteristics we are examining. Crime data can be a variable and is most often our dependent variable in regression analysis (i.e., the phenomenon that is the subject of our study). Population data, data on living conditions, data on the distribution of liquor stores, data on the distribution of schools, and data on the location of parking lots all can be variables but usually for crime analysis are explanatory (or, another term often used, "independent") variables in a regression model. Explanatory variables in a regression model are tested to examine if each is related to the dependent variable. That is, does the explanatory variable help explain the presence of the dependent variable?

variance (and error terms): A measure of how data is spread out and is calculated from measuring how far off a set of values are from their average value. In regression analysis, variance of the error terms describes how much an observed value differs from its expected value. For example, if we fit a trend line between a series of x,y values on a scatter plot, the variance in the error term refers to how far an observation is from this trend line (the expected value).

variance inflation factor (VIF) (see also *multicollinearity* and *multicollinearity condition number*): Quantifies the severity of multicollinearity in a regression model. It is particularly useful in determining which explanatory variables (significant and not significant) to remove from a model once multicollinearity is identified. In many software applications of regression analysis (e.g., the ArcGIS GWR tool), the VIF is calculated for each explanatory variable and indicates whether the variable is not contributing to the model. If the VIF value for an explanatory variable is greater than 7.5, it suggests the variable is redundant in the model and should be removed. It can be useful for additionally determining explanatory variables to remove that are not significant in a regression model. This also applies when an explanatory variable is significant and has a VIF value greater than 7.5. This case is likely because the variable offers a similar explanation to another (statistically significant) variable, and therefore one (the variable with the highest VIF) should be removed.

Further resources

Following are web links to the data sources referred to in this chapter and links to the GeoDa software and stand-alone GWR software created by the technique's inventors.

World Bank Open Data database: https://data.worldbank.org.

International Labor Organization, ILOSTAT database: https://www.ilo.org/global/statistics-and-databases/lang--en/index.htm.

United Nations (UN) database: https://data.un.org.

GeoDa: https://geodacenter.github.io.

Multiscale geographically weighted regression software: https://sgsup.asu.edu/sparc/mgwr.

References

Anselin, L. 1980. "Estimation Methods for Spatial Autoregressive Structures." Regional Science Dissertation and Monograph Series No. 8, Cornell University, Ithaca, NY.

Anselin, L. 1988. Spatial Econometrics: Methods and Models. Dordrecht: Kluwer.

Anselin, L. 2005. *Exploring Spatial Data with GeoDa: A Workbook*. University of Illinois: Center for Spatially Integrated Social Science.

Anselin, L., and D. A. Griffith. 1988. "Do Spatial Effects Really Matter in Regression Analysis?" Papers, Regional Science Association 65:11–34.

Austin, M. P., and J. A. Meyers. 1996. "Current Approaches to Modeling the Environmental Niche of Eucalypts: Implication for Management of Forest Biodiversity. *Forest Ecology and Management* 85:95–106.

Babor, T., R. Caetano, S. Casswell, G. Edwards, N. Giesbrecht, K. Graham, J. Grube, P. Gruenewald, L. Hill, H. Holder, R. Homel, E. Osterberg, J. Rehm, R. Room, and I. Rossow. 2003. Alcohol : *No Ordinary Commodity: Research and Public Policy*. Oxford: Oxford University Press.

Cahill, M., and G. Mulligan. 2007. "Using Geographically Weighted Regression to Explore Local Crime Patterns." *Social Science Computer Review* 25 (2): 174–93.

Can, A. 1990. "The Measurement of Neighborhood Dynamics in Urban House Prices." *Economic Geography* 66 (3): 254–72.

Ching-Chi, H., and M. D. Pugh. 1993. "Poverty, Income Inequality, and Violent Crime: A Meta-Analysis of Recent Aggregate Data Studies." *Criminal Justice Review* 18 (2): 182–202.

Dubin, R. A. 1992. "Spatial Autocorrelation and Neighborhood Quality." *Regional Science and Urban Economics* 22:433–52.

Dubin, R. A. 1998. "Spatial Autocorrelation: A Primer." *Journal of Housing Economics* 7:304–32.

Durbin, J. 1960. "Estimation of Parameters in Time-Series Regression Models." *Journal of the Royal Statistical Society B* 22:139–53.

Fischer, M. M. 1997. "Computational Neural Networks: A New Paradigm for Spatial Analysis." *Environment and Planning A* 29:1873–91.

Foody, G. M. 2003. "Geographical Weighting as a Further Refinement to Regression Modeling: An Example Focused on the NDVI-Rainfall Relationship." *Remote Sensing of Environment* 88 (3): 283–93.

Fotheringham, A. S., C. Brunsdon, and M. Charlton. 2002. *Geographically Weighted Regression: The Analysis of Spatially Varying Relationships*. Chichester: Wiley.

Fox, J. C., P. K. Ades, and H. Bi. 2001. "Stochastic Structure and Individual-Tree Growth Models." *Forestry Ecology Management* 154:261–76.

Gao, X., Y. Asami, and C.-J.F. Chung,. 2006. "An Empirical Evaluation of Spatial Regression Models." *Computers and Geosciences* 32:1040–51.

Graham, K., and R. Homel. 2008. *Raising the Bar: Preventing Aggression in and around Bars, Pubs, and Clubs*. Crime Science Series. Cullompton: Willan.

Gray, P. 2000. "Repeat Victimization in the Asian Community: A Study of Domestic Burglary." *Crime Prevention and Community Safety* 2:53–65.

Griffith, D. A. 2008. "Spatial-Filtering-Based Contributions to a Critique of Geographically Weighted Regression (GWR)." *Environment and Planning A* 40:2751–69.

Guisan, A., T. C. Edwards, and T. Hastie. 2002. "Generalized Linear and Generalized Additive Models in Studies of Species Distribution: Setting the Scene." *Ecological Modeling* 157:89–100.

Haiganoush, P. K., N. G. Rappaport, and D. L. Wood. 1997. "Regression Methods for Spatially Correlated Data: An Example Using Beetle Attacks in a Seed Orchard." *Forest Science* 43 (1): 71–77.

Hastie, T., and R. Tibshirani. 1990. *Generalized Additive Models*. London, England: Chapman and Hall.

LeSage, J. P. 1997. "Bayesian Estimation of Spatial Autoregressive Models." *International Regional Science Review* 20:1 and 2:113–29.

Littell, R. C., G. A. Milliken W. W. Stroup, and R. D. Wolfinger. 1996. *SAS System for Mixed Models*. Cary, N.C: SAS Institute.

Loftin, C., and S. K. Ward. 1983. "A Spatial Autocorrelation Model of the Effects of Population Density on Fertility." *American Journal of Sociology* 48 (1): 121–28.

Maguire, M., and M. Hopkins. 2003. "Data Analysis for Problem-Solving: Alcohol and City Center Violence." In *Crime Reduction and Problem-Oriented Policing*, edited by K. Bullock and N. Tilley. Cullompton, England: Willan.

McCreith, S., and S. Parkinson. 2004. *Crimes Against Students: Emerging Lessons for Reducing Student Victimization*. London, England: Home Office.

Messner, S. F., and L. Anselin. 2004. "Spatial Analyses of Homicide with Areal Data." In *Spatially Integrated Social Science*, edited by M. F. Goodchild and D. G. Janelle. New York, NY: Oxford University Press.

Mohebbi, M., R. Wolfe, D. Jolley, A. B. Forbes, M. Mahmoodi, and R. C. Burton. 2011. "The Spatial Distribution of Esophageal and Gastric Cancer in Caspian Region of Iran: An Ecological Analysis of Diet and Socioeconomic Influences." *International Journal of Health Geographics* 10:13.

Nakaya, T., A. S. Fotheringham, M. Charlton, and C. Brunsdon., 2005. "Geographically Weighted Poisson Regression for Disease Associative Mapping." *Statistics in Medicine* 24:2695–2717.

Roncek, D. W. 1981. "Dangerous Places: Crime and Residential Environment." *Social Forces* 60 (1): 74–96.

Rosenshein, L., L. Scott, and M. Pratt. 2011. *Finding a Meaningful Model.* Redlands, CA: Esri.

Rossmo, K. 2000. *Geographic Profiling.* Boca Raton, FL: CRC Press.

Schabenberger, O., and F. J. Pierce. 2002. *Contemporary Statistical Models for the Plant and Soil Sciences.* New York, NY: CRC Press.

Shin, M., and A. L. Goel. 2000. "Empirical Data Modeling in Software Engineering Using Radial Basis Functions." *IEEE transactions on Software Engineering* 26 (6): 1–10.

Voss, P. R., D. D. Long, R. B. Hammer, and S. Friedman. 2006. "County Child Poverty Rates in the US: A Spatial Regression Approach." *Population Research and Policy Review* 25:369–91.

Waller, L. A., L. Zhu, C. A. Gotway, D. M. Gorman, and P. J. Gruenewald. 2007. "Quantifying Geographic Variations in Associations between Alcohol Distribution and Violence: A Comparison of Geographically Weighted Regression and Spatially Varying Coefficient Models." *Stochastic Environmental Research and Risk Assessment* 21 (5): 573–88.

Wang, Q., J. Ni, and J. Tenhunen. 2005. "Application of a Geographically Weighted Regression Analysis to Estimate Net Primary Production of Chinese Forest Ecosystems." *Global Ecology and Biogeography* 14 (4): 379–93.

Wheeler, D., and C. Calder. 2007. "An Assessment of Coefficient Accuracy in Linear Regression Models with Spatially Varying Coefficients." *Journal of Geographical Systems* 9:145–66.

Wheeler, D., and L. Waller. 2009. "Comparing Spatially Varying Coefficient Models: A Case Study Examining Violent Crime Rates and Their Relationships to Alcohol Outlets and Illegal Drug Arrests." *Journal of Geographical Systems* 11 (1): 1–22.

Wheeler, D., and M. Tiefelsdorf. 2005. "Multicollinearity and Correlation among Local Regression Coefficients in Geographically Weighted Regression." *Journal of Geographical Systems* 7:161–18.

Zhang, L., H. Bi, P. Cheng, and C. J. Davis. 2004. "Modeling Spatial variations in Tree Diameter-Height Relationships." *Forest Ecology and Management* 189:317–29.

Zhang, L., and H. Shi. 2004. "Local Modeling of Tree Growth by Geographically Weighted Regression." *Forest Science* 50 (2): 225–44.

Zhang, L., J. H. Gove, and L. Heath, L. 2005. "Spatial Residual Analysis of Six Modeling Techniques." *Ecological Modeling* 186:154–77.

Zhang, L., Z. Ma, and L. Guo. 2009. "Spatial Autocorrelation and Heterogeneity in the Relationships between Tree Variables." *Forest Science* 55 (6): 533–48.

CHAPTER 9

Mapping and analyzing crime risk

Objectives

Crime risk refers to the possibility of crime happening. In this chapter, we introduce the crime risk framework to approach the task of mapping and analyzing crime risk. The crime risk framework involves three temporal settings: immediate risk, near risk, and enduring risk. We suggest that different data and different spatial analysis techniques are required for each one, informed by the presence of different spatiotemporal patterns of crime. Many of the techniques discussed draw from techniques we already have referred to in this book. They include considering the application of the patterning principles of repeat and near repeat victimization (chapter 5), hot spot analysis (chapter 3), and the application of spatial analysis techniques using multiple variables as inputs (such as spatial regression analysis, in chapter 8). In this chapter, we explore the application of these techniques to determine crime risk, which in turn informs different types of police and public safety responses.

Key learning points

- Understand the value of mapping and analyzing crime risk for different temporal settings rather than thinking that a single method is optimal for determining crime risk.
- Understand how to generate crime risk zones for identifying immediate risk, based on the patterning principles of repeat and near repeat crimes.
- Recognize how spatial patterns can inform the transition from immediate risk to near risk and from near risk to enduring risk.
- Consider the application of composite mapping methods, such as spatial regression analysis and risk terrain modeling, to identify the factors that contribute to enduring risk.

Introduction

Risk relates to the possibility of something bad happening. We purposely use the term *risk* in this chapter, rather than *prediction* or *forecasting*, because it is more appropriate in the application of crime and conceptually aids better understanding of crime. Prediction typically gives a sense of exactness and certainty to what will happen. Risk refers to the *possibility* of something happening, which overall is a better way to think about the future occurrence of crime. Risk and forecasting both require the use of data from previous events to calculate what may happen in the future. Forecasting refers to things that are good or bad. Risk naturally makes us associate that what it is attached to is something bad. Crime is bad! Crime risk relates to the possibility of crime happening.

In this chapter, we organize the concept of crime risk into a crime risk framework, arranged across three temporal settings: immediate, near, and enduring. If we want to be accurate in determining the risk of crime, we must recognize the necessity of using different techniques for different temporal settings of risk. That is, what is good at determining immediate risk is likely not to be as good at determining the risk of crime in the near future or a risk that may endure for some time; additionally, the methods we use to determine near risk are likely to be particular for this temporal setting rather than also being suitable for determining immediate crime risk and enduring crime risk. Determining enduring risk is also likely to require a different approach than for the other two temporal settings. Principally, we refer to three types of spatial analysis techniques for determining crime risk, one for each of the three temporal settings: the analysis of repeat and near repeat victimization patterns for determining immediate risk, hot spot mapping for near risk, and composite mapping for enduring risk. Composite mapping involves using variables other than, or in addition to, crime data to measure risk. We refer to spatial regression analysis for determining enduring risk but also describe the use of risk terrain mapping to determine risk for this temporal setting. Other techniques also are likely to be suitable for determining crime risk, and we encourage you to experiment with them, but we recommend that to do so, you will consider these techniques based on the temporal settings we define in the crime risk framework.

The crime risk framework

If we can determine where there is a risk of crime, resources can be targeted to prevent crime, reduce victimization, or catch an offender. However, the data and techniques we can use to judge the risk of crime are likely to be different for different temporal settings. In chapter 5, we discussed the heightened risk following an initial incident, which is observed in the patterns of repeat victimization and near repeat victimization. In chapter 3, we described how hot spot mapping techniques provide a good indication of where crime is likely to take place in the future. And in chapter 8, we described how spatial regression techniques can identify how explanatory variables are related to the occurrence of crime, and how changes in these variables can influence changes in crime. All three sets of techniques can be used to determine some form of crime risk (i.e., the possibility of crime happening) but evaluate that risk in different ways.

The crime risk framework structures how we map and analyze crime risk. The framework organizes our consideration of crime risk into immediate, near, and enduring risk. Rather than being led to believe that we can predict, forecast, or estimate the risk of crime using a single technique or software tool, thinking in terms of the crime risk framework helps us recognize that different data and different techniques are better at determining crime risk for different temporal settings. By organizing crime risk into these three temporal settings of immediate, near, and enduring, we can better determine the most suitable type of policing and public safety response activities for addressing the risk of crime.

Determining crime risk for different temporal settings: A weather forecasting analogy

Data on current and recent weather conditions is perhaps the best predictor of what the weather is likely to be in the immediate future (i.e., for the rest of today and tomorrow). For example, think about what the weather has been like over the last eight hours. Over the next eight hours, it is likely to be similar. In this, we have used data about the weather from the recent past to forecast the weather in the immediate future. If we had used data about the weather for the same time of day but on a day four weeks ago, our forecast on what the weather is likely to be over the next eight hours likely would be less accurate. Perhaps, the same is true for crime. We know from research on repeat and near repeat patterns that a heightened risk of crime immediately follows a recent incident. Although it is not definite that a crime will occur, the crime risk to the same target or nearby targets associated with a recent incident indicates a heightened risk that crime may happen to those same or nearby targets.

To forecast what the weather may be like next week, data in addition to recent conditions probably must be used. A high-pressure system, for example, is moving in, suggesting that weather conditions are likely to be stable over the next week. To consider next week's crime risk patterns, knowing that crime previously has clustered in certain places gives a good indication that this also is where we are likely to observe crime to cluster again soon. That is, our high crime risk areas next week are likely to be our hot spots from last week.

To forecast what the weather may be like next year, data other than that on recent conditions and from just the last week or so must be used. We also may include data other than weather data to improve our longer-term forecasts. This data may include information on ocean currents, pollution, and volcanic events. Similarly, for determining crime risks that endure, we would use not only crime data but also data from other sources, such as on population, land use, and socioeconomic conditions.

The analytical methods used to forecast what the weather may be like tomorrow are different from the methods used to forecast the weather outlook for next week. For example, if I wanted to determine if rain was imminent, I could review the current weather radar map to see if a rain cloud was approaching my location (or I could look out the window to see in which direction that dark cloud was moving). But I wouldn't use the same weather radar map to forecast the weather outlook for next week. We also would use a different method to forecast what the weather may be like next year. So, when we think about crime risk, it makes sense that we also should use different data, and different methods for determining crime risk for different temporal settings.

In the sections that follow, we illustrate why different data and techniques are required to inform different temporal settings of crime risk. Here, we illustrate how this data and different techniques are organized across the crime risk framework so that we go into the next sections with an initial appreciation of the differences in inputs and outputs resulting from their use.

Table 9.1. The crime risk framework—a temporal framework for mapping and analyzing crime risk

Temporal period	Input data	Spatial patterns	Types of responses
 Immediate risk	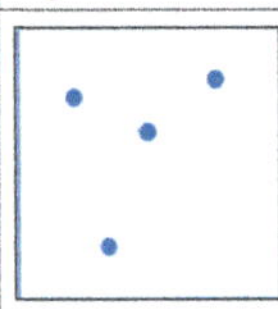 Recent incidents of crime	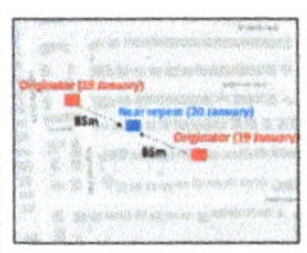 Repeat and near repeat victimization	• Targeted operational tactical response • Targeted offender supervision
 Near risk	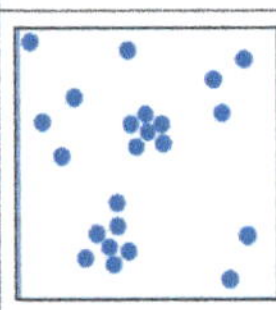 Incidents of crime	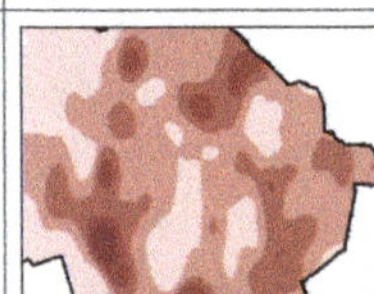 Crime hot spots	• Targeted operational strategy • Problem-solving crime prevention initiatives
 Enduring risk	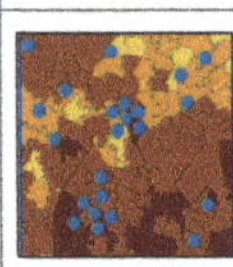 Incidents of crime Explanatory variables	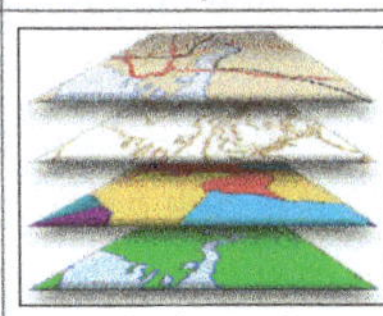 Spatial relationships	• Strategic interventions and changes in policy

Table 9.1 shows the crime risk framework organized into the three temporal settings. The framework states that patterns relating to repeat and near repeat victimization can inform immediate crime risk. In turn, this immediate risk of crime can be responded to with targeted operational police responses to the areas or targets at highest risk. This intervention could use offender management resources to help disrupt and deter the activity of those suspected to be involved in the commission of crime in the area of high risk and encouraging people who live or frequent this area to carry out practical crime prevention measures to minimize their risk of victimization. It also may

involve asking these high-risk targets to be extra vigilant over the next few days and to report any suspicious activity.

Crime risk for the near future can be determined using hot spot mapping. In turn, police and public safety activity should be focused on addressing the conditions that make an area conducive to crime. This action could involve implementing a hot spot policing program if the crime risk is associated with street crime or introducing projects to ensure the problem of crime risk in this area is better understood and addressed.

An assessment of enduring crime risk can be determined using techniques that consider the relationship between crime and a set of explanatory variables. The variables significantly correlated to this type of modeling then can be used to inform the direction of strategic interventions. Activity focused on addressing the influence these variables have on crime could help bring long-term reductions in crime risk.

Each of the three types of patterns or techniques described above for determining crime risk is naturally suited to an individual temporal setting, and each is proven to contribute to our understanding of crime and the risk of being a victim of crime in the future. These techniques may not be the only ones, but they set a foundation that encourages other techniques to be used if their performance in accurately determining crime risk is better than those we initially propose in the crime risk framework. For example, in some situations, the analysis of risky facilities for informing near crime risk may improve hot spot analysis or complement it. In the sections that follow, we describe how each pattern or technique that we initially propose within the crime risk framework can inform crime risk.

The prediction accuracy index

The prediction accuracy index (PAI) (Chainey, Tompson, and Uhlig 2008) is one of the most used measures for comparing the prediction (and risk) performance of crime mapping output. The PAI is calculated by dividing the *hit rate percentage* (the percentage of crime events committed in the areas in which crimes were predicted to occur) by the *area percentage* of the predicted areas (the proportion of the area in which crime is predicted to occur, in relation to the whole study area). The equation (9.1) for the PAI is as follows:

$$\frac{\left(\frac{n}{N}\right)*100}{\left(\frac{a}{A}\right)*100} = \frac{HitRate}{AreaPercentage} = \text{prediction accuracy index,} \qquad (9.1)$$

where *n* refers to the number of *crimes* that were committed in the areas in which crimes were predicted to occur (or the areas determined as having the highest crime risk), *N* is the number of crimes committed in the study area, *a* refers to the area (e.g., km^2) of the areas in which crimes were predicted to occur, and A is the area (e.g., km^2) of the study area. The higher the PAI, the better the predictive performance of the mapping output.

For example, let's say we wanted to compare the PAI of the Gi* hot spot mapping technique for different types of crime—residential burglary, theft of vehicles, and violent assaults. The study area is 112.7 km^2 in size. Table 9.2 shows the area covered by the hot spots generated by the Gi* statistic and the number of crimes that subsequently took place in these hot spots. The PAI involves dividing the hit rate percentage by the percentage of the area covered by the hot spots. The PAI was greatest for violent assaults, suggesting the Gi* hot spot map of violent assaults was better in predicting where assaults subsequently would take place than the other Gi* maps for burglary and theft from vehicles.

Table 9.2. PAI results of Gi* hot spot maps for different types of crime

	Area of Gi* hot spots	Gi* hot spots area percentage	Crimes in study area	Crime in Gi* hot spots ($p < .05$)	Hit rate percentage	PAI
Residential burglary	2.70	2.4%	130	22	16.9%	7.1
Theft from vehicles	2.37	2.1%	189	35	18.5%	8.8
Violent assaults	1.13	1.0%	154	56	36.4%	36.4

The PAI also can be used to compare the predictive performance of mapping output produced by different techniques (e.g., compare Gi* outputs to KDE outputs to risk terrain modeling). The PAI is applicable to any study area, any crime point data, and any analysis technique that aims to predict, forecast, or calculate where there is a risk of crime.

Immediate crime risk

In chapter 5, we described how recent incidents can inform the heightened risk of future incidents. This heightened risk of crime occurring after a recent incident is displayed in the patterns of repeat victimization and near repeat victimization. The heightened risk is greatest at the location in which the previous incident occurred, declining spatially and temporally from this originator incident. Research also has shown that a large proportion of these repeats and near repeats take place outside crime hot spots (Chainey et al. 2018). Therefore, hot spot analysis, perhaps, is less suited for estimating immediate crime risk, and instead, an approach that uses the patterning observed in repeat and near repeat victimization can provide a better measure of this immediate crime risk.

Repeat and near repeat victimization—a refresher

Repeat victimization is the pattern described for the heightened risk resulting after an initial victimization. This heightened risk is greatest within a few days of the initial incident, and then declines rapidly over time.

Near repeat victimization is the pattern described for incidents that take place within a short distance and short time of an originator incident. The heightened risk of crime declines with distance from the location of the originator and declines over time.

Patterns of repeat victimization and near repeat victimization have been observed for a wide range of crimes, including burglary, vehicle crime, street robbery, robbery from a convenience store, shooting, bicycle theft, assault, and (in the case of near repeats) homicide. Chapter 5 covers repeat and near repeat victimization in more detail.

The first stage in using repeat victimization and near repeat victimization patterns to determine immediate crime risk is to measure the extent to which repeats and near repeats are present in the data being analyzed. The presence of patterns of repeats and near repeats suggests these patterns can be used to determine immediate crime risk. Table 9.3 shows the extent of repeat and near repeat victimization for residential burglary for several cities. The results from these cities show that, in most places, repeats can account for up to a third of all burglaries, and near repeats (defined as being within 200 meters and seven days) can account for an additional 12 to 25 percent of all burglaries. In the city of Belo Horizonte, Brazil, the study showed little evidence of burglary repeats and near repeats. In situations in which very few repeats or near repeats are present, hot spot analysis is likely to be the most suitable technique for determining immediate crime risk. We will elaborate on this further in the next section. In most situations, a repeat and near repeat pattern is likely to be present.

Table 9.3. (a) The extent of burglary repeat and (b) near repeat victimization as reported from a range of studies conducted by police agencies in the US, UK, Japan, Sweden, the Netherlands, Brazil, and New Zealand

a

Location of study and source	Proportion of burglary that was repeat victimization
Belo Horizonte, Brazil: 2013 (Chainey and de Silva 2016)	5%
Birmingham, England: 2011 (Chainey and de Silva 2016)	30%
England and Wales: 2012–13 (Chainey and de Silva 2016)	14%
Japan: 1989 (Farrell and Bouloukos 2001)	22%
Newcastle, England: 2010 (Chainey and de Silva 2016)	15%
Netherlands: 1996 (Farrell and Bouloukos 2001)	21%
Northumberland County, England: 2009 (Chainey et al. 2018)	7%
South Auckland, New Zealand: 2014 (Chainey et al. 2018)	10%
Sweden: 1996 (Farrell and Bouloukos 2001)	13%
USA: 1996 (Farrell and Bouloukos 2001)	33%

b

Measure of near in space and near in time to originator incident, and location of study	Proportion of burglary that was near repeat victimization
Within 200 m and 7 days Within 200 m and 14 days Herefordshire, UK (Chainey and Silva 2016)	14% 23%
Within 300 m and 10 days Kettering, UK (Chainey and Silva 2016)	16%
Within 200 m and 7 days Newcastle, UK (Chainey and Silva 2016)	23%
Within 100 m and 1 day Within 100 m and 7 days Within 200 m and 7 days Auckland, New Zealand (Chainey et al. 2018)	1% 5% 12%
Within 100 m and 1 day Within 100 m and 7 days Within 200 m and 7 days Wellington, New Zealand (Chainey et al. 2018)	2% 5% 12%
Within 100 m and 7 days Within 200 m and 7 days Within 300 m and 7 days Belo Horizonte, Brazil (Chainey and Silva 2016)	2% 6% 13%

In chapter 5, we described the use of the Near Repeat Calculator for measuring the presence of repeat and near repeat victimization. This calculator also can measure the extent of repeats and near repeats, but so too can an ArcGIS tool—the Repeat and Near Repeat Classificatio*n* tool. The Repeat and Near Repeat Classification tool is simple to use, requiring the user only to enter the parameters of the near repeat spatial and temporal bandwidths (e.g., within 200 meters and seven days).

In this chapter, we again use the district of Newcastle, England, because it is a study area you now probably are familiar with from other chapters, so you can place the results from this chapter into the context of the results shown elsewhere in the book. The crime data used was residential burglary and assaults for a one-year period. Table 9.4 lists the Repeat and Near Repeat Classification results for Newcastle and shows that repeats (i.e., burglaries against the same property) and near repeats (defined as being within 200 meters of a previous burglary), within seven days and 14 days of a previous incident, accounted for 18 percent and 27 percent of all burglaries, respectively. For assaults in Newcastle, repeats (defined as being at the same location) and near repeats (within 200 meters) within seven days and 14 days of a previous incident accounted for 40 percent and 47 percent of all assaults, respectively.

Let us now consider these results in terms of crime risk. The research evidence on repeats and near repeats (refer to chapter 5, where we cite this) would lead us to suggest that heightened crime risk surrounds recent incidents soon after they occur. Using the results from Newcastle as an example, in the case of residential burglary, over the next week, nearly one in five burglaries that now will take place are likely to take place within 200 meters of a previous incident. In the case of assaults, over the next week, 40 percent of all assaults are likely to take place within 200 meters of a previous incident. Research on repeats shows that the probability of being victimized again declines quickly over time, and for near repeats, also declines spatially from the originator incident. We can suggest, therefore, that crime risk is present around an incident when it occurs in our study area of Newcastle, with this risk decaying spatially and temporally. That is, the risk of crime is present around a recent incident but only for the immediate period after the incident has occurred.

Table 9.4. The extent of burglary and assault repeats and near repeats in Newcastle, England

	Within seven days	**Within 14 days**
Burglary repeats	23 (1.8%)	31 (2.4%)
Burglary near repeats (within 200m)	211 (16.2%)	318 (24.4%)
Assault repeats	202 (10.9%)	260 (14.0%)
Assault near repeats (within 200m)	533 (28.7%)	620 (33.4%)

Operationalizing the patterning principles of repeat and near repeat victimization into a tool that shows areas of immediate crime risk involves creating buffers around each recent incident on the basis of the spatial extent of the crime risk while also taking into account how the crime risk decays over time. This can be attempted manually, or semiautomatically, as was done by police crime analysts in Manchester, England (Fielding and Jones 2012).

The ArcGIS Generating Crime Risk Zones tool automates this process. The tool requires the user to specify the date from which recent incidents have occurred (referred to as the *initial processing date*), and then risk zones are created. So, for example, if we wanted to determine the crime risk for today, we would use data starting from yesterday to determine the immediate crime risk. The tool also requires specification of the spatial range of influence of crime risk, which relates to the size of the risk zones around each incident. This value refers to the maximum distance in which the incident will influence the risk of future incidents. The results from measuring the extent of near repeat victimization can help decide this value, or it can be based on the area around an incident considered to be practical for resource targeting and deployment (e.g., the area for which additional patrols are to be targeted). So, for example, using the results from Newcastle, and considering the practical deployment of police resources to areas in Newcastle, a sensible value for this example would be 200 meters. The user also is required to enter the value for the spatial half distance. This is the distance from an incident in which the risk of crime is estimated to be half that for another incident occurring. The *spatial half distance* is used to calculate the rate of spatial decay in the risk of crime from the incident that has occurred. This value is calculated by the Repeat and Near Repeat Classification tool. For the Newcastle burglary and assaults data, the spatial half distance was 108 meters and 123 meters, respectively.

Crime risk also declines temporally. That is, more recent incidents have a greater weighting on crime risk than the risk associated with incidents that occurred a short period before. To account for this, the Crime Risk Zones tool requires the user to enter a temporal range of influence and a temporal half-life. The temporal range of influence refers to the number of days before the initial processing date that incidents are estimated to have an influence on the risk of future incidents taking place. Incidents that took place between the initial processing date and the value for the temporal range of influence are used for creating the crime risk zones. For example, if this value was set to three, only incidents for the initial processing date and the two days before would be used for creating crime risk zones. Incidents that took place three days ago would have less of an influence on the risk of future incidents than those that took place on the initial processing date. The temporal half-life refers to the number of days before the initial processing date that the risk of crime is estimated to be half that for another incident occurring. This value is used to calculate the rate of decay in the temporal risk of repeat and near repeat incidents. Like the spatial half distance, temporal half-life is generated when running the Repeat and Near Repeat Classification tool. For the Newcastle burglary and assault data, the temporal half-life, based on a temporal band of seven days, was four days and three days, respectively. The user also can specify the risk calculation method as being either cumulative (in which the value of each cell is the sum of any crime risk zone cells that overlap) or maximum (in which the value of each cell is the maximum value of any crime risk zone cells that overlap). The tool also creates polygonal output of the crime risk zones, which can be used to define the boundaries of the crime risk zones and grade areas with each zone by its risk level (e.g., if three risk ranges are used, the areas closest to the center of the zone could be referred to as *primary risk*, radiating out to areas of secondary and tertiary risk).

Figure 9.1 shows crime risk zones generated for residential burglary and assaults using the scenario that today is May 20. The crime risk zones relate to immediate crime risk, using 200-meter spatial bands, a seven-day temporal band, and the spatial half-distance and temporal half-life values specified above. Each crime risk zone is positioned over an incident that took place within the last seven days, with the risk of crime decaying spatially from the center of each risk zone. Each crime risk zone also decays temporally from yesterday's date to seven days ago, with incidents that happened yesterday having more influence on today's crime risk than incidents that took place on the days before. If our previous measures of the extent of repeats and near repeats provide an accurate indication of the proportion of crime that would take place today in each crime risk zone, we would expect about one in five burglaries and about four in every 10 assaults to take place in the crime risk zones. The process is then repeated the next day to generate up-to-date crime risk zones.

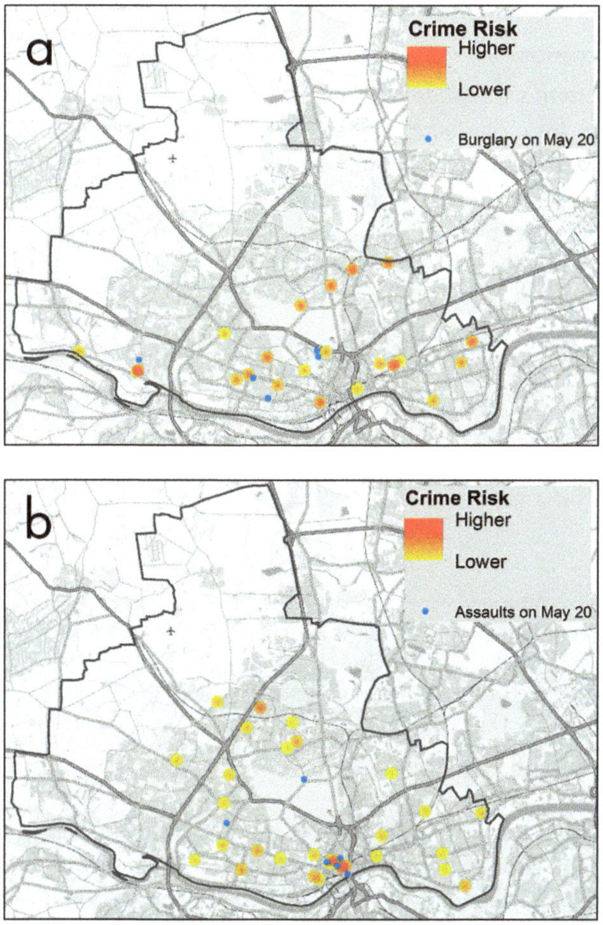

Figure 9.1. Crime risk zone maps for (a) residential burglary and (b) assaults, based on the scenario that today is May 20.

On May 20, there were five burglaries in Newcastle. One took place in a crime risk zone, and three of the others took place just outside the border of a crime risk zone (figure 9.1a). On May 20, there were six assaults in Newcastle (figure 9.1b). Four took place in the area identified as the highest crime risk zone. Two assaults took place in areas not identified as crime risk zones. These mapping outputs and results not only illustrate how crime risk zones can identify immediate risk but also are indicative of their accuracy.

To test the accuracy of the crime risk zones for your study area, perform a similar test to that shown for Newcastle, but repeat the test several times for at least 10 randomly selected dates. You'll use data for about a one-year retrospective period, selecting dates within this period as "today," and then calculating how many crimes took place in the crime risk zones. You should find that the extent of repeats and near repeats calculated in the initial stage provides a good indication of the number of crimes that then occur in the crime risk zone you generate.

How accurate are the crime risk zones?

In a test of the accuracy of crime risk zones, crime data was used for Naperville, Illinois, USA. Using a dataset of 541 incidents, for a one-year period, 77 incidents were repeats that took place within 14 days of a previous incident. Using a spatial band of 1,300 feet and a temporal band of 14 days, 183 incidents were near repeats (table 9.5). These results indicate that we would expect about 48 percent of all future incidents to take place within 1,300 feet and 14 days of a previous incident.

Table 9.5. (a) Parameters for repeat and near repeat classification, and (b) the extent of repeats and near repeats in Naperville, Illinois, USA

a

Parameter	Parameter value
Spatial bands (ft)	1,300
Temporal bands (days)	14
Estimated spatial half distance (ft)	472
Estimated temporal half-life (days)	4

b

	Within 14 days	
	Number of incidents	Proportion of all incidents
Repeats	77	14.2%
Near repeats (<= 1,300 ft)	183	33.8%

A total of 28 dates were randomly selected from the one-year period. Crime risk zones were created for each date, using a spatial range of 1,300 feet, a spatial half distance of 472 feet, a temporal range of 14 days, and a temporal half-life of four days. For each crime risk zone, three risk ranges were created, referring to areas of primary risk, secondary risk, and tertiary risk. The number of incidents that took place on the day after each of the 28 dates, and the number of incidents within the Crime Risk Zones (by risk range) on these "day after" dates was calculated.

Some 49 percent of the incidents that took place on the day after were in crime risk zones (table 9.6). This result illustrates the strength of the accuracy of crime risk zones in determining crime risk. The result also illustrates how the initial measure of the extent of repeats and near repeats provides an accurate indicator of the proportion of crime likely to take place in the crime risk zones.

A total of 5 percent of all the crimes were in the primary risk areas, compared with 16 percent in the secondary risk area and 28 percent in the tertiary risk area. The primary areas, though, were much smaller in size than the secondary and tertiary areas, with the tertiary areas the largest of the three. An additional calculation was performed using the prediction accuracy index, which accounts for differences in the size of the three crime risk zones. The higher the PAI, the better the performance of the crime risk zone in its estimation of risk. The PAI was highest for the primary crime risk zone areas and the lowest for the tertiary areas, matching what was expected of the primary risk areas being the areas of highest crime risk.

Table 9.6. PAI values for primary, secondary, and tertiary crime risk zones in Naperville

Incidents that took place on the next day following 28 randomly selected dates	**Hit rate**		**Area**		**PAI**
	n	**%**	**Miles²**	**%**	
Study area	76	100%	41.45	100%	NA
Crime risk zone—primary risk	4	5%	0.01	0.03%	152
Crime risk zone—secondary risk	12	16%	0.32	0.8%	21
Crime risk zone—tertiary risk	21	28%	3.19	7.7%	4
Total	**37**	**49%**	**3.52**	**8.5%**	**n/a**

Near crime risk

Crime risk zones use the patterning principles of repeat and near repeat victimization to determine where crime may occur in the immediate future. This immediate future represents a time setting that is typically not long enough to allow for clusters of crime to form. For example, taking today as May 20 in a study area, figure 9.2a shows the number of crimes that subsequently took place. Figure 9.2b shows the number of crimes that then took place between May 20 and May 23. In both figures, 9.2a and 9.2b, clusters have yet to form, and the spatial distribution of these patterns appears to be random. However, we know that the possibility of repeat and near repeat victimization patterns could be contained within these crime points, which in turn can inform immediate crime risk (i.e., the areas of high crime risk on May 20–23. By May 27, clusters of crime have begun to form (figure 9.2c). This pattern suggests that as we approach the *near future*, techniques that identify where clusters are likely to form would be a better measure for determining the risk of crime than continuing to use patterning that relates to repeat and near repeat victimization.

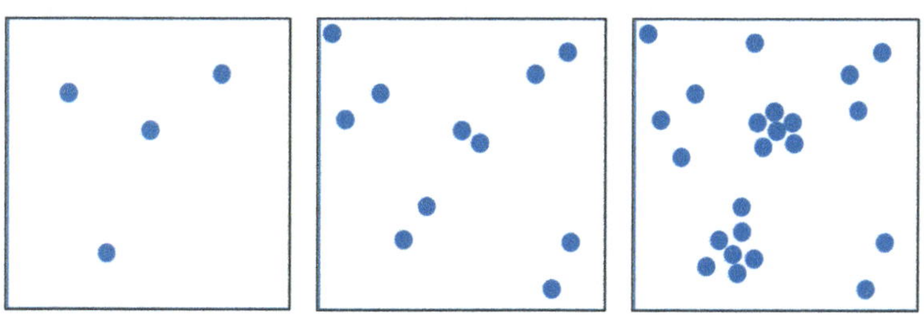

(a) May 20 **(b) May 20–23** **(c) May 20–27**

Figure 9.2. From random spatial patterns (a) and (b) to (c) the clustering of crime over an eight-day period.

Table 9.7 shows the results from a study that compared the PAI for crime risk zones to the PAI for Gi* hot spot maps, for different periods. Data on residential burglary over the previous 14 days was used to create the crime risk zones and Gi* hot spot maps. The PAI was used to compare the performance of the two techniques in relation to the crimes that then took place on each day after this 14-day retrospective period. So, on day 1, the PAI was calculated for the crime risk zones and Gi* hot spot maps in relation to the crime on day 1. On day 2, the PAI was calculated for the crime risk zones and Gi* hot spot maps in relation to the crime on day 1 to day 2, and so on, to day 8, for which the PAI was calculated for the crime risk zones and Gi* hot spot maps in relation to the crimes on day 1 to day 8. The table also includes the results from the nearest neighbor index test to determine whether clustering was present in the crimes on day 1, day 1 to day 2, day 1 to day 3, and so on, to day 1 to day 8.

Table 9.7. PAI results for the eight days after the date when crime risk zones and Gi* hot spot maps were created, using 14 days of retrospective crime data

	Day 1	Day 2	Day 3	Day 4	Day 5	Day 6	Day 7	Day 8
PAI: Crime risk zones	7.6	6.3	5.9	5.3	5.1	4.7	4.1	3.6
PAI: Gi* hot spot maps	3.2	4.1	4.3	4.6	5.0	5.2	8.9	9.8
NNI	3	3	2.1	1.3*	1.1*	0.8*	0.7**	0.7**

Note: NNI results refer to the number of crimes that have taken place by day listed.
* p < .05; ** p < .01; *** p < .001

The results show the crime risk zones performed better than the Gi* hot spot maps in determining where crime took place on day 1 to day 5, but from day 6, the Gi* hot spot maps began to perform better than the crime risk zones. The results from this study illustrate the difference in the performance of different techniques for determining areas of crime risk with respect to different prospective periods. The results from this study illustrate the transition from the immediate future to the near future was on about day 6, when the Gi* hot spot maps began to perform better than using the patterning principles of repeats and near repeats to determine crime risk. The NNI results are indicative of this transition, showing that it was at the point when there was a sufficient amount of crime to begin to form clusters that the cluster mapping technique (Gi* hot spot maps) began to perform better than the crime risk zones. The spatial pattern of crime began to be observed in different ways, transitioning from the use of recent incidents and the patterning principles of repeats and near repeats to inform crime risk to using hot spot maps to determine the areas of crime risk.

Knowing that areas of high crime risk in the near future are likely to be the areas in which crime has previously clustered helps decide the types of policing and public safety responses that are better suited for countering the clustering of crime. These types of responses are likely to be operational strategies that the police can deploy, such as hot spot policing, rather than reactive tactical responses to suppress an immediate crime risk. Problem-oriented policing interventions also are suited to this category of response, involving examining more deeply the patterns of crime in the areas of high crime concentration and devising problem-solving responses to counter situational factors that make these areas conducive to crime.

Predictive mapping software and mixed methods

As a reader of this book, you are conscious, I am sure, of the developments in predictive mapping software and solutions, and the claims that their producers have made. For me, a fundamental issue in many of these applications is that they do not recognize that different techniques are required to predict different periods of the future. Predicting the immediate future and predicting the near future require different types of spatial analysis because of the differences in the patterns that crime follows. For instance, determining where hot spots previously have occurred may not be the best technique for determining a risk of crime in the immediate future. To determine areas of crime risk in the immediate future, and before the point that crime begins to cluster, we need to examine whether patterns of repeats and near repeats are present, and if so, we can use them to inform immediate crime risk. But using the patterning principles of repeats and near repeats is likely to be useful only for determining crime risk in the immediate future. As crime begins to cluster, for example, after seven to 10 days, determining the areas of crime risk is likely to be better informed by good hot spot analysis. Many predictive mapping solutions mix these two methods together, and often include other methods as well, undermining the actual predictive potential that each technique can offer on its own in predicting crime or determining crime risk for the temporal period of interest.

There is often a temptation to add other data, such as land-use data, to retrospective crime data to determine areas of crime risk. In principle, there is no problem in adding extra data, but only if it improves the estimation of crime risk. To date, few studies have measured fairly and reliably the contribution that additional data brings to retrospective crime data for determining crime risk and whether this added contribution improves the measure of crime risk when considering where crime will occur in the next few days, weeks, or beyond. One study (Drawve 2016), which was fair and consistent in the use of each technique, concluded that the use of retrospective crime data for producing KDE hot spot maps was more accurate in determining where crime occurred than using other types of spatial data to determine crime risk. However, other types of spatial data can be useful for understanding the underlying conditions contributing to the risk of crime, and perhaps are better suited for determining crime risks that are longer and more enduring.

Enduring crime risk

In almost all cases, crime, at some point, forms into spatial clusters. When crime does not form clusters, it is either distributed uniformly or is random, which can make deciding where to deploy resources extremely difficult. In my 25 years of experience in examining crime data, I never have observed either a pattern of complete uniformity or a persistence of spatial randomness in any crime pattern. Before the advent of crime clusters, what may appear as being a spatially random pattern (e.g., has an NNI of 1) may contain patterns relating to repeat victimization and near repeat victimization. As illustrated in a previous section of this chapter, the patterning principles of repeats and near repeats then can inform the location of an immediate crime risk, before crime begins to form into clusters. This pattern can inform tactical police deployment (e.g., where should the police be deployed in the next shift?). Knowing where crime clusters can form can inform where to organize a strategic response, such as a hot spot policing program that involves the regular deployment of police patrols to areas of high crime concentration. Clusters of crime often can persist, and neighborhoods or other areas of our cities often continually endure the experience of more crime than other areas.

In chapter 8, we described how spatial regression analysis techniques can identify variables that help explain the distribution of crime, and in doing so help identify the underlying causes of crime risk. The vulnerable localities index (VLI) (Chainey 2008) and risk terrain modeling (RTM) (Caplan, Kennedy, and Miller 2011) are other techniques that use other data than, or in replacement of, crime data to help determine the underlying conditions contributing to crime risk. Here, we refer to these techniques as *composite mapping techniques*, as they involve using more than one type of data to measure risk. When examining the enduring risk of crime, it is important to analyze the underlying factors, conditions, and causes contributing to this crime risk. To do so requires us to use techniques such as spatial regression, the VLI, and RTM.

A comprehensive description of spatial regression analysis already is provided in chapter 8. Here, we refer to one example from chapter 8 and place the use of spatial regression analysis into the structure of the crime risk framework for informing enduring crime risk. We know from several examples in this book that examine the spatial distribution of assaults in the district of Newcastle that these assaults cluster. In turn, this clustering could result in a hot spot policing program being implemented in the main hot spots—thus fitting the near risk temporal category of the crime risk framework. We also could use techniques such as the Space Time Cube and Emerging Hot Spot Analysis tools described in chapter 6 to identify whether there were any enduring hot spots of assaults. In places in which high levels of crime persist, even after the introduction of a hot spot policing program, an analysis involving the use of data on factors that may explain the persistence of assaults could contribute to understanding the enduring crime risk of assaults.

In chapter 8, we illustrated that the distribution of licensed premises and patterns in income deprivation were related significantly to the distribution of assaults in Newcastle. Figure 9.3 shows the distribution of assaults, licensed premises, and income deprivation in Newcastle, the areas in which the relationship between these variables was greatest (using GWR), and a map of the standardized residuals from the GWR model. In crime risk terms, licensed premises and income deprivation can be considered factors that significantly contribute to the underlying conditions of the risk of assault. So, to address the enduring risk of assault, results from the spatial regression analysis would suggest the need for a strategic response that considered how changes in the licensing of premises for the sale of alcohol, changes in the spatial distribution of licensed premises, or restrictions on the sale of alcohol could address the underlying factors that relate to their association with assaults in Newcastle. Additionally, programs that aimed to alleviate income deprivation in areas in which it was mostly associated with assaults or targeting programs in these areas that aimed to reduce alcohol misuse by those involved in committing assaults could be other strategic interventions to help address the enduring crime risk of assaults.

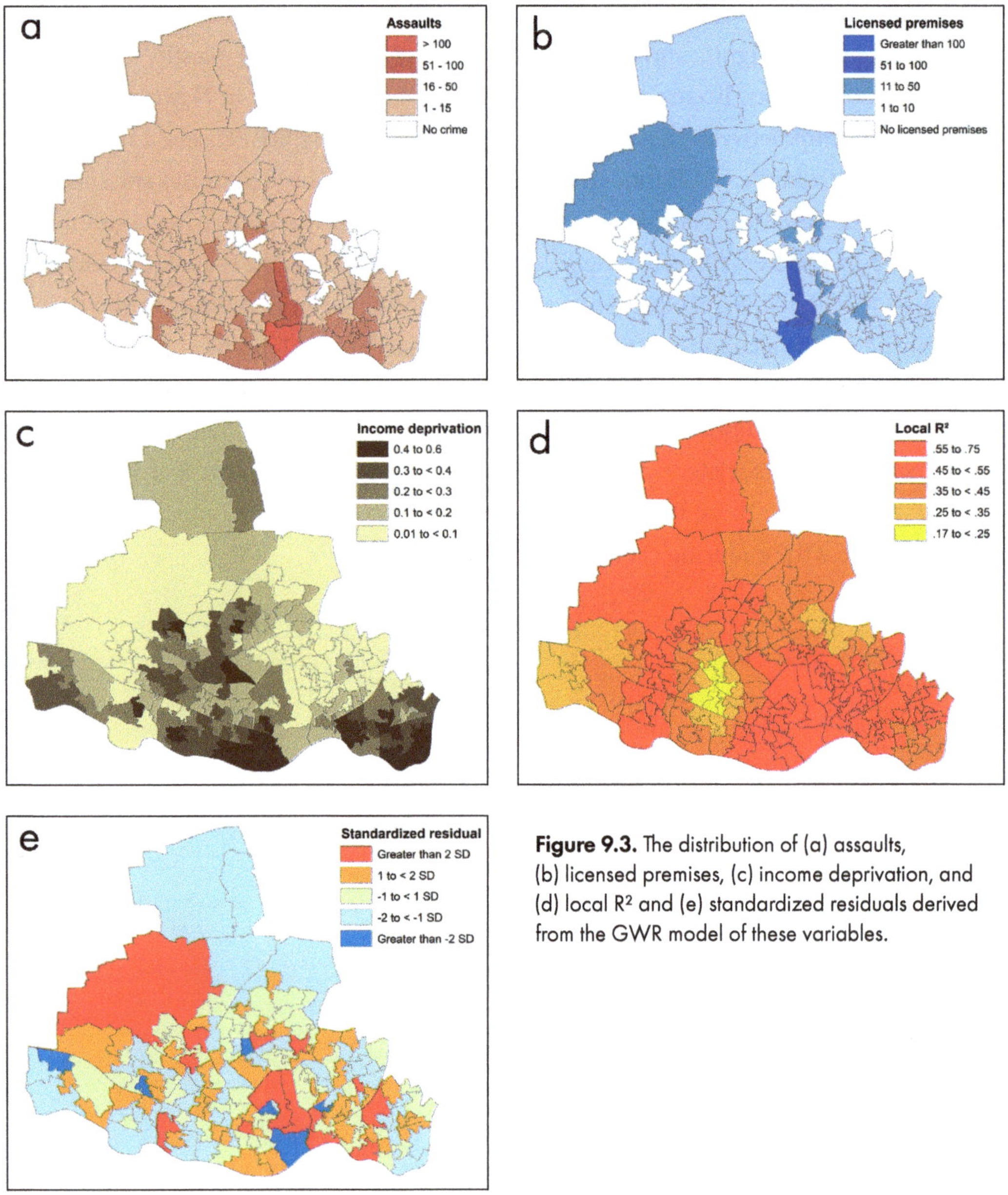

Figure 9.3. The distribution of (a) assaults, (b) licensed premises, (c) income deprivation, and (d) local R^2 and (e) standardized residuals derived from the GWR model of these variables.

Crime prediction and perpetuating bias

When predicting crime, estimating risk, or forecasting crime, care must be taken to ensure the results are not biased by the data inputs used to generate the results. Data must be selected that is evidenced to show its value in accurately predicting crime. For example, many peer-reviewed studies have shown that where street robberies previously have concentrated is where street robberies are likely to concentrate in the future. However, as has been shown in this chapter, these predictions are less accurate for the immediate future but more accurate for near future events. In turn, this temporal reality then dictates the type of response to apply—that is, implementing solutions that aim to counter the concentrated clustering of crime rather than implementing a tactical response plan that targets individuals. We also need to be confident in the source of our data inputs. If recorded crime data on street robberies is poorly recorded and not an accurate reflection of victimization across the population, we shouldn't use it. To predict, estimate risk, or forecast crime, the first thing we must do is qualify the data inputs we plan to use as being accurate in predicting crime, and determine the settings when they are not accurate. We also must consider whether adding other data as an input to our prediction calculation adds to the accuracy of the prediction, or whether it does the same thing as data we already carefully have chosen to include. We should not include data that contributes the same as other data because it will begin to bias our results. Also, if our data is not an accurate reflection of victimization across the population, it also may bias our results.

The same principles apply to the technique or technology solution we use to make our predictions, estimates of risk, or forecasts. If we don't understand how a technique or technology solution works but still use it, how can we be confident our results are accurate? Also, if the technique or technology has not been subject to the rigors of peer review, how can we trust it does what is claimed? Many peer-reviewed studies have shown that repeat victimization and near repeat victimization patterns can be valuable predictors of crime—for most types of crime, in most settings. When the patterning principles of repeats and near repeats are built into a technology solution, we can be confident most of the time in the predictions that result. For the types of crime and settings where repeat victimization and near repeat victimization are not reliable predictors of crime, we would not use them. If we did, we are likely to generate results that are inaccurate. We need to ensure we understand how the technique or technology uses our data to make the predictions so that we can be confident our results are accurate for the purpose intended.

If we find ourselves in a situation in which we are using data inputs for predicting crime by simply thinking they would be interesting variables to include, and we are uncertain how the technology solution works, we are unlikely to be able to qualify how each input contributes to the results. This could result in biasing attention to certain places and people. Absolutely, we must target where resources are deployed, but when doing so, we must be confident that we understand the inclusion of data inputs in any model of prediction and understand how the technology works so we can be confident that our predictions, estimation of risk, or crime forecasts are appropriate, while also protecting and safeguarding the civil rights and liberties of all people.

Risk terrain modeling is another type of composite mapping technique. RTM involves the analysis of several variables related to the crime type being studied, and from this modeling, generating a composite crime risk map. The method involves using the variables significantly related to crime to calculate weights that relate to the contribution that each one makes to the risk of crime. These weights then are summed for each geographic unit of the study area to indicate areas of high to low crime risk. Here, we do not go into detail on the use of RTM because it requires the purchase of specialized software, which can be acquired only by attendance in the RTM training course operated by researchers from Rutgers University. We direct you to the details provided in the "Further Resources" section at the end of this chapter to find out more about RTM. However, many examples of the application of RTM are available, including its use for examining assaults (Kennedy, Kaplan, and Piza 2016), residential burglary (Andresen and Hodgkinson 2018), robbery against tourists (Drawve et al. 2020), gun crime (Drawve, Moak, and Berthelot 2016), gang violence (Valasik 2018), assaults against police officers (Caplan et al. 2014), illicit drug markets (Escudero and Ramírez 2018), the likelihood of recidivism (Drawve, Caplan, and Ostermann 2019), and terrorist threats (Marchment, Gill, and Morrison 2019). We also refer you to several other resources on the use of RTM to consider its application in determining crime risk (Drawve 2016; Kocher and Leitner 2015; Ohyama and Amemiya 2018).

Chapter summary

Crime risk refers to the possibility of crime happening. Determining the risk of crime, however, is sensitive to temporal settings—the immediate risk, near risk, and enduring risk. To be accurate in determining crime risk for each of these temporal settings requires consideration of different data and different spatial analysis techniques, informed by the presence of different spatiotemporal patterns of crime. In turn, the type of responses then to implement to address crime risk will be different, informed by a better understanding of the factors, conditions, and causes that create the risk of crime. The crime risk framework provides a means of better structuring how we approach the task of mapping and analyzing crime risk, so that we make better choices in the data and techniques we use for determining crime risk, and in turn be better placed to respond to the risk of crime.

Further resources

Predictive mapping using the patterning principles of repeat and near repeat victimization:

UCL JDI briefing note, https://discovery.ucl.ac.uk/id/eprint/1344080.

UK Channel 4 News article, https://www.channel4.com/news/predictive-policing-tackling-crime-before-it-happens.

Risk Terrain Modeling: http://www.rutgerscps.org/rtm.html.

References

Andresen, M. A., and T. Hodgkinson. 2018. "Predicting Property Crime Risk: An Application of Risk Terrain Modeling in Vancouver, Canada." *European Journal of Criminal Policy Research* 24:373–92.

Caplan, J. M., L. W. Kennedy, and J. Miller. 2011. "Risk Terrain Modeling: Brokering Criminological Theory and GIS Methods for Crime Forecasting." *Justice Quarterly* 28 (2): 360–81.

Caplan, J. M., P. Marotta, E. Piza, and L. W. Kennedy. 2014. "Spatial Risk Factors of Felonious Battery to Police Officers." *Policing: An International Journal* 37 (4): 823–38.

Chainey, S. P. 2008. "Identifying Priority Neighborhoods Using the Vulnerable Localities Index." *Policing: A Journal of Policy and Practice* 2 (2): 196–209.

Chainey, S. P., and B. de Silva. 2016. "Examining the Extent of Repeat and Near Repeat Victimization of Domestic Burglaries in Belo Horizonte, Brazil." *Crime Science* 5 (1): 1-10.

Chainey, S. P., S. J. Curtis-Ham, R. M. Evans, and G. J. Burns. 2018. "Examining the Extent to which Repeat and Near Repeat Patterns Can Prevent Crime." *Policing: An International Journal of Police Strategies and Management* 41 (5): 608–22.

Chainey, S. P., L. Tompson, and S. Uhlig. 2008. "The Utility of Hotspot Mapping for Predicting Spatial Patterns of Crime." *Security Journal* 21:1–2.

Drawve, G. 2016. "A Metric Comparison of Predictive Hot Spot Techniques and RTM." *Justice Quarterly* 33 (3): 369–97.

Drawve, G., J. Caplan, L. W. Kennedy, and J. Sarkos. 2020. "Risk of Robbery in a Tourist Destination: A Monthly Examination of Atlantic City, New Jersey." *Journal of Place Management and Development*. Advanced access version.

Drawve, G., J. Caplan, and M. Ostermann. 2019. "Utilizing a Risk of Crime Measure for Recidivism Research: Moving Environmental Corrections Forward." *Crime and Delinquency* 65 (5): 606–29.

Drawve, G., S. C. Moak, and E. R. Berthelot. 2016. "Predictability of Gun Crimes: A Comparison of Hot Spot and Risk Terrain Modeling Techniques." *Policing and Society* 26 (3): 312–31.

Escudero, J. A., and B. Ramírez. 2018. "Risk Terrain Modeling for Monitoring Illicit Drug Markets across Bogota, Colombia." *Crime Science* vol. 7, no. 3.

Farrell, G., and A. Bouloukos. 2001. "International Overview: A Cross-National Comparison of Rates of Repeat Victimization." In *Repeat Victimization*, edited by G. Farrell and K. Pease. Crime Prevention Studies, vol. 12. New York: Criminal Justice Press.

Fielding, M., and V. Jones. 2012. "Disrupting the Optimal Forager: Predictive Risk Mapping and Domestic Burglary Reduction in Trafford, Greater Manchester." *International Journal of Police Science and Management* 14 (1): 30–41.

Kennedy, L. W., J. M. Caplan, and E. L. Piza. 2016. "Vulnerability and Exposure to Crime: Applying Risk Terrain Modeling to the Study of Assault in Chicago." *Applied Spatial Analysis* 9:529–48.

Kocher, M., and M. Leitner. 2015. "Forecasting of Crime Events Applying Risk Terrain Modeling." *Journal for Geographic Information Science* 1:30–40.

Marchment, Z., P. Gill, and J. Morrison. 2019. "Risk Factors for Violent Dissident Republican Incidents in Belfast: A Comparison of Bombings and Bomb Hoaxes." *Journal of Quantitative Criminology* (online-first article).

Ohyama, T. and M. Amemiya. 2018. "Applying Crime Prediction Techniques to Japan: A Comparison Between Risk Terrain Modeling and Other Methods." *European Journal of Criminal Policy Research* 24:469–87.

Valasik, M. 2018. "Gang Violence Predictability: Using Risk Terrain Modeling to Study Gang Homicides and Gang Assaults in East Los Angeles." *Journal of Criminal Justice* 58:10–21.

CHAPTER 10

Measuring performance and evaluating impact

Objectives

In this chapter, we describe several techniques for measuring how crime has changed over time as a result of crime interventions. Many techniques measure the impact of interventions—our focus in this chapter is on those that measure the change in crime as a result of geographically targeted interventions. Many police agencies operate CompStat-like processes to measure performance. These performance-monitoring settings usually track simple numeric and percentage changes in crime. In this chapter, we describe simple ways to measure crime over time that support performance-monitoring requirements, but then build on these by describing how to measure the impact of police and public safety interventions more robustly. We discuss the value of using control groups (or comparison areas), describe techniques that can help evaluate the impact of a targeted intervention more robustly, and explain how we can measure for displacement.

Key learning points

- Consider how to use simple performance measures that help monitor changes in crime. These measures include calculating the percentage change in crime and considering comparable periods for evaluation.
- Understand the value in using control groups to measure effectively the impact of crime.
- Understand the value in using randomized control trials (RCTs) to measure the impact of an intervention, the practical challenges in conducting an RCT, and the use of alternative approaches for robust evaluation. Alternatives include the use of quasi-experimental approaches and the weighted displacement difference test.

Introduction

A key question we want to answer is, *What is the impact that an intervention(s) has had in reducing crime?* Additionally, we should be aiming to answer this question: *How do we know that the reduction in crime is due to the intervention we implemented rather than being due to something else?* The simplest way to answer the first question is to calculate the change in crime before and after the intervention. For reasons that we discuss in the next section, although measuring the change in crime over time in this simple manner is quick and easy to do, it does not allow us to answer the second question. To determine whether an observed reduction in crime is due to the intervention implemented, we must use more sophisticated approaches.

Numerous techniques measure how crime changes over time. In this book, we already have described several, such as those for identifying diminishing hot spots of crime and emerging problem areas. In this chapter, the focus is on techniques for measuring how crime changes over time as a result of the activities and interventions we have designed and implemented to reduce crime. Numerous techniques also can be used to evaluate the impact of interventions, too many to cover in this chapter. In this chapter, we focus on practical techniques that will help monitor performance (e.g., to measure progress against a key performance indicator), and evaluate the impact of geographically targeted interventions. We illustrate how simple measures of performance help monitor the change in crime (e.g., for CompStat-style performance monitoring purposes) but do not allow you to conclude that we can be certain the change was due to an intervention implemented over the period of monitoring.

Using control groups can help us determine whether the reduction in crime is because of the activities we implemented. In this chapter, we describe how control groups can be created and used for measuring the impact of geographically targeted interventions. Randomized control trials are the gold standard in evaluation. RCTs help ensure that the evaluation findings that measure the impact of an intervention are robust. RCTs, however, can be difficult to employ and perhaps are not necessary for performance measurement scenarios such as CompStat-style settings. With this in mind, we describe how to use other techniques to evaluate effectively the impact of geographically targeted interventions. These measures include using multiple control groups, statistical modeling techniques, and impact measurement techniques, such as the weighted displacement quotient and weighted displacement difference test, which can be calibrated easily by following the advice we provide in this chapter.

The importance of evaluation

Too often, evaluation is treated as an afterthought, if it is done at all. Policing typically deals with the here and now, seeking to focus attention on the current issue, addressing it, and then moving on to the next one. But if we don't evaluate what we did to address it, how do we know if it worked? How can we be confident that our efforts were not a complete waste of time? Often in the world of policing and law enforcement, we assume that everything we do works. In turn, this often orients the police toward determining success from the mere existence of an initiative or by the number of tasks completed rather than measuring whether whatever we did worked. A pat on the back is given for doing something to reduce crime without questioning whether what was done worked and assuming that everything that is done provides some benefit.

Many things done by police and public safety agencies don't work and, in some instances, make things worse. To improve a profession, we must learn from what has worked for us, what has worked for others, and learn from our mistakes.

The evidence-based policing movement has encouraged improvement in the evaluation of interventions and activities and has helped develop a better knowledge base of what works and what doesn't in policing and public safety. This approach includes not only what works in reducing crime, but also has helped identify how practices and processes in policing (e.g., the criminal investigation process) work best, work badly, and can be improved. The following list includes several of my go-to websites on what works in policing and public safety.

Evaluation helps us learn lessons on what worked. Evaluation also helps us pilot new ideas, test them, and refine them so they are effective. Evaluation helps us be more accountable (by showing whether the resources we used had an impact) and can help us make better use of limited budgets (by focusing on the things that work rather than repeating old mistakes).

What works? Three useful websites with information on what works in policing and public safety:

Crime Solutions.gov, https://www.crimesolutions.gov/.
UK College of Policing What Works in Crime Reduction Toolkit,
https://whatworks.college.police.uk/toolkit/Pages/Welcome.aspx.
Center for Problem-Oriented Policing, https://popcenter.asu.edu/.

Measuring performance—simple statistics

The simplest way to measure the change in crime, whether for a CompStat-style performance-monitoring purpose or to measure the impact of a geographically targeted intervention, is to compare crime levels for a before period to the most recent period (e.g., the period after the intervention began). As an example, let us imagine that a crime prevention intervention to reduce street robberies and thefts began on July 1 and operated for six months. We would need to compare the levels of crime for periods of the same length, so we could compare the six months before the intervention and the six months after the implementation to the six months during which the intervention was active. Table 10.1 lists statistics for these periods, showing that during the intervention, robberies and thefts declined by 828 crimes, equivalent to a 9.6 percent reduction. During the six months after the intervention, robberies and thefts increased by 99 crimes, equivalent to a 1.3 percent increase. The conclusions you initially could draw from these statistics are that there was a good reduction in crime while the intervention was operating, but after the intervention, crime levels increased a little. However, the crime levels during the post-intervention period were more than 8 percent lower than they were before the intervention, suggesting a reasonable degree of sustainability in the reductions achieved after the intervention was implemented.

Table 10.1. Street robberies and thefts for the six-month periods before, during, and after an intervention

	Before intervention	During intervention	Post intervention
Recorded crime	8,659	7,831	7,930
Percentage change	---	−9.6%	1.3%

CompStat and performance monitoring

CompStat is an operational management process implemented by many police agencies across the world. Although police agencies may have changed the name they use for this process, and may operate it in a slightly different format from the original processes implemented by the New York City Police Department, four central principles are typically in play (Schick 2004):

- Timely and accurate intelligence
- Effective tactics
- Rapid deployment
- Relentless follow-up and assessment

The CompStat process aims to provide actionable information and intelligence to operational commanders on a regular basis, such as data on crime levels for the last few days and weeks, and to monitor whether the activities they have implemented are having a positive effect in reducing crime and improving public safety. The operational commanders are held accountable for what they do, with follow-up on the activities they implemented and a review of crime statistics the next time a meeting is held.

A challenge that many operational commanders face is determining the *effective tactics* to use. When there's a lack of effective evaluation in police agencies, determining what works from what does not and not giving time to review the base of evidence often leaves operational commanders relying on their experience ("*We did it before. I am sure it worked, so let's do it again*"). Effective evaluation helps generate evidence on tactics that are effective.

Operational dashboards have become an increasingly popular tool for collating and presenting information to support CompStat and other performance-monitoring processes. An example of how to use this type of operational dashboard for showing statistics, trends, and hot spots is shown here: https://www.arcgis.com/apps/opsdashboard/index.html#/35966e2ccd464a3c98a2ba6ae20891a9.

An operational dashboard from the Rochester Police Department (New York, USA) on shootings provides an excellent crime-specific example of how this approach can measure changes in these types of crime over time: https://modelpd.maps.arcgis.com/apps/opsdashboard/index.html#/415ef13fe15e48ed848b66c52b915efb.

A problem with the before and post-intervention periods used in table 10.1 is that they refer to different times of the year from when the intervention was active. Instead, it is better to compare the period when the intervention was operating to the same periods in the year before and the year after the intervention (i.e., compare crime levels for July 1 to December 31 for each year). This is particularly important if the type of crime being analyzed is sensitive to seasonality.

Table 10.2 shows robbery and theft statistics for the period during which the intervention operated (July to December), for the same period in the previous year, and for the same period in the year after the intervention. The statistics show that during the intervention, robberies and thefts dropped by 628 crimes, a 7.4 percent reduction. In the following year, for the same period, robberies and thefts dropped by 1,296 crimes (in comparison with the same period before the intervention was active), equivalent to an 8.5 percent reduction. The conclusions you could draw from these statistics are that a good reduction in crime occurred during the period the intervention was operating, but in the year that followed when no intervention was in place, robberies and thefts dropped even more. From this decline, we then may question whether it was the intervention that caused the reduction. That is, if we had done nothing, would we still have experienced a reduction of about 7–8 percent, like that experienced in the year that followed?

Table 10.2. Street robberies and thefts compared over six-month periods

	Same period, previous year	**During intervention**	**Same period, following year**
Recorded crime	8,459	7,831	7,163
Percentage change	---	-7.4%	-8.5%

Thinking about evaluation before the implementation of an intervention

When thinking about evaluation, it is important to consider how you are going to evaluate the intervention or activity for a period before it commences. You will need a baseline to compare against when you measure the impact of the intervention at a future point. The data you use for the baseline already may be collected (e.g., recorded crime data), making your evaluation reasonably straightforward. In some situations, data may need to be collected before the intervention commences (e.g., conducting a victimization survey or a survey of current operating practices).

Even when data is available before the intervention on how it is going to be measured, it is important that the data is checked to ensure the content is sufficient. For example, if the intervention aims to reduce thefts and robberies against schoolchildren, does the recorded crime data contain enough information to determine accurately whether the crime was committed against a schoolchild? Also, does any information currently exist on the level of crime reported by schoolchildren? An awareness campaign that supports the intervention to reduce thefts and robberies against schoolchildren may result in an improvement in the reporting of these crimes by schoolchildren themselves. So, if you had not considered the latter in your evaluation, and the results from your impact analysis show there has been no change in the recorded number of thefts and robberies against schoolchildren, rather than determining the intervention was unsuccessful, perhaps instead it *did* reduce thefts but at the same time increased the confidence schoolchildren had in reporting these crimes. That is, the reporting of crime by schoolchildren increased, and offset the reduction in crimes against schoolchildren that were a result of the new intervention.

The most robust way to evaluate an intervention or activity is to set up experimental conditions that compare what happened in your treatment group to what happened to a control group. The treatment group could be the areas in which the intervention is targeted, and the control group could be other areas to compare with the intervention.

The example illustrated in tables 10.1 and 10.2 is a useful example of some of the problems, and potentially wrong conclusions, we can reach when we use only simple before and after statistics to measure the impact of interventions or activities. In addition, coarse aggregations of data to large periods may hide some of the detail in the change in crime over time. Figure 10.1 shows the change in robberies and thefts, by month, for the full 30-month time span for the periods used in table 10.2. During the before-intervention period, crime levels in general remained at a steady level (albeit with some fluctuations month to month). During the intervention, a large reduction in crime was experienced in the first month of the intervention (comparing statistics for June to July), remained at approximately the same level for three months, but then began to increase a little over the last two months of this period. During the after-intervention period, although crime levels were lower than during the intervention, a large reduction was observed in the first month of the after-intervention period. From this first month, the crime trend during the after-intervention period was, in general, an increasing one.

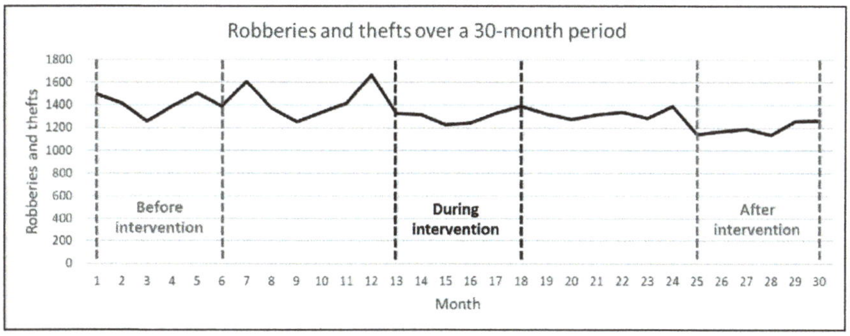

Figure 10.1. Changes in robberies and thefts over a 30-month period.

This example refers to an area in a large city in which a robbery and theft reduction intervention was targeted. Simple before and after statistics can provide insight into the impact an intervention has had but do not allow you to conclude whether it was the intervention itself that was responsible for the reductions. Did the intervention work? Possibly, but we cannot be certain. Should we replicate the intervention in the future? Possibly, but we cannot be certain it will work, because we don't really know whether it worked the first time we implemented it.

Crime displacement

Crime displacement is the movement of crime from its place or position. Crime displacement comes in many forms: temporal displacement (from one period of time to another, such as by the time of day); target displacement (the offender changes their offending by committing crimes against different targets); tactical displacement (the offender changes how they offend by using different tactics); displacement by crime type (the offender changes their offending by, for example, committing burglaries from businesses rather than burglaries from houses); and geographic displacement (the movement of crime from one location to another location).

The results from many evaluations of geographically targeted interventions indicate that crime does not get displaced simply from one location to another (Braga, Papachristos, and Hureau 2014; Weisburd and Telep 2014). Most of these studies have observed that crime levels decline in areas targeted by an intervention and may increase a little in the other areas that did not receive the intervention, but that the overall net effect is a reduction in crime across the whole area.

Crime displacement is always a threat but is not inevitable. Consider the following—our analysis identifies an area that is a crime hot spot. We decide to focus a crime reduction intervention in the area covered by the hot spot. The reason why the hot spot exists is because it is where the most favorable conditions for crime exist. The intervention is implemented, crime declines in the hot spot, and it does not increase in the surrounding areas. Why? In the surrounding areas, the same favorable conditions for crime do not exist. If they did, the surrounding area most likely also would have been a hot spot before the intervention.

But what about the offenders who previously committed crime in the hot spot? They just don't become, suddenly overnight as a result of our intervention, good, law-abiding citizens. We have limited their opportunities to commit crime by taking away the easiest opportunities present in the hot spot. They are likely to continue to commit crime, but now less crime than they did before. Our intervention in the crime hot spot effectively has removed the easy opportunities they previously were taking advantage of.

In some cases, the evidence suggests that not only is crime reduced in the area targeted by an intervention, but it also reduces a little in areas that immediately surround where the intervention is deployed. This is known as a *diffusion of benefits effect,* and research has found it to be just as likely to occur as displacement (Braga et al. 2012; Bowers et al. 2011).

Using control groups

The benefits

Control groups help us determine whether what we did (i.e., an intervention or activity) is responsible for the reduction in crime. By targeting interventions to certain geographic areas, we can compare the change in crime in these areas (our treatment areas) to similar areas (our control areas). By doing so, we can observe whether the reduction in crime was observed only in our treatment areas, and not in our control areas. This would lead us to be more confident that what we did in our treatment areas resulted in reducing crime because the same reductions were not observed in our control areas. Control areas also provide a means of measuring whether a geographic displacement of crime occurred, or whether there has been a diffusion of benefit effect from our treatment areas to other areas. In the next section, we describe techniques for measuring displacement and the diffusion of benefit.

Hierarchy of evaluation reliability (and difficulty in using)

How we create and use control groups influences the robustness of the evaluation we can perform. This hierarchy is displayed in figure 10.2. Our control groups should be as similar as possible to our treatment areas. That is, similar in size, population, land use, street layout, and levels of crime before the implementation of any intervention. When targeting interventions to reduce crime, it is likely the area we are targeting has a much higher level of crime than other areas (hence, why we are targeting it), adding an initial complication to consider. We also must be careful in choosing control areas that are not going to be *contaminated* by the intervention in our treatment areas. This procedure most likely involves selecting control areas that are far enough away from our treatment areas so that the activities associated with the intervention influence only our treatment areas.

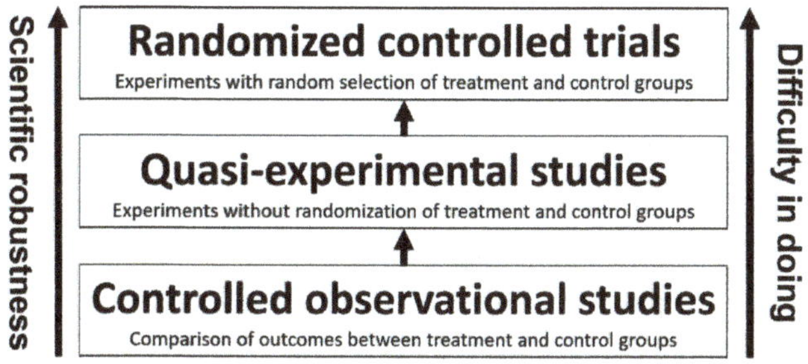

Figure 10.2. The hierarchy of evaluation reliability for studies using control groups.

Determining control areas

Determining control areas involves three main approaches: user selection (based on user knowledge and experience), selection using quantitative analysis, and randomized selection.

- User selection: relies on knowledge of the study area to choose suitable areas as controls. This approach usually means our evaluation will be under the *controlled observational studies* category of evaluation techniques. Our evaluation will be more prone to bias (because of our personal selection of the control areas), and hence less scientifically robust but likely easier to do. Figure 10.3a provides an illustration, showing control areas chosen on the basis of what were believed to be similar areas to the treatment area. Each treatment area has been paired with an area considered to be a suitable area for comparison. Another example would be to choose a neighborhood as the treatment area and then the rest of the city as the control.
- Quantitative analysis: uses quantitative techniques to determine control areas. This approach could involve the analysis of crime data for the study area to help identify the areas that are most like the treatment areas (i.e., similar in crime levels before the intervention). For example, if the intervention is deployed to several neighborhoods in a city, the controls could be neighborhoods in a similar city nearby. The analysis also could include reviewing data on the population and land use in the treatment and control areas to help ensure they are similar. Figure 10.3b provides an illustration of this approach. The treatment areas are the same as those shown in figure 10.3a but were paired with comparable areas after an analysis of crime, land use, and population across the whole study area. Using this approach, the evaluation likely would fall under the *quasi-experimental studies* category.
- Randomized selection: involves identifying the potential areas to be treated, and then randomly assigning each area to a treatment or control group. Only those areas assigned

to the treatment group receive the intervention, whereas the others remain as controls for the duration of the intervention. Figure 10.3c illustrates this approach by showing 12 areas that potentially could receive the intervention, but where six have been selected randomly as treatment areas and six as control areas. This approach means our evaluation would fall under the *randomized controlled trials* category of evaluation techniques.

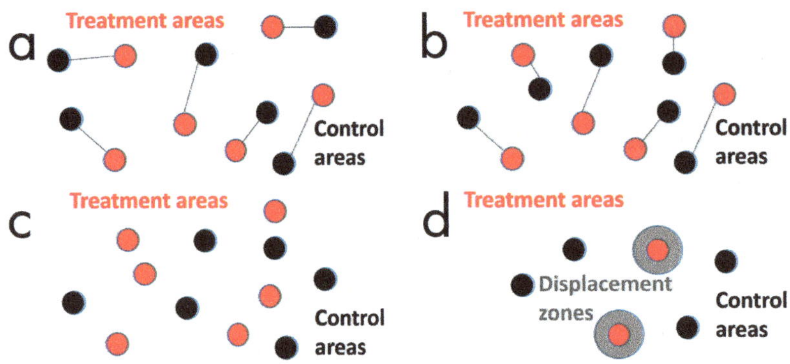

Figure 10.3. Selection of treatment and control areas based on (a) user selection, (b) quantitative analysis selection of control areas (linked to relevant treatment areas), (c) random selection of treatment and control areas, and (d) user selection of displacement zones and control areas.

The random selection of treatment areas and control areas means that the results of an evaluation examining the change in crime between these two groups will be more reliable and less prone to bias. However, the randomized selection approach may be more difficult to implement because it requires the control areas, which previously have been identified as potential treatment areas (experiencing high levels of crime), to be liable to experience continually the high levels of crime that initially identified them for possible treatment. It is important to recognize that control areas continue to receive the attention that all areas received before the intervention (i.e., business continues as usual in the control areas), and that it is only the additional resources associated with the intervention being targeted to the treatment areas that the control areas do not receive. By setting up treatment and control areas in this manner, we will know, with greater certainty, if the intervention works to reduce crime. If the intervention reduced crime, we then could expand the intervention to the areas initially selected as control areas.

Control areas must be far enough away from any treatment areas to ensure they are not influenced by the intervention implemented in the treatment areas. Although we should attempt to ensure the intervention is applied only within the boundaries of each treatment area, if the control area immediately neighbors the treatment area, there is a risk of *contamination* to the control area from the treatment area. However, as we often are concerned about geographic crime displacement, to areas close to the areas in which interventions are targeted, we also may want to measure whether local displacement has occurred. To do so may require us to define another group of control areas, which we shall call *displacement zones*—areas that neighbor or surround a treatment area. Usually, a combination of user selection and quantitative analysis approaches is required to help us select these displacement zones. Our evaluation then can involve comparing our treatment areas with control areas and displacement zones to provide a comprehensive assessment of the impact our intervention has had on crime. Figure 10.3d provides an illustration, showing a displacement zone surrounding a treatment area, and control areas positioned away from the treatment areas.

The influence of the number of areas on the reliability of the results

The number of areas assigned to treatment and control areas affects the reliability of the findings from the evaluation. RCTs and quasi-experimental studies typically require many treatment and control areas to ensure the results are robust. If the effect of the intervention is small, and the number of areas in the treatment and control groups is also small, the results could imply that the intervention did not have a statistically significant effect on reducing crime. As the number of areas in each treatment and control group increases, the same evaluation approach may demonstrate that the intervention had a statistically significant effect, even if the reduction in crime was small.

Evaluating the impact

In this section, we describe several techniques for evaluating the impact of geographically targeted interventions: a randomized control trial approach, which includes a measure of odds ratios; a quasi-experimental approach, which includes a difference-in-differences statistical technique; and the use of the weighted displacement quotient and weighted displacement difference test applied to both a large and small number of treatment and control groups.

Randomized control trial

Selection of treatment and control areas

We use the Philadelphia Foot Patrol Experiment (Ratcliffe et al. 2011) to illustrate an example of an RCT. The Philadelphia Foot Patrol Experiment was a violence reduction hot spot policing program targeted to 60 areas in the city of Philadelphia. Initially, 120 potential treatment areas were identified. A three-step process, illustrated in figure 10.4, was used to determine treatment and control areas. The process was as follows:

1. For each of the 120 areas, the number of recorded violent crimes (in each area) for the previous three-year period (2006–08) was calculated but giving greater weight to the most recent year: crimes recorded in 2008 were weighted as 1.0; crimes from 2007 as 0.5,

and crimes from 2006 as 0.25. The 120 areas were then ranked in descending order by the weighted number of violent crimes in each area.

2. Each area was then paired with the next area in the ranked list. For example, the area ranked first was paired with the area ranked second, the area ranked third was paired with the area ranked fourth, and so on until the final areas (119th and 120th) were paired. This process created 60 pairs, each with similar levels of crime.
3. For each pair, a random selection process was used to assign one area to the treatment group and one to the control group. This assignment was performed simply by the toss of a coin. If the coin landed on heads, the first area listed in the pair was assigned to the treatment group. If the coin landed on tails, the second area was assigned to the treatment group. This coin toss was repeated for all 60 pairs, creating a list of 60 treatment areas and 60 control areas.

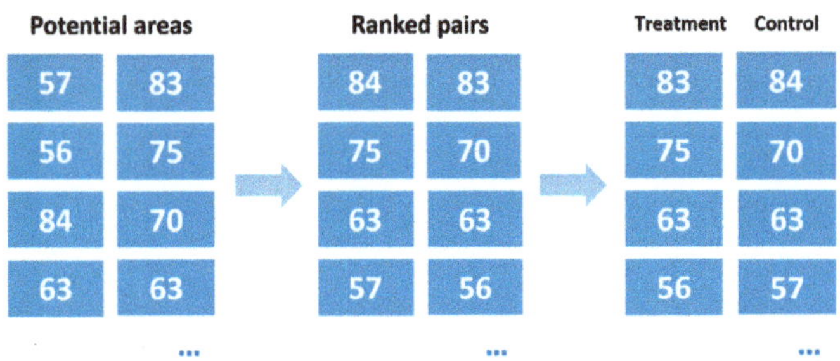

Figure 10.4. Random selection of treatment and control areas, based on initial ranking of levels of violent crime in each area (showing details for eight areas that potentially could receive the intervention).

Techniques used in the evaluation: Change in number of crimes, the odds ratio, percentile groupings, and the F-test

The evaluation of the Philadelphia Foot Patrol Experiment involved several techniques for comparing the change in crime in the treatment areas to the change in crime in the control areas. The intervention operated for three months and compared the three months during the operational period of the intervention to the three months before its implementation. During the intervention, the number of violent crimes in the control area declined by 53 offenses but increased in the control areas by 31 offenses (table 10.3). Thus, the overall effect of the intervention, when comparing treatment and control areas, was a reduction of 22 offenses, equivalent to a 3.4 percent reduction. However, this approach does not consider the differences in the level of crime in the two areas for the before period and can result in an overestimation or underestimation of the effect of the intervention. One way to correct this issue is to use an inverted odds ratio to calculate the reduction in crime between the two areas, as follows:

$$\text{Odds ratio} = 1/((a \times d)/(b \times c)), \qquad (10.1)$$

where a is the count of crimes in the treatment areas before the intervention, b is the count of crimes in the treatment areas during the intervention, c is the count of crimes in the control areas before the intervention, and d is the count of crimes in the control areas during the intervention. Substitution of the crime data from table 10.3 into equation (10.1) gives:

$$\text{Odds ratio} = 1/(359 \times 327/306 \times 296).$$

This calculation produces an odds ratio of 0.77, which is equivalent to a percentage change in the target areas relative to the control areas of a reduction of 23 percent.

Table 10.3. Violent crime in treatment and control areas for the Philadelphia Foot Patrol Experiment

Areas	Before intervention	During intervention	Change in crime (%)
Treatment	359	306	−53 (14.8%)
Control	296	327	31 (10.5%)
Treatment and control	655	633	22 (3.4%)

Testing for the statistical significance of this change in crime can be a challenge. The simplest way to test for significance in this scenario is to calculate a *change score*, based on the difference between the before period and the period during which the intervention operated. However, this approach is not appropriate because it does not account for the differences in levels of crime in the before period for the treatment areas and control areas, across the ranked pairs of treatment and control areas. For instance, although the total number of violent crimes for all areas in the treatment groups was similar before the intervention to the total for all areas in the control groups, several treatment and control areas experienced much lower levels of violent crime than other treatment and control areas before the intervention. Effective evaluation must account for differences in the level of crime between treatment and control areas before the intervention. A second option would be to use independent sample t-tests to determine whether the change in crime between the before and during periods was significantly different for the treatment and control areas. However, this approach fails to adequately control for short-term changes in crime (as in the case of the Philadelphia Foot Patrol Experiment, which examined changes associated with an intervention that lasted only three months).

A suitable approach in these circumstances would be to examine the data further to see if there were differences in the change in crime for treatment and control areas that had similar levels of crime before the intervention. The Philadelphia Foot Patrol Experiment evaluation used this approach, involving a comparison of changes in crime among the pairs with the highest levels of violence, and separately comparing changes in crime among pairs with the lowest levels of violent crime. This approach involved organizing the pairs into five percentile groupings, based on violent crime levels for the before period, corresponding to the 20th, 40th, 60th, 80th, and 90th percentiles (e.g., for the 20th percentile, comparing only the lowest-ranked 12 pairs of treatment areas and control areas, and then for the 90th percentile, comparing only the top-ranked six pairs of treatment areas and control areas). The differences in the number of violent crimes for the treatment and control areas then were assessed to determine whether the intervention had a significant impact, relative to the levels of crime before the intervention. To do so, an *F-test* statistical test was applied to compare the variation between the means for the treatment and control groups for each percentile. The results (table 10.4) show that for areas organized into the 20th and 40th percentile groupings (i.e., the treatment and control areas in which violent crime levels were low), there were no differences between treatment and control groups before and during the intervention. For the 60th, 80th, and 90th percentile groups, treatment areas experienced less violent crime than their control area counterparts, with this finding being significant to at least $p = .05$.

Table 10.4. Differences in violent crime in treatment and control areas, by percentile group

Percentile (average number of violent crimes before intervention)	Treatment area—violent crime during intervention	Control area—violent crime during intervention	Difference (Treatment area—control area)	F	Significance
20th (3.0)	4.449	3.936	0.513	0.538	0.47
40th (4.5)	4.667	4.719	−0.052	0.007	0.94
60th (6.0)	5.104	6.286	−1.182	3.919	0.05
80th (8.0)	5.540	7.852	−2.312	10.148	0.002
90th (11.0)	6.195	10.202	−4.007	13.705	< 0.001

Note: Based on levels of violent crime before the intervention.

Measuring impact using harm

Rather than using the number of crimes, the impact of an intervention can be measured instead based on the change in harm. Harm refers to the consequences of criminal activities (Von Hirsch and Jareborg 1991; Ratcliffe 2015), with each crime incident weighted according to its consequential impact. Rather than treating each crime event equally, each crime event is weighted by its harmful effect. Most harm studies use sentencing guidelines to calculate the harm weights, which involves multiplying each crime incident by the number of days an offender is imprisoned (Curtis-Ham and Walton 2017; Sherman, Neyroud, and Neyroud 2016).

Using a harm index rather than only the number of crimes to measure the change in crime is most useful when an evaluation involves examining the change in a variety of types of crime. For example, if the impact of an intervention was measured in relation to the change in robberies and thefts, robberies in which a weapon was used in the commission of the crime would have a greater weight than minor theft offenses. If thefts increased by the same number that robberies declined, the impact in relation to harm would indicate a reduction. The Danish Harm Index produced by Andersen and Mueller-Johnson (2018) is a good example of how the change in crime can appear different when a harm measure is used compared with a change in only the number of crimes.

Quasi-experimental evaluation

Selection of treatment and control areas

The evaluation of a hot spot policing program in Montevideo, Uruguay (Chainey, Serrano-Berthet, and Veneri 2020), provides an illustration of a quasi-experimental evaluation. PADO (Programa de Alta Dedicación Operativa, the name of the Montevideo hot spot policing program) was implemented in April 2016 and was one of the first hot spot policing programs in Latin America. Patrols were predominantly foot patrols (with each patrol consisting of two or three police officers) and operated between the hours of 17:00 and 01:00 on weekdays and Saturdays (informed by a temporal analysis of robbery patterns). PADO used existing police resources but in a more dedicated and targeted manner.

At the time PADO was implemented, little thought was given to how to evaluate the program. A total of 120 areas were identified for the hot spot policing program, each area consisting of between one and five street segments. In total, PADO covered 736 street segments and 649 street junctions in Montevideo (table 10.5). No control areas were determined initially, but two groups of controls were subsequently created to support the evaluation of the hot spot policing program: control area 1, consisting of street segments and street junctions within 300 meters of a PADO patrol street segment or junction; and control area 2, consisting of all other street segments and street junctions in the city. PADO's evaluation approach followed a quasi-experimental design, in that the treatment areas were predetermined (based on the condition that they were hot spots of crime), and the control areas were selected based on a criterion (rather than random selection) that would permit their comparability to the treatment areas. For instance, street segments and junctions that control area 1 comprised would allow for the measurement of any local displacement effects, and street segments and junctions in control area 2 would allow for comparison in the change of crime across the rest of the city. A difficulty this approach creates for robust evaluation is that, from the outset, the treatment and control groups are unlikely to have equivalent baselines (i.e., they are likely to have experienced different levels of crime before the program's implementation), which in this case is highly likely because the treatment areas were hot spots, experiencing high crime levels.

Table 10.5. Units of analysis for the evaluation of PADO—treatment street segments and junctions and control areas

	PADO areas (treatment)	Control areas 1	Control areas 2	Total
Distance (meters) from PADO areas	---	1–300	> 300	
Street segments (% of all segments)	736 (2.5%)	5,561 (19.3%)	22,550 (78.2%)	**28,847 (100%)**
Street junctions (% of all junctions)	649 (3.3%)	3,915 (19.6%)	15,365 (77.1%)	**19,929 (100%)**

Techniques used in the evaluation: Change in number of crimes and the difference-in-differences panel data model

Rather than using only a before and after comparison of crime levels, a difference-in-differences approach was used in the evaluation of PADO to measure the impact of the program. A difference-in-differences approach can help measure the impact of an intervention when control areas are not determined at the outset of the intervention. A difference-in-differences approach involves examining the change in outcome (e.g., the change in crime) in a group of nonparticipants (e.g., control areas in which PADO was not deployed) to estimate the change in outcome if the program did not exist. The approach then compares the differential effect between areas in which PADO was deployed and the control areas by determining whether the change in crime was statistically significant. Some knowledge of statistical modeling is required to use this difference-in-differences approach. In the next section, we describe software that applies this method but, in this section, refer to the specific equations used in this evaluation and computed using standard statistical modeling software.

Recorded crime data on robbery for the period January 2013 to December 2016 was used to examine the impact of PADO. Before the start of the program in April 2016, Montevideo had experienced consecutive annual increases in robberies for over a decade. For example, in the two years before the implementation of PADO (in 2014 and 2015), the increase in robberies against the previous year was 14 percent and 10 percent, respectively. The first step of the evaluation (after the creation of the control areas) required assigning the robbery data to one of the three groups of areas: a group representing the treatment areas (street segments and street junctions covered by the PADO hot spot policing foot patrols) and the two control groups. Each robbery record then was aggregated into monthly crime counts for the 48 one-month periods from January 2013 to December 2016 for each street segment and street junction. This data then was used as the input data for the evaluation models. Each street segment or junction was coded with data representing the number of robberies that took place at each location for each month across the three-year period before the implementation of PADO, and for each of the nine months from April to December 2016 when PADO was active.

Table 10.6 provides before and after robbery statistics for the treatment areas and control areas. The table shows that the total number of robberies in the PADO treatment areas increased by 37 percent in the year before the implementation of PADO, compared with increases of 8 percent and 5 percent in control areas 1 and 2, respectively. In the year PADO was implemented, robberies declined in the treatment areas by 31 percent, declined by 2 percent in the areas immediately surrounding where PADO was deployed (i.e., in control area 1), and increased by 1.5 percent in all other areas of the city (i.e., in control area 2). In total, after increases in robbery of over 10 percent in Montevideo for the two years before the implementation of PADO, in 2016 robberies declined by 5.7 percent across the city—the first decrease in more than a decade. These results provide an initial indication that PADO was influential in the reduction in robbery. The use of a more sophisticated evaluation technique can provide greater certainty that it is likely the reductions were because of PADO.

Table 10.6. Robbery in the PADO treatment areas and control areas in Montevideo, Uruguay (2013–16)

Areas	Robbery	2013	2014	2015	2016
PADO treatment areas	*n*	1,958	2,486	3,403	2,338
	% change	---	27.0%	36.9%	−31.3%
	Average per month (per unit of analysis)	1.41	1.79	2.46	1.69
Control area 1 (1–300 meters)	*n*	2,501	2,971	3,215	3.147
	% change	---	18.8%	8.2%	−2.1%
	Average per month (per unit of analysis)	0.26	0.31	0.34	0.33
Control area 2 (> 300 meters)	*n*	9,118	10,045	10,517	10,676
	% change	---	10.2%	4.7%	1.5%
	Average per month (per unit of analysis)	0.24	0.27	0.28	0.28
Total	***n***	**13,577**	**15,502**	**17,135**	**16,161**
	% change	**---**	**14.2%**	**10.5%**	**-5.7%**
	Average per month (per unit of analysis)	**0.28**	**0.32**	**0.35**	**0.33**

To further evaluate the impact of PADO on robbery in Montevideo, a difference-in-differences negative binomial panel data model was created to compare between treatment and control areas, in equation (10.2):

$$Y_{it} = \alpha + \beta_1 D_{PADO} + \beta_2 T_{Period} + \beta_3 (T_{Period} * D_{PADO}) + \gamma_{year} X_{year} + \sum_{month=2}^{12} \theta_{months} D_{month} \qquad (10.2)$$

The model included a dummy variable to indicate whether the unit of analysis (a street segment or street junction) was in the treatment group ($D_{PADO} = 1$) or in a control group ($D_{PADO} = 0$). The variable T_{PERIOD} in the model identified whether the data was from the period before the implementation of PADO ($T_{PERIOD} = 0$) or for the implementation period of PADO ($T_{PERIOD} = 1$). The use of the model represented in equation (10.2) means that difference-in-differences estimates could be calculated by comparing robberies between areas (e.g., treatment areas in comparison with control 1 areas and treatment areas in comparison with control 2 areas), with respect to the different periods (i.e., the periods before PADO and the period PADO was operating). The main value of interest in the model is for β_3. β_3 is a coefficient value associated with the interaction of D_{PADO} and T_{PERIOD}. If β_3 is less than 1, it indicates a decrease in the rate at which the robberies occurred in the treatment areas, in comparison with the control area it is compared to; if β_3 is greater than 1, the coefficient would indicate the treatment areas experienced an increase in the robbery rate, in comparison with the control area it is compared to. Yearly trend (γ_{year}) and seasonal trend (θ_{month}) variables also were included in the model as controls. The β_3 value is presented as an *incident rate ratio* (IRR) and can be interpreted as the rate at which the dependent variable occurs while controlling for other factors. For example, a β_3 value of 0.9, generated by comparing treatment areas with a control area, would suggest that the treatment (the PADO hot spot police patrols) was associated with a 10 percent decrease in the rate at which crime occurs, in comparison with the control area.

The model represented by equation (10.2) was run four times:

- Model 1 compared the change in robberies between treatment areas and the control 1 areas.
- Model 2 compared the change in robberies between treatment areas and the control 2 areas.
- Model 3 compared the change in robberies between treatment areas and all control areas (i.e., combining control 1 and control 2 areas to create a single area).
- Model 4 compared differences between control area 1 and control area 2 to further examine displacement effects.

When PADO areas are compared with the control 1 areas using the model represented in equation (10.2), the model generates a statistically significant IRR of 0.789 ($p < .01$). This result is shown in table 10.7 as the β_3 value of 0.789. This is equivalent to a 21.1 percent reduction in the rate at which robberies occurred in the treatment areas compared with neighboring areas. When PADO areas are compared with control 2 areas, the model generates a statistically significant IRR of 0.769 ($p < .01$), equivalent to a 23.1 percent reduction in robbery in PADO areas compared with the rest of Montevideo. When comparing PADO areas to all non-PADO areas of Montevideo (i.e., control areas 1 and 2 combined), there was a statistically significant IRR of 0.774 ($p < .01$), equivalent to a 22.6 percent reduction in the rate of robberies. Table 10.7 also shows that the IRR of 0.974 between control area 1 and control area 2 was not significant, indicating there was no evidence of geographic displacement from PADO treatment areas to any non-PADO areas. If anything, the result from the fourth model indicates that areas immediately surrounding PADO areas marginally benefited from the intervention, as previously indicated in the reduction of robberies in control area 1 of 2.1 percent (shown in table 10.6). Although the IRR of 0.974 from the fourth model was not significant, it indicates a reduction of 2.6 percent in the rate of robberies in control 1 areas in comparison with control 2 areas.

Table 10.7. A comparison of changes in PADO treatment areas to control areas in Montevjdeo following equation (10.2)

	Model 1: PADO vs. Control 1	Model 2: PADO vs. Control 2	Model 3: PADO vs. all controls	Model 4: Control 1 vs. Control 2
IRRs	0.789*** (0.0311)	0.769*** (0.0258)	0.774*** (0.0255)	0.974 (0.0269)
Equivalent percent change in the rate at which robberies occurred	21.1%	23.1%	22.6%	2.6%
Observations (number of units times 48 months)	150,192	544,944	667,104	639,072

Note: Standard errors are shown in parentheses. The coefficients are expressed as incident rate ratios (IRR).
*$p < .1$, **$p < .05$, ***$p < .01$

Before the implementation of PADO, robberies had been increasing year over year. The simple before and after statistical results that compared the levels of crime between the PADO treatment areas and control areas provided a useful indication that the hot spot policing program had a good impact on reducing robberies. By using two sets of controls that represented areas that neighbor the PADO patrol areas and that represented all other parts of the city of Montevideo, the change in crime in the PADO areas could be compared against change in crime in the control areas, and displacement effects could be measured. These results indicated that robbery declined by over 30 percent in the PADO areas (see table 10.6), marginally declined by 2 percent in neighboring areas to which PADO was deployed, and increased slightly (by 1.5 percent) across all other parts of the city. The difference-in-differences technique provides a means of examining differences in the change in crime between areas and determining whether this change is statistically significant. The results from the difference-in-differences models provide further indication that PADO had a significant impact in reducing crime and that no significant geographic displacement occurred. Collectively, this example illustrates the use of control areas, outside an RFT, and how the use of two approaches (descriptive statistics and statistical modeling) can help us evaluate the impact of a geographically targeted intervention.

Technique limitations

The methods used in this evaluation are not without their faults. Although the difference-in-differences technique helps provide a robust means for determining whether treatment areas have experienced differences in crime trends in comparison with control areas, the differences in crime levels and changes in crime in treatment areas and control areas before the implementation of an intervention could mean that the true effect of the program was underemphasized. Robberies had been significantly increasing in the areas in which PADO was to be deployed before its implementation. In the two control areas, robbery levels remained much the same over the duration of the study period, albeit reducing slightly in 2016. For an evaluation to be fully robust, the trends in crime in the areas to be treated and the trends in crime in the areas that form controls should be of the same direction and magnitude before the implementation of the intervention. In practice, this arrangement can be difficult because of the interventions for reducing crime naturally being targeted to the areas in which crime is highest, or in which crime is increasing the most. Limitations such as these can be difficult to avoid, but they require consideration when interpreting the results.

Process evaluation: How did the intervention work?

This chapter focuses on evaluation techniques that measure the *impact* of an intervention. That is, did the intervention have an impact on crime levels? A *process* evaluation aims to determine *how the impact was achieved.* This type of evaluation seeks to answer such questions as:

- What resources were required to achieve the impact?
- Did the activities go as expected?
- Were the activities implemented in the sequence planned?
- What alterations were made to the implementation of the intervention?
- Who did not cooperate in implementing the intervention, and why?

Although an impact evaluation can indicate whether the intervention was successful in reducing crime, the process evaluation importantly helps identify how this impact was achieved. For instance, if you identified an example from practice on what worked for a police agency to reduce a crime problem you are also trying to reduce, you would want to know *how* they achieved the reductions in crime. Often, process evaluations are overlooked in studies on crime reduction. For instance, does hot spot policing work by deterring would-be offenders through increased police activity (e.g., increase in stop and frisk and in arrests) or through improved community engagement? How many patrols should be allocated to a hot spot policing program to have optimal impact, and what is the optimal frequency and duration of patrol visits to the hot spots? To what extent does the management supervision of patrols influence the impact of a hot spot policing program? Much is known about the impact of hot spot policing, but much less is known about how it achieves these impacts. The same typically applies to all other interventions that have (and do not have) an impact on crime.

Most evaluations of interventions focus on impact, but to determine how the impact was achieved, it is also important to conduct a process evaluation. Further information on process evaluation is available in a free guide called *Assessing Responses to Problems*, at https://popcenter.asu.edu/content/assessing-responses-problems-did-it-work-page-2.

Weighted displacement quotient and the weighted displacement difference test

In this section, we describe the WDQ and WDD test and illustrate their use when working with a large sample of treatment and control areas, as well as a small sample of these areas. The WDQ (Bowers and Johnson 2003) is a simple measure for determining whether displacement has occurred as a result of a geographically targeted intervention. The WDD test (Wheeler and Ratcliffe 2018) complements the WDQ by helping evaluate whether a geographically targeted intervention has resulted in a statistically significant reduction in crime while controlling for changes in crime in control areas (such as the displacement of crime to these control areas). Both the WDQ and WDD require crime data for before and after implementation of the intervention and can be calculated simply in a spreadsheet. Like the other approaches and techniques described in other sections of this chapter, the WDQ and WDD assume the treatment and control areas to have similar levels of crime before implementation of the intervention for the results to be robust. As already discussed, this assumption often is not possible when evaluating geographically targeted interventions and will be considered later in this section when reviewing examples of the application of the WDQ and WDD.

The WDQ requires data for three areas—treatment areas, displacement zones, and control areas. The WDD ideally requires data for four types of areas: treatment areas, control areas for the treatment areas, displacement zones, and control areas for the displacement zones (e.g., buffer areas around the control areas for the treatment areas). Data may not be available for the control areas for both the treatment areas and displacement zones, nor for the control areas for the displacement zones. Examples in this section illustrate results for the full range of circumstances. Counts of crime are required for the WDD. Counts and rates of crime can be used for the WDQ.

Calculating the WDQ

The WDQ involves comparing a success measure to a buffer displacement measure (BDM). The success measure (SM) can be calculated in equation (10.3) as follows:

$$SM = T1/Ct1 - T0/Ct0, \tag{10.3}$$

where T0 and T1 refer to the level of crime in the treatment areas before (T0) and after (T1) the implementation of the intervention, respectively, and Ct0 and Ct1 refer to the level of crime in the control areas for the treatment areas before and after the implementation of the intervention. If the success measure is negative, the intervention was a success (i.e., crime declined in the treatment area relative to the change in crime in the control area). If the success measure is positive, the intervention was not successful (i.e., crime has increased in the treatment area relative to the change in crime in the control area).

The second stage involves calculating the buffer displacement measure. The BDM indicates whether crime displacement has occurred, or whether there has been a diffusion of benefit to the displacement zone. The BDM can be calculated in equation (10.4) as follows:

$$BDM = D1/Ct1 - D0/Ct0, \qquad (10.4)$$

where D0 and D1 refer to the level of crime in the displacement zones before and after the implementation of the intervention, respectively. If the BDM is positive, crime displacement has occurred to the displacement zones. If the BDM is negative, a diffusion of benefit has occurred to these displacement zones. The WDQ then is calculated as a ratio of the BDM and SM in equation (10.5), as follows:

$$WDQ = BDM / SM \qquad (10.5)$$

If the WDQ is less than −1, it suggests that a large amount of displacement of crime has occurred, and is more than the direct impact that the intervention has had in the treatment areas. If the WDQ is between 0 and −1, displacement has occurred, but it is less than the direct impact of the intervention in the treatment areas. If the WDQ is greater than 0 but less than 1, it suggests a diffusion of benefit effect in the displacement zones, but it is less than the direct impact of the intervention in the treatment areas. If the WDQ is greater than 1, it suggests that a diffusion of benefit effect has been experienced in the displacement zones, and this diffusion effect is greater than the direct impact of the intervention in the treatment areas. Table 10.8 lists the full range of WDQ values and their descriptive interpretation. The WDQ does not indicate whether the changes in crime are statistically significant, and provides only a descriptive indication of the impact of the intervention.

Table 10.8. WDQ interpretation guide

WDQ	Diffusion/displacement	Overall impact of intervention
WDQ greater than 1	Diffusion to displacement zones greater than impact in treatment areas	Positive impact of the intervention
WDQ close to 1	Diffusion to displacement zones equal to impact in treatment areas	Positive impact of the intervention
WDQ less than 1, greater than 0	Diffusion to displacement zones but less than impact in treatment areas	Positive impact of the intervention
WDQ equals 0	No displacement or diffusion	Positive impact of the intervention
WDQ less than 0, greater than −1	Displacement to displacement zones but less than impact in treatment areas	Positive net impact of the intervention
WDQ close to −1	Displacement to displacement zones equal to impact in treatment areas	No net benefit of the intervention
WDQ less than −1	Displacement to displacement zones greater than impact in treatment areas	Negative net benefit of the intervention

Calculating the WDD

The WDD is a measure of the comparison of the differences in the change in crime in the treatment areas and in their control areas, compared with the differences in the change in crime in the displacement zones and their control areas, given in equation (10.6):

$$WDD = (\Delta T - \Delta Ct) + (\Delta D - \Delta Cd), \qquad (10.6)$$

where ΔT is the change in crime in the treatment areas, ΔCt is the change in crime in the control areas of the treatment areas, ΔD is the change in crime in the displacement zones, and ΔCd is the change in crime in the control areas of the displacement zones. The WDD can measure whether the change in crime is statistically significant, calculated by using the free spreadsheet file provided by Wheeler and Ratcliffe (2018). This calculation includes the standard error and z-score for the input data to determine the level of statistical significance.

Application of WDQ and WDD to a hypothetical example

To illustrate how the WDQ and WDD work, we begin with a hypothetical example using the data shown in table 10.9. In the treatment area, crime declined by 40 offenses. In the control areas and displacement zones, there were small changes in crime. We call this *scenario 1*.

Table 10.9. Counts of crime in treatment areas, displacement zones, and control areas

	Before the intervention	After (or during) the intervention
Treatment areas (T)	100	60
Control areas for the treatment areas (Ct)	100	105
Displacement zones (D)	100	95
Control areas for the displacement zones (Cd)	100	105

Taking the data from table 10.9 and inserting it into the WDQ equations—equations (10.3), (10.4), and (10.5)—gives:

$$SM = 60/105 - 100/100$$
$$SM = -0.43$$
$$BDM = 95/105 - 100/100$$
$$BDM = -0.10$$
$$WDQ = -0.10/-0.43$$
$$WDQ = 0.22$$

This WDQ result suggests the intervention has had a positive impact in reducing crime and that there has been a diffusion of benefit to displacement zones, but it has been less than the impact in treatment areas.

Taking the data from table 10.9 and inserting it into the WDD equation—equation (10.6)—gives:

$$WDD = ((60 - 100) - (105 - 100)) + ((95 - 100) - (105 - 100))$$
$$WDD = -55$$

The WDD result was significant to $p < .05$, and the standard error for the WDD was 27.7. These WDD results suggest the reduction in crime was statistically significant, after controlling for changes in crime in displacement zones and control areas.

Let's say, using the hypothetical example here, that the change in the displacement zones was an increase of 40 offenses compared with the before period (changing from 100 to 140). We will call this *scenario 2*. This would produce a WDQ of −0.78, suggesting the displacement to displacement zones had been less than reductions in the treatment areas, so the intervention still had a positive net impact. The increase of 40 offenses in the displacement zones would produce a WDD of −10 [((60 − 100) − (105 − 100)) + ((140 − 100) − (105 − 100))] and a p-value of .36. This calculation would indicate, however, that the overall crime reduction experienced from the intervention was not significant.

Let us now say the change in the displacement zone was a reduction of 20 offenses (*scenario 3*). This would produce a WDQ of 0.56, suggesting the intervention had a positive impact in reducing crime, and that there was a diffusion of benefit to displacement zones, but it was less than the impact in treatment areas. The WDD for this scenario would be −70 [((60 − 100) − (105 − 100)) + ((80 − 100) − (105 − 100))] and produce a p-value of .005. This would indicate the crime reduction was significant to $p < .01$, supported by the diffusion of benefit effect in the displacement zones.

Let us now consider that we had data for only the control areas of our treatment zones (*scenario 4*). We still can use the WDD but cannot use the WDQ because we do not have data to assess whether local displacement has occurred (i.e., data relating to variables D0 and D1). We can, though, calculate the success measure component of the WDQ. Using the data from the previous example (scenario 3), it generates an SM measure of −0.43, suggesting the intervention has been successful. The WDD for this scenario would be −45 ($p < .01$) [((60 − 100) − (105 − 100)) + ((0 − 0) − (0 − 0))]. That is, the crime reduction in the treatment areas has been significant, while controlling for change in crime in the treatment's control areas.

Let us next consider another scenario in which we had data for displacement zones but did not have data for control areas for the displacement zone (*scenario 5*). Using the data from scenario 3, we know it would produce a WDQ of 0.56 (a positive impact in reducing crime, with some diffusion of benefit to displacement zones). If we used the WDD test, the content of the WDD equation would look like this: [((60 − 100) − (105 − 100)) + ((80 − 100) − (0 − 0))]. This would produce a WDD of −65 and a p-value of .003. However, in these circumstances, Wheeler and Ratcliffe (2018) recommend a modification of the WDD that involves multiplying by two the change in the control areas for the treatment zones to compensate for no data for the controls of the displacement zones. Hence, equation (10.7) would be:

$$\textit{Alternative}\ \text{WDD} = (\Delta \text{T} - 2 \times \Delta \text{Ct}) + (\Delta \text{D}) \qquad (10.7)$$

The change in the formula for the WDD also requires a change in the calculation of the z-score and p-value, substituting the new value for the WDD into these equations provided in the spreadsheet from Wheeler and Ratcliffe (2018). This alternative WDD technique, applied to the situation in which control areas for displacement zones are not available, would produce a WDD of −70 ($p < .01$) [((60 − 100) − 2 × (105 − 100)) + (80 − 100)] and a p-value of .001. That is, crime has declined significantly over the period when the intervention was implemented.

These examples illustrate the application of the WDQ and the flexibility in the use of the WDD when the full range of displacement zones and control zones is not available. As the purpose of the techniques is to examine whether a reduction in crime in treatment areas has been offset by displacement, it is highly advised to include data for displacement zones. The examples also illustrate how the two measures complement each other: the WDQ provides an indication of the impact of an intervention and the extent to which displacement has inhibited this impact, or whether a diffusion of benefit has contributed to the intervention's overall impact; whereas the WDD helps calculate whether the reduction in crime, while controlling for changes in crime in control areas, was statistically significant.

Application of WDQ and WDD to real examples

Let us now examine the use of the WDQ and WDD for the two real examples previously presented in this chapter: the PADO hot spot policing program in Montevideo and the Philadelphia Foot Patrol Experiment. Both these examples involved many units of analysis for the treatment and control areas. We then illustrate an application of the WDQ and WDD for an example with a single treatment area, a displacement zone, and separate control areas (as per the illustration in figure 10.3d).

Application to large data samples: Montevideo PADO and Philadelphia Foot Patrol Experiment

The evaluation of PADO in Montevideo involved comparing changes in crime in treatment areas to areas immediately surrounding treatment areas, and to a control area that consisted of all other streets in the city. The areas immediately surrounding treatment areas are, effectively, *displacement zones*. The number of robberies in each of these areas is shown in table 10.10. In this example, there were no control areas for the displacement zones. For the WDQ and WDD to work robustly, the number of crimes in each area before the intervention should be similar. In this case, the number of robberies in the displacement zones before the implementation of PADO is like the number of robberies in the treatment zones. The number of robberies in the control areas was higher (because of the large number of units for analysis of this area). In the example that follows, we illustrate the results from the WDQ and WDD when this number is not adjusted and compare these results when adjusting for the number of robberies in control areas. The control areas experienced three times the number of robberies than in the treatment areas. To adjust for this, a third of the original control areas were randomly selected from the original full sample of control areas. This produced an *alternative* control area, which accounted for 3,485 robberies before the intervention began and 3,508 robberies while PADO was operational.

Table 10.10. Counts of crime in treatment, displacement, and control areas for PADO in Montevideo, Uruguay, and the Philadelphia Foot Patrol Experiment

	Number of crimes before the intervention	Number of crimes after the implementation of the intervention
PADO Montevideo		
Treatment areas	3,403	2,338
Displacement zones	3,215	3,147
Control areas	10,517	10,676
Alternative control areas	3,485	3,508
Philadelphia Foot Patrol Experiment		
Treatment areas	359	306
Control areas	296	327

Table 10.11 lists the WDQ and WDD results using the data from PADO. The WDQ results suggest that for both scenarios in the use of control areas, there has been a positive net impact of the intervention, and that no displacement or diffusion of benefit effect to displacement zones has occurred. Both WDD results, for both scenarios in the use of control areas, show the reduction in crime during the PADO operational period was statistically significant. Although there is little difference in the results for the two different control areas, the results from the adjusted control areas are more robust. However, the example illustrates little difference in the conclusions on the impact of PADO between the use of the original control areas and the ones that were adjusted to ensure the baseline count of crime in control areas was similar to that recorded in the PADO treatment areas.

Table 10.11. WDQ and WDD results for PADO and the Philadelphia Foot Patrol Experiment

	WDQ	WDD	Z-score	Alternative WDD	Alternative WDD z-score
PADO Montevideo (using original control areas)	0.10	−1,292***	-7.08	−1,451***	-7.95
PADO Montevideo (using adjusted control areas)	0.08	−1,156***	-8.37	−1,179***	-8.53
Philadelphia Foot Patrol Experiment	n/a	-84***	-2.34	n/a	n/a

*** $p < .01$; ** $p < .05$; * $p < .1$

Table 10.11 also shows the data from the Philadelphia Foot Patrol Experiment. For this intervention, only data for the treatment areas and controls of the treatment areas were recorded, meaning that the WDQ could not be calculated. The WDD results for this program show the reduction in crime was statistically significant.

Application to a small data sample: A problem-oriented policing project in Montevideo

The final example we use is based on an intervention to reduce crime in a district in the city of Montevideo. In this district, the police recently had adopted problem-oriented principles in their neighborhood policing work and wanted to pilot them for an intervention to reduce thefts of vehicles in a hot spot they had identified. The neighborhood police team created a displacement zone around their treatment area and two control areas within their district that were similar in size and land use to the treatment area. The displacement zone partially crossed the border into a busy commercial area of a neighboring police district in which theft of vehicle offenses were also high (but which the police working in the treatment area could have no direct impact on influencing because it was a different jurisdictional area). Figure 10.5 shows the treatment area, displacement zone, and control areas. Table 10.12 lists the number of theft of vehicle offenses in the treatment and control areas, and displacement zones for before and after the introduction of the intervention. The table also lists data in each of the areas for other types of thefts and robberies for the purposes of examining whether displacement occurred to these other crime types in the geographic areas of analysis. The intervention operated for six months, and police compared data with the before period on the basis of the same six months in the previous year.

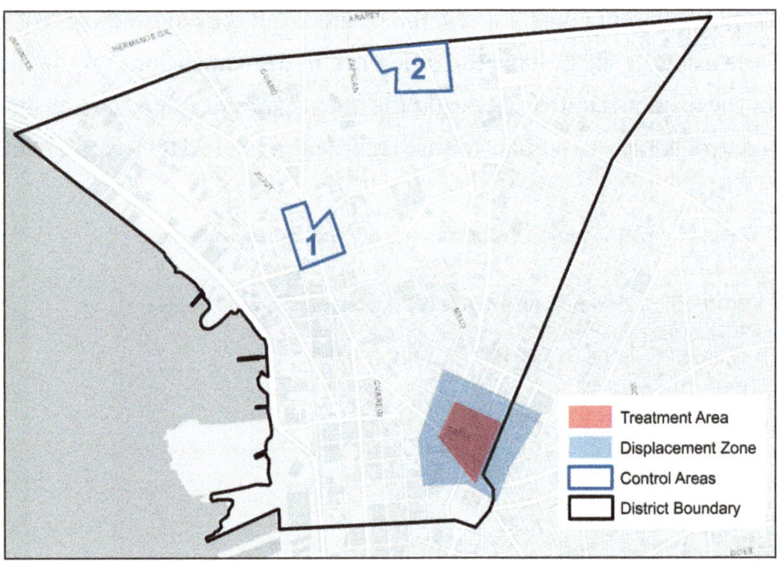

Figure 10.5. Treatment area, displacement zone, and control areas for a problem-oriented policing intervention in a police district in Montevideo, Uruguay.

Table 10.12. Counts of crime in a problem-oriented policing intervention in a police district in Montevideo, Uruguay

	Before the intervention	After the implementation of the intervention
Theft of vehicles		
Treatment area	26	4
Control area 1	4	5
Control area 2	9	11
Displacement zone	39	8
Robberies and other thefts		
Treatment area	106	64
Control area 1	42	44
Control area 2	57	40
Displacement zone	171	55

For the purposes of calculating the WDQ and WDD, the numbers of crimes in each of the two control areas were combined so the control areas for the treatment areas had crime levels that were more similar to those in the treatment area before the intervention. Table 10.13 shows the WDQ and WDD results. Both WDD and alternative WDD results were generated. The WDQ for thefts from vehicles was 1.43, indicating that the intervention had a positive impact and that there also had been a diffusion of benefit effect to the displacement zone that was even greater than the impact of the intervention in the treatment area. Both the treatment area and displacement zone experienced large reductions in thefts from vehicles, with little change in the treatment area's control areas. The WDD results showed the reductions in thefts of vehicles had been significant, once controlling for changes in the control areas. The results also showed a significant diffusion of benefit effect to robberies and other thefts—these crimes declined significantly in the treatment area once controlling for the other control areas, with these benefits also diffusing significantly into the displacement zone. So, not only was the POP intervention significant in reducing thefts of vehicles in the targeted treatment area, it also resulted in a significant reduction in robberies and other thefts in the treatment area, with these effects diffusing not only to neighboring areas of the treatment area in the police's own jurisdiction but also benefiting their police colleagues in the neighboring jurisdiction.

Table 10.13. WDQ and WDD results for a POP intervention in Montevideo, Uruguay

	WDQ	WDD	Z-score	Alternative WDD	Alternative WDD Z-score
Theft of vehicles	1.43	−56***	−5.44	−59***	−5.73
Robberies and other thefts against a person	3.47	−143***	−5.94	−128***	−5.32

*** $p < .01$; ** $p < .05$; * $p < 0.1$

When conducting an evaluation on the change in crime between treatment areas and control areas (including displacement zones), it is important to ensure the baseline levels of crime before the implementation of the intervention are similar. This precaution helps ensure robustness in the results of the evaluation. Also, the trend in crime for each of those areas should be similar. That is, if crime is increasing, this increasing trend should be observed in the treatment areas, control areas, and any displacement zones before the intervention. This trend also ensures further robustness in the results. Of course, this trend may not always be the case, and in circumstances in which the level of crime and the trend in crime in treatment areas and control areas are different before the implementation of an initiative, this difference then should be considered in the interpretation of the results.

Chapter summary

This chapter has focused on techniques for measuring the impact of geographically targeted interventions. A fundamental question we often want to answer is what impact has an intervention had in reducing crime? Although the use of before and after statistics is useful in helping us monitor the impact of activities (such as in CompStat-style performance management scenarios), they do not allow us to answer a second key question, on whether the reduction in crime is because of the activities we performed. The use of treatment areas and control areas places us in a better position to answer the second question.

In this chapter, we have described the use of randomized control trials—the gold standard of evaluation approaches. RCTs are not possible in many scenarios in policing and public safety, so other approaches are necessary instead. In this chapter, we have described the use of quasi-experimental and controlled observational studies and several statistical techniques. The statistical techniques have varied in their level of sophistication. Our aim is to include techniques that allow for easy replication and illustrate the variety of settings for their use.

Further resources

Assessing Responses to Problems: the Problem-Oriented Policing Center's tool guide on assessing responses is a good introductory guide on how to design an evaluation, https://popcenter.asu.edu/content/assessing-responses-problems-did-it-work-page-2.

Impact Evaluation Resources Hub: the Inter-American Development Bank's Impact Evaluation Resources Hub contains many useful tools for conducting evaluations and a free practice guide on evaluation techniques, https://publications.iadb.org/en/impact-evaluation-practice-second-edition.

References

Andersen, H. A., and K. Mueller-Johnson. 2018. "The Danish Crime Harm Index: How It Works and Why It Matters." *Cambridge Journal of Evidence Based Policing* 2:52–69.

Bowers, K. J., and S. D. Johnson. 2003. "Measuring the Geographical Displacement and Diffusion of Benefit Effects of Crime Prevention Activity." *Journal of Quantitative Criminology* 19:275–301.

Bowers, K. J., S. D. Johnson, R. T. Guerette, L. Summers, and S. Poynton. 2011. "Spatial Displacement and Diffusion of Benefits among Geographically Focused Policing Initiatives: A Meta-Analytical Review." *Journal of Experimental Criminology* 7 (4}: 347–74.

Braga, A. A., A. V. Papachristos, and D. Hureau. 2012. "Hot Spots Policing Effects on Crime." *Campbell Systematic Reviews* 8 (8): 1–96.

Braga, A. A., A. V. Papachristos, and D. Hureau, D. 2014. "The Effects of Hot Spots Policing on Crime: An Updated Systematic Review and Meta-Analysis." *Justice Quarterly* 31 (4): 633–63.

Chainey, S. P., R. Serrano–Berthet, and F. Veneri. 2020. "The Impact of a Hot Spot Policing Program in Montevideo, Uruguay: An Evaluation Using a Quasi-experimental Difference-in-Difference Negative Binomial Approach." *Police Practice and Research.*

Curtis-Ham, S., and D. Walton. 2017. "The New Zealand Crime Harm Index: Quantifying Harm Using Sentencing Data." *Policing: A Journal of Policy and Practice.*

Ratcliffe, J. H. 2015. "Harm-Focused Policing." *Ideas in American Policing* 19:1–12.

Ratcliffe, J. H., T. Taniguchi, E. R. Groff, and J. D. Wood. 2011. "The Philadelphia Foot Patrol Experiment: A Randomized Controlled Trial of Police Patrol Effectiveness in Violent Crime Hotspots." *Criminology* 49 (3): 795–831.

Schick, W. 2004. "CompStat in the Los Angeles Police Department." *Police Chief* 71 (1}: 17–23.

Sherman, L., P. W. Neyroud, and E. Neyroud. 2016. "The Cambridge Crime Harm Index: Measuring Total Harm from Crime Based on Sentencing Guidelines." *Policing: A Journal of Policy and Practice* 10 (3): 171–83.

Von Hirsch, A., and N. Jareborg. 1991. "Gauging Criminal Harm: A Living-Standard Analysis." *Oxford Journal of Legal Studies* 11:1.

Weisburd, D., and C. W. Telep. 2014. "Hot Spots Policing: What We Know and What We Need to Know." *Journal of Contemporary Criminal Justice* 30 (2): 200–220.

Wheeler, A. P., and J. H. Ratcliffe. 2018. "A Simple Weighted Displacement Difference Test to Evaluate Place-Based Crime Interventions." *Crime Science* 7:11.

Index

A

A. Costello, 15, 16–17, 31–32
acute temporal category, 75, 76
address points, 44
adjusted R^2 statistic, 167, 220
adolescent exuberance, 6
Aggregation Shape Type, 119
Akaike correlation coefficient, 51
Akaike Information Criterion (AIC), 172, 218
ambush style of attack, 138
analysis variable, 119
anchor points, 24, 32, 34, 145, 146, 148, 150–52, 153
Andresen, M., 65, 66, 67, 79–80, 124
aoristic data clocks, 71–73
ArcGIS: for applying FDR correction, 60–61; data clock tool, 69; Emerging Hot Spot Analysis tool, 114, 115–17, 119, 124, 240; for importing spatial lag regression analysis, 170; for OLS regression, 167, 173, 189; Generating Crime Risk Zones tool, 233, 234; GWR tool, 222; Repeat and Near Repeat Classification tool, 104, 232; Space Time Cube tool, 114, 115–17, 119, 124, 240
area percentage, 229–30
areas of crime occurrence, 15–16
Ashby, M., 72
assaults: GWR regression of violent, example of, 200–207; sexual, in Bath, England, 12, 24–25, 143–53, 157; spatial error models of, 212–17; spatial lag and spatial error models of, 215–17
attack methods, 137–38
Auckland, New Zealand, 96, 231, 232
Australia, 10, 96
Automated Fingerprint Identification System, 149
available targets, 10, 11, 12
average nearest neighbor test, 27–31
awareness space, 14–16, 18, 32, 40, 135–38, 153, 155

B

bandwidth: concentric zones using, 4; for near repeat analysis, 101–5, 232; for proportion of micro-places, 42, 45; in GWR modeling, 195–96, 205; in KDE technique, 49–52, 54, 56, 61; spatial, 42, 101–4, 195, 232; temporal, 101–4, 232
bandwidth size: automated calculation of, in GWR modeling process, 195; optimization routines, 51–52
bank robbery, 96, 150, 153
Banksy, 142
Barwell, C., 151
baseline, in linkage analysis, 138
Bath, England, 12, 24–25, 143–53, 157
bias, perpetuating, 243
binary regression, 164
bins, 117–18, 119
Bonferroni correction, 59–61
boost account: near repeat victimization, 101; repeat victimization, 97
border security, 142
bounded rationality, 12
Brantingham, P. J., 2, 7, 14–15
Brantingham, P. L., 2, 7, 14–15
Breusch-Pagan test, 174, 218
Burgess, E., 3–4
Burgess model, 3–5
burglary: distribution of, 26, 27–29; GWR regression of residential, example of, 190–200; Kirkholt Burglary Prevention Project, 96; least-effort principle, 16–17, 31; residential, 190–200, 212–14; spatial error models of, 212–14

C

Cambridge Crime Harm Index, 53
Canter, D., 145, 147
capable guardians, 8–9, 11
Cartographic School, 3
causation, correlation and, 211–12
cell size, in KDE technique, 49–52, 54, 56
census blocks, 22, 121, 166

center of gravity, 23–24, 153
central tendency, patterns of, 22, 23, 31, 143
centrographic statistics, 23–25; spatial mean, 23–24; standard deviational ellipse, 25; weighted spatial mean, 24–25
centrographic techniques, 143–45
centroid, 23–24, 25, 49, 56, 58
centroid point, 25–26
Chainey, S. P., 3, 14–15, 51, 52
change score, 259
Chicago School, 3–7; meso-level explanations of crime, 7; social disorganization and collective efficacy, 5–6; sociocultural triggers of crime and models for urban development, 3–5
child abduction, 150
Chula Vista, California, 86
circle model, 4
circle theory, 145–46
Clarke, R. V., 10
clustered spatial concentration, 75, 76
clustering: nearest neighbor index (NNI), 27–31; of residuals, 188; spatial clustering techniques, 37, 38, 41, 61 (See also hot spot analysis)
coefficients: Akaike correlation, 51; Gini, 48; regression, 166, 172; values of, interpreting, 209–10
cognitive maps, 14
Cohen, L. E., 9, 10
cold cases, 142
collective efficacy (or social efficacy), 5–6, 39
commercial robbery, 96
commuter, 137–38
commuters zone (zone V), 4
comparative case analysis. See linkage analysis
composite mapping techniques, 240
CompStat, 39, 247, 248, 249–51, 276
concealable targets, 10
concentric (or zonal) model of urban development, 3–5
concentric zones, 3–4
consecutive cold spot, 118
consecutive hot spot, 117
contamination to control area, 256
control groups, 254–57; benefits of using, 254; control areas, determining, 255–56; hierarchy of evaluation reliability, 254–55; influence of the number of areas on the reliability of the results, 256
controllers, 9–10
co-offending, 153
correlation and causation, 211–12
Costello, A., 15, 16–17, 31–32
counterterrorism, 142
CRAVED, 10–12; to address property crime problem, 11; characteristics of targets prone to theft, 10
crime: four dimensions of, 2; journey to, 22, 31–35; meso-level explanations of, 7; persistence of, in certain areas, 5
crime attractors, 17
crime circles, journey to, 146–47
crime concentration at micro-place level: calculation of crime concentration levels across the micro-place units, 44–45; checklist, 47; Gini coefficients, 48; Gi* statistic, 58–62; hot spot mapping using spatial analysis techniques, 49–62; joining geographically referenced crime data to micro-place units, 44; kernel density estimation (KDE), 49–56; local Moran's I, 56–58; Lorenz curves, 48; mapping of high crime concentration micro-places, 46–47; measuring and mapping, 41–48; preparing micro-place geographic unit data, 42–43
Crime Concentration Dispersion Index (CCDI), 114, 124–28
crime controllers, 9–10
crime data: geographically referenced, 44, 47; measuring and mapping crime concentration, 42–44; micro-place geographic unit data, 42–44; regression diagnostics, 171–74; retrospective, 30, 79, 235, 238, 239; risky facilities, 87; spatial error regression, 180; spatial weights matrix, 180; WDQ and WDD test, 267, 272–75
crime displacement, 253–54
crime generators, 17
crime locations, 155–57
crime opportunity, 10
crime pattern theory, 13–18, 134–35; advances in, 7; awareness space in, 14–16, 32, 40, 135, 136–38, 155; overview of, 13–18; theoretical principles of, 14
crime prediction, 77, 95, 96, 101, 226, 243
crime prediction and perpetuating bias, 243
Crime Prevention through Environmental Design (CPTED), 7
crime problem analysis triangle, 9–10
crime risk, mapping and analyzing, 225–45; crime prediction and perpetuating bias, 243; crime risk framework, 226–30; enduring crime risk, 240–44; for different temporal settings, 227–28; immediate crime risk, 230–37; near crime risk, 237–39
crime risk framework, 226–30
crime risk zones, 234–38
criminal decision-making, 12–13, 134
criminal event decision, 12
criminal geographic targeting (CGT) equation, 133–34, 148, 150, 151, 154, 156
criminal groups, 153
criminal involvement decision, 12
criminal occurrence space, creation of, 15
cross validation method, 51

D

daily routines, 40
Danish Harm Index, 261
data clocks, 68–75; aoristic, 71–73; overnight concentrations of crime, 70; profiles for different areas, 74–75
Davies, A., 33–34
defensible space, 7
dependent variables, 163, 221
deployment, 38, 102, 127, 233, 240, 250
diagnostic tests, 186–89. See also model respecification process
diffused temporal category, 76
diminishing cold spot, 118
diminishing hot spot, 117–18
Dirichlet process mixture (DPM) model, 158
disease prevention, 142
dispersed spatial concentration, 75, 76
Dispersion Calculator, 121–28
displacement zones, 257
disposable targets, 10, 11
distance band, 56
distance decay, 31–32
Distance Interval, 119
DNA, 139, 142, 149, 151
Dragnet, 147–48
drivers of crime, 13
Duque de Caxias, 80, 121–23, 126–28

E

Eck, J., 9
ecology, 142
ECRI, 148
effort, in offender decision-making, 12–13
Emerging Hot Spot Analysis tool, 114, 115–17, 119, 124, 240
emerging problem areas, 120–24
enduring crime risk, 240–44
enjoyable targets, 10
environmental criminology, 2, 7–8, 83, 133–34
epidemiology, 142, 158
error, defined, 218
error terms, 163–64, 166, 174–75, 180, 218, 221
Euclidean distance, 32–34
evaluation. See also performance measures: before implementation of intervention, 252; hierarchy of evaluation reliability, 254–55; of impact, 257–76; importance of, 249; quasi-experimental evaluation, 261–76; randomized control trial, 257–61
evidence-based policing movement, 249
Excel Analysis Toolpak, 167
excuses, in offender decision-making, 12–13
explanatory variables, 163, 221
exploratory approach, 165

F

factory zone, 4
false discovery rate (FDR), 59–60
family-wise error rate (FWER), 59
Felson, M., 9, 10
First Law of Geography, 12, 22, 25
first-order spatial effects, 221
flag account, 97
focused temporal category, 75, 76
forecasting, 226, 227, 243
Frisbie, D. W., 31
F-test, 218, 260

G

gas, making off without paying for, 86–94. See also risky facilities
Geary's C, 27, 31
Generating Crime Risk Zones tool, 233, 234
geocoding, 7, 148
GeoDa, 170, 172–73, 175, 177, 180, 212, 215, 222
geographical analysis of crime, 1–19; Cartographic School, 3; Chicago School, 3–7; GIS School, 7–18
geographically referenced crime data, 44, 47
geographically targeted interventions, evaluating impact of, 257–76
geographically weighted regression (GWR), 185–212; bandwidth size, 195; correlation and causation, 211–12; data and regression approaches, 186; definition and use, 165, 185–86; diagnostic tests and model respecification process, 186–89; model bias, 188; model performance, 186–87; model significance, 188; multicollinearity, 187; OLS model specification in preparation for GWR analysis, 189; outliers, 189; residential burglary, example of, 190–200; results, for inference, 211; results, interpreting and using, 208–11; spatial clustering of residuals, 188; value of, 189; violent assaults, example of, 200–207; what-if scenarios, 211
geographic information systems (GIS): development of, 7; software, 41, 49, 55, 98, 170
geographic profilers, 133, 140–41
geographic profiling analysis, 131–59; analysts, 133–34; applied to co-offending and criminal groups, 153; in other applications, 142; linkage analysis, 138–41; offender hunting methods, 137–38; professional geographic profilers, 133, 140–41; questions to ask in, 140–41; spatial analysis techniques for, 143–58;

theoretical principles underpinning, 134–36; to inform investigative strategies, 141–42; training, 134
Geographic Profiling Analysis training program, 150
geographic units: Crime Concentration Dispersion Index, 125; geographically weighted regression (GWR), 186, 195–96, 201–3, 207; Gi* statistic, 58, 59, 60; identifying emerging problem areas, 121; identifying persistent spatial patterns of crime, 117, 119; Local Moran's I, 56, 60; mapping hot spots of crime, 41–43; risk terrain modeling (RTM), 244; S Index, 80; spatial lag regression, 170, 171, 179; spatial weights matrix, 171; temporal stability index, 79
geoprofiling_w_map.R, 158
Getis-Ord Gi* statistic. See Gi* statistic
Gini coefficients, 48
GIS School, 7–18
Gi* statistic, 58–62; compared to KDE map, 61; identifying significant clusters of crime, 62; statistical significance, 58, 59, 60, 117; using uncorrected procedure, 60
GitHub, 158
goodness-of-fit line, 167
guardianship, 9, 14
Guerry, A.-M., 3

H

handlers, 9–10
harm, mapping, 53
harm index, 260–61
harm studies, 53, 260
Harries, K., 7
HDTV, 51
heterogeneity, 41, 42, 218, 221
heteroscedasticity, 174–75, 178, 179, 181–85, 188, 191, 203–5, 218
hierarchy of evaluation reliability, 254–55
high-crime areas, 4, 25–26, 56, 114
high crime concentration micro-places, mapping of, 46–47
historical cold spot, 118
historical hot spot, 118
hit rate percentage, 229–30
homicide, 34–35, 132, 133, 147–50
homogeneity index, 77
hot point, 75, 76
hot spot analysis, 37–62; hot spot mapping using spatial analysis techniques, 48–62; measuring and mapping crime concentration at micro-place level, 41–48; neighborhood characteristics, 39; purpose of, 38–39; situational characteristics, 40; spatial concentration, 75–76; theoretical principles behind, 39–40
hot spot mapping using spatial analysis techniques, 49–62; Gi* statistic, 58–62; kernel density estimation (KDE), 49–56; local Moran's I, 56–58
hot spot maps: across network, 55; determining what's hot, 56; Gi* statistic compared to KDE map, 61; Gi* statistic identifying significant clusters of crime, 62; Gi* statistic using uncorrected procedure, 60; for identifying key strategic areas of focus, 54; harm hot spots, 53; KDE cell size and its effect on results of, 51; Local Moran's I maps, 57; of crime distribution, 49; using different bandwidths, 52; using same bandwidth, 50
hot spot matrix, 75–76
hot spot policing program, 229, 240, 257, 261, 265, 266, 272
hunter, 137–38
hypothesis-driven approach, 165
hypothesis test, 219

I

ILOSTAT database, 166
immediate crime risk, 230–37
impact, harm for measuring, 260–61
impact of geographically targeted interventions, evaluating, 257–76; quasi-experimental evaluation, 261–76; randomized control trial, 257–61
independent variables. See explanatory variables
infectious sources of disease, 142
inference, GWR results for, 211
innermost zone (zone I), 3–4
intelligence assessments, 38–39
intelligence products, 38–39
intensifying cold spot, 118
intensifying hot spot, 117
intensity value, 25–26, 27
intercept, 166
International Labor Organization Department of Statistics, 166
iron law of troublesome places, 85
Irvine, California, 139
Irvine Chair Burglar, 139

J

Jarque-Bera test, 174, 176, 177, 181, 185, 188, 189, 191–93, 204, 219
Jenks (natural breaks) classification process, 52, 54–55, 99
joint F statistic, 188, 218, 219
joint Wald statistic, 185, 188, 191, 193, 194, 203, 204
journey to crime, 22, 31–35; distance decay, 31–32; geographic profiling analysis, 135–36; measuring,

32–33; principle of least effort, 31–32; returning to scene of the crime, 34–35
journey to crime circles, 146–47
juvenile delinquency, 4, 5–6

K

kernel, 49
kernel density estimation (KDE), 49–56. See also hot spot maps; bandwidth, 49–52, 54, 56, 61; expressing in mathematical terms, 49; parameters, 50–52
kernel density values, 49–51, 54
kidnapping, 149, 150
Kirkholt Burglary Prevention Project, 96
Knox, P. L., 5
Koenker-Bassett test, 174, 187–88, 191, 192, 193, 194, 203, 204, 219
Koenker statistic, 174, 187–88, 191, 192, 193, 194, 203, 204, 219

L

Lagrange multiplier, 176–77, 185
Lagrange multiplier statistics, 176, 212, 215
Lambda, 183
land parcels, 44
law of crime concentration, 42
least-effort principle, 14, 16–17, 18, 31–32, 134, 135
LeBeau, J. L., 7, 24
legal dimension of crime, 2
likelihood ratio test, 178, 183–85, 213, 216
likely offenders, 8–9
limited rationality, 12
linear regression, 164, 166
linkage analysis, 138–41; baseline, establishing, 138; errors in, 139; successful, principles for, 138–39
local Moran's I, 56–58, 60
locations of crime, 155–57
logistic regression, 164
log likelihood, 172, 174, 175, 177, 178, 181–85, 219
log likelihood ratio, 219
loop, 3–4
Lorenz curves, 48

M

Manhattan distance, 32–34
maximum, 43
Mayhew, H., 3
McKay, H., 3, 4
mean center, 23–24
mean integrated squared error, 51
MeasuringU, 60
memory-processing capacity of computers, 51
Merseyside, England, 86, 96
meso-level explanations of crime, 7, 17
metrics on micro-place units, 43
micro-grid cells, 42–44, 46, 47, 49, 57, 61, 80
micro-level explanations of crime, 17
micro-place geographic unit data, 42–43
micro-place identity number, 44, 87–88
micro-place level, measuring and mapping crime concentration at, 41–48, 62, 124; calculation of crime concentration levels across the micro-place units, 44–45; checklist, 47; Gini coefficients, 48; joining of geographically referenced crime data to micro-place units, 44; Lorenz curves, 48; mapping of high crime concentration micro-places, 46–47; preparing micro-place geographic unit data, 42–43
Microsoft Excel, 44, 68, 87–88, 98, 115, 123, 167
minimum, 43
minimum bounding rectangle (MBR), 51–52
model bias, 188
model performance, 186–87
model respecification process, 186–89; in preparation for GWR analysis, 189; model bias, 188; model performance, 186–87; model significance, 188; multicollinearity, 187; OLS model specification in preparation for GWR analysis, 189; outliers, 189; spatial clustering of residuals, 188; value of, 189
model significance, 188
modifiable areal unit problem (MAUP), 41
modus operandi, 97, 101, 105, 139, 150
Montevideo, Uruguay, 38, 43, 261–65, 272–76
Moran's I, 25–27
multicollinearity, 174, 177, 181, 187, 190, 196, 219, 222
multicollinearity condition number, 174
multinomial logistic regression, 164
murder, 34–35, 132, 133, 147–50

N

National Crime Agency, UK, 149, 151
natural breaks (Jenks) classification process, 52, 54–55, 99
natural log transformation, 188, 191, 192, 203, 205, 212, 215, 220
near crime risk, 237–39
nearest neighbor index (NNI), 27–31
Near Repeat Calculator, 101, 103, 104, 106, 232
near repeat chains, 109
near repeat victimization, 100–109; analysis, example of, 107–9; analysis, usefulness of, 100; defining spatial and temporal bandwidths for, 101–4; mapping,

104–5; measuring, 101–6; Near Repeat Calculator to produce output file of, 106; reasons for, 101
negative spatial autocorrelation, 25, 26, 27, 31, 56
neighborhood: canvassing by the police, 142; characteristics, 39; spatial weights matrix for defining, 171
Newcastle, England, 26, 27–28, 79, 98–99, 171–15, 177, 178, 180–84, 186, 190, 194–95, 197, 199–216, 231, 232–35, 240–41
new cold spot, 118
new hot spot, 117
New York City, 35, 43
New York City Police Department, 250
nodes, 14
Non-Contributory Dispersion Index (NCDI), 114, 121–22, 126–27
nonstationarity, 188, 191, 193, 194, 203–5, 218, 219, 220, 221
no pattern detected, 118
null hypothesis, 59, 219, 220

O

odds ratio, 258–59
offender decision-making, 12–13, 134
offender dimension of crime, 2
offender hunting methods, 137–38
offender registries, 141
offender's awareness space, 14–16, 18, 32, 40, 135–38, 153, 155
offending behavior, 8, 18, 22, 31, 77, 208
Offense Dispersion Index (ODI), 114, 121–23, 126–27
operational briefing of police patrols, 38
operational dashboards, 251
Operation Lynx, 149–50, 151
opportunistic offenses, 155
opportunity-based criminal behavior, 32
optimal foraging theory, 101
optimization routines, 51–52
ordinal logistic regression, 164
ordinary least squares (OLS), 166–68, 186–89; geographically weighted regression, 185–212; Lagrange multiplier, 176–77; model bias, 188; model performance, 186–87; model significance, 188; model specification and results, 190–94; model specification in preparation for GWR analysis, 189; multicollinearity, 187; outliers, 189; regression results and spatial autocorrelation, 174–75; regression results using spatial weights matrix, 175–77; spatial clustering of residuals, 188; spatial error regression, 180–85; spatial lag and spatial error models of burglary and assaults, 212–17; spatial lag regression, 170–79; spatial relationships, examining, 168–70; value of, 189
originator offenses, 103–6, 109
oscillating cold spot, 118
oscillating hot spot, 118
outliers, 155, 189
overnight concentrations of crime, 70

P

PADO (Programa de Alta Dedicación Operativa), 261–66, 272–73
parameter estimates, 163, 170, 179, 180, 186, 202, 208, 213–15, 217
Pareto, V., 85
pathways, 14
patrol saturation and surveillance, 141
patterns of central tendency, 22, 23, 31, 143
peak profile area, 150–51, 152, 154, 156
percentile groupings, 259–60
performance measures, 247–57. See also evaluation; CompStat and performance monitoring, 250–51; control areas, determining, 255–57; control groups, 254–57; crime displacement, 253–54; impact, harm for measuring, 260–61; influence of the number of areas on reliability of results, 257; process evaluation, 266; simple statistics, 249–54
perpetuating bias, 240
persistent cold spot, 118
persistent hot spot, 117
persistent spatial patterns of crime, identifying, 115–20
Philadelphia Foot Patrol Experiment, 257–60, 272–73
place management, 9, 14
place managers, 9–10
planned criminal behavior, 32
poacher, 137–38
point objects, 42, 171
Poisson regression, 164, 201–3
polygon objects, 42
positive spatial autocorrelation, 25–27, 56
prediction accuracy index (PAI), 229–30
predictive mapping software and mixed methods, 239
prevention, 142
professional geographic profilers, 133, 140–41
property at greater risk of being stolen, 10–12
property gazetteers, 44
provocations, in offender decision-making, 12–13
pseudo R^2 value, 177
public awareness campaigns, 142
p-values, 167, 172, 187, 219, 220

Q

quartic function, 49

quasi-experimental evaluation, 261–76; selection of treatment and control areas, 261–62; technique limitations, 266; techniques used in the evaluation, 262–65; WDQ and WDD test, 267–75
Queen Mary University, London (QMUL), 158
Quetelet, A., 3

R

randomized control trial, 257–61; selection of treatment and control areas, 257–58; techniques used in evaluation, 258–60
rape, 33–34, 132, 133, 145–46, 149, 150, 151
raptor style of attack, 137
Ratcliffe, J. H., 3, 14–15, 51, 58
rational-choice perspective, 12–13, 134
regression coefficient, 166, 172
regression diagnostics, data and, 171–74
regression model respecification, 179
removable targets, 10, 11
Repeat and Near Repeat Classification tool, 104, 232
repeat victimization, 95–99. See also near repeat victimization; example of, 107–9; findings of analysis, 96; for different crimes, 96; mapping, 98–99; measuring, 97–98; reasons for, 97; research findings, 96; usefulness of analysis, 95
residential burglary, 190–200, 212–14
residential zone (IV), 4
residuals, 168, 188, 191–93, 197, 205, 206, 241, 242
resource targeting, 81, 83, 113, 124, 126, 129, 233
retrospective crime data, 30, 79, 235, 238, 239
returning to scene of the crime, 34–35
rewards, in offender decision-making, 12–13
R geographic profiling package, 158
Rgeoprofile package, 158
Rigel, 148–53
Rio de Janeiro, 43, 80, 121–24, 126–28
risk, defined, 226. See also crime risk, mapping and analyzing
risks, in offender decision-making, 12–13
risk terrain modeling (RTM), 240, 244
risky facilities, analyzing, 86–94; calculating statistics and visualizing results in a chart, 90; distribution of crime across facilities, 86–91; figuring contribution of each facility, 89; mapping results, 90–91; number of crimes at each facility, 87–88; ranking facilities, 88–89; review of crime data, and data on facilities, 87; revision of analysis after considering a denominator, 91–94; risk factors associated with facility, analysis of, 94; type of crime and type of facility, 87
robbery: bank, 96, 150, 153; commercial, 96; repeat victimization, 96; street, 13, 33–34, 38, 68, 76, 100, 108–9, 150, 231
Rochester Police Department, 251
Rossmo, K., 31, 33–34, 132, 149
routine-activity perspective, 7, 8–12, 134
R^2 statistic, 167, 220

S

Sampson, R. J., 6
San Diego, California, 24, 143
São João de Meriti, 121–23, 126–28
search to crime, 135–36
serial murderers and rapists, 147–48
serial rapist in San Diego, California, 24, 143
serious acquisitive crime (SAC) offenders, 174, 190–200, 208, 212, 213
sex offenders, 149, 152
sexual assault series in Bath, England, 12, 24–25, 143–53, 157. See also geographic profiling analysis
sharks, predatory attacks of, 142
Shaw, C., 3, 4
S Index, 79–80
situational characteristics, 40
social capital, 6, 7
social disorganization, 5–6
social disorganization theory, 5–6
social efficacy (or collective efficacy), 6, 39
sociocultural triggers of crime, 3–5
software: for applying S Index, 79; for geographic profiling, 143; for spatial lag regression, 170; for spatial regression analysis, 172; for treating zero-inflated models, 172; GIS, 41, 49, 55, 98, 170; GWR, 189, 195, 203; predictive mapping software and mixed methods, 239; spreadsheet data software tool, 44, 68, 87–88, 98
Son of Sam, 35
Space Time Cube tool, 114, 115–17, 119, 124, 240
spatial analysis techniques for geographic profiling analysis, 143–58; centrographic techniques, 143–45; circle theory, 145–46; Dragnet, 147–48; excluding and including locations, 155–57; journey to crime circles, 146–47; misleading results, 153–55; R geographic profiling package, 158; Rigel, 148–53
spatial autocorrelation, 25–27, 174–75
spatial behavior of offenders, 16–17, 133, 135, 136, 137, 140, 143, 145, 156
spatial clustering of residuals, 188
spatial concentration, 47, 58, 75–76
spatial contiguity, 171
spatial decision-making behavior, 140, 145, 153, 155
spatial dependence, 169, 177, 178, 180, 183, 220
spatial dimension of crime, 2
spatial effects, 163–65, 172, 175, 177, 179, 220, 221

spatial error models, 212–17; residential burglary example, 212–14; violent assaults example, 215–17
spatial error regression, 180–85; data and spatial weights matrix, 180; definition and use, 180; introducing another test for model performance and model specification, 185; OLS regression results using spatial weights matrix, 181–82; results, 182–85
spatial half distance, 233–34, 236
spatial heterogeneity, 41, 42, 218, 221
spatial homogeneity, 77, 78, 221
spatial lag regression, 170–79; data and regression diagnostics, 171–74; definition and use, 170; interpretation and regression model respecification, 179; OLS regression results, 174–77; results, 177–78; spatial autocorrelation, 174–75; spatial error models of burglary and assaults and, 212–17; spatial weights matrix, 175–77
spatial mean, 23–24, 43
spatial outliers, 155
spatial patterns of crime, 21–31, 113–29; centrographic statistics, 23–25; Crime Concentration Dispersion Index (CCDI), 114, 124–28; emerging problem areas, Dispersion Calculator for identifying, 120–24; nearest neighbor index (NNI), 27–31; output from Emerging Hot Spot Analysis tool, 117–19; output from Space Time Cube tool, 119; persistent, identifying, 115–20; S Index, 79–80; spatial autocorrelation techniques, 25–27; stability of, 77–81; statistical measures for, 23–31; temporal stability index (TSI), 77–78
spatial regression analysis, 161–222; geographically weighted regression (GWR), 185–212; methods, types of, 164; ordinary least squares (OLS), 166–68; regression methods, 164; spatial error regression, 180–85; spatial lag and spatial error models of burglary and assaults, 212–17; spatial lag regression, 170–79; spatial relationships, examining, 165, 168–70
spatial scales, 7, 207
spatial weights matrix, 180–82; data and, 180; OLS regression results using, 181–82
spatiotemporal analysis techniques, 7
spatiotemporal patterns of crime, 7, 83, 84, 100, 107
sporadic cold spot, 118
sporadic hot spot, 118
spreadsheet data software tool, 44, 68, 87–88, 98
stability of spatial patterns of crime, 77–81; S Index, 79–80; temporal stability index (TSI), 77–78
stalker form of attack, 137–38
standard deviation, 25, 26, 27, 58, 197
standard deviational ellipse, 24, 25
standard error, 167, 168, 172, 197–99, 202, 205–6, 221
standardized residuals, 197, 205, 206, 241, 242
stationarity, 218, 221
statistical measures, descriptions of, 218–22
statistical measures for spatial patterns of crime, 23–31; centrographic statistics, 23–25; nearest neighbor index (NNI), 27–31; spatial autocorrelation techniques, 25–27
statistical significance: of Geary's C, 27; of Gi* statistic, 58, 59, 60, 117; of Jarque-Bera statistic, 188, 189, 219; of joint F statistic and joint Wald statistic, 205, 219; of Koenker statistic, 187, 219; of model significance, 188; of Moran's I, 26; of nearest neighbor index, 27; of parameter estimates, 163, 186; of performance measures, 259, 269; of p-value, 187, 220
street junctions, 41, 42–43, 46, 261, 262
street robbery, 13, 33–34, 38, 68, 76, 100, 108–9, 150, 231
street route distance, 32–34
street segments, 42–47, 55, 62, 261–62
suitable targets, 8–9, 10, 11, 135, 137, 152
suspect prioritization, 141
Sutcliffe, P., 24
Sydney, Australia, 10

T

targeted police response plans, 141
temporal analysis of crime, 65–82; hot spot matrix, 75–76; stability of spatial patterns of crime, analyzing, 77–81; temporal concentrations of crime, analyzing, 67–75
temporal categories, 75–76
temporal settings, 226–28
temporal stability index (TSI), 77–78
terms, descriptions of, 218–22
tests, descriptions of, 218–22
theft, characteristics of targets prone to, 10
theoretical distribution, 26, 27
theoretical principles of geographic profiling analysis, 134–36; crime pattern theory, 134–36; least-effort principle, 135; rational-choice perspective, 134; routine-activity perspective, 134
threshold distance, 56
time step, 117–19
Tobler, W., 25
trapper, 137–38
troller, 137–38
t-statistics, 167, 205, 221
Type I errors, 59
Type II errors, 59

U

unique identity code, 87
unique identity numbers, 44, 87–88

urban development, models of, 3–5
urban geography model of city expansion, 5

V

valuable targets, 10, 11
variables, 163, 221
variance, 166, 221
variance inflation factor (VIF), 186–87, 219, 222
variance of the error term, 166
vehicle crime, 10–11, 38, 231; theft from vehicles, 29, 30, 174, 230; vehicle thefts, 11, 28, 33, 34, 150
victim dimension of crime, 2
victimization: near repeat, 100–109; repeat, 95–99
violent assaults, 215–17
vulnerable localities index (VLI), 240

W

weighted displacement difference (WDD) test, 267–75; application of, to hypothetical example, 269–72; application of, to real examples, 272–75; calculating, 269; large data samples, 272–73; small data samples, 274–75
weighted displacement quotient (WDQ), 267–75; application of, to hypothetical example, 269–72; application of, to real examples, 272–75; calculating, 267–69; large data samples, 272–73; small data samples, 274–75
weighted spatial mean, 24–25
Weisburd, 38, 42, 43
Wellington, New Zealand, 99, 103, 107–10
what-if scenarios, 161, 211
Wiles, P., 15, 16–17, 31–32
World Bank Open Data database, 166

Y

Yorkshire Ripper, 24

Z

zonal (or concentric) model of urban development, 3–5
zone in transition (zone II), 4
zone of workingmen's homes, 4
z-score, 58–59, 60–61, 178, 183, 269, 271, 273, 275